Returning to Scientific Practice

This book is the result of a collective study of the philosophy of scientific practice (PSP), which began around 2002 and is still ongoing. There is an apparently increasing interest in scientific practice, influenced by the historicistic philosophy of science and the sociology of scientific knowledge (SSK). Prof. WU Tong and his research group believe that it is necessary for PSP to turn from a theory-dominant position to practice dominance. PSP has also brought forward the possibility of reinterpreting the epistemic status of local knowledge in the Chinese tradition, which provides the most significant motivation to participate in this study.

In this book, we have selected three main cases – namely, Chinese medicine, Fengshui, and Ethnobotany – to examine the effect of PSP. The aim of our collective studies is not merely a theoretical construction of PSP, but also to consider the various applications of PSP, especially in re-interpreting and demonstrating the variety of local knowledge from traditional China, which seems to be a genuine contribution to the international enterprise of philosophy of science, particularly when made by Chinese scholars.

Xu Zhu, Associate Professor in Department of Philosophy, ECNU. Research interests includes Epistemology, Philosophy of Action, Wittgenstein, and Philosophy of Social Science. The author of *Understanding the Social: from Normativity to Mechanism* (monograph published in Chinese) and several papers published both in Chinese and in English.

Wu Tong, Professor of Philosophy of Science and Technology (tenured) in the STS Centre, Tsinghua University. Research interests include philosophy of scientific practice, philosophy of system and complexity, and local knowledge. Author of *The Melody of Growth: the Self-Organizing Evolution of Science* (monograph published in Chinese) and many influential papers on Chinese philosophy of science and technology.

China Perspectives

The *China Perspectives* series focuses on translating and publishing works by leading Chinese scholars, writing about both global topics and China-related themes. It covers Humanities & Social Sciences, Education, Media and Psychology, as well as many interdisciplinary themes.

This is the first time any of these books have been published in English for international readers. The series aims to put forward a Chinese perspective, give insights into cutting-edge academic thinking in China, and inspire researchers globally.

Titles in philosophy currently include:

Explanation, Laws, and Causation
Wei Wang

Confucius and the Modern World
Chen Lai

Secret Subversion I
Mou Zongsan, Kant, and Original Confucianism
Tang Wenming

Secret Subversion II
Mou Zongsan, Kant, and Original Confucianism
Tang Wenming

Central Issues of Philosophy of Science
Wei Wang

Returning to Scientific Practice
A New Reflection on Philosophy of Science
Xu Zhu and Wu Tong

A Research on the Metaphysics of Tao
Kai Zheng

For more information, please visit https://www.routledge.com/series/CPH

Returning to Scientific Practice
A New Reflection on Philosophy of Science

Xu Zhu and Wu Tong

Routledge
Taylor & Francis Group
LONDON AND NEW YORK

清华大学出版社
TSINGHUA UNIVERSITY PRESS

First published 2019 by Routledge

2 Park Square, Milton Park, Abingdon, Oxon, OX14 4RN
605 Third Avenue, New York, NY 10017

Routledge is an imprint of the Taylor & Francis Group, an informa business

First issued in paperback 2020

Copyright © 2019 Xu Zhu and Wu Tong

The right of Xu Zhu and Wu Tong to be identified as author of this work has been asserted by them in accordance with sections 77 and 78 of the Copyright, Designs and Patents Act 1988.

All rights reserved. No part of this book may be reprinted or reproduced or utilised in any form or by any electronic, mechanical, or other means, now known or hereafter invented, including photocopying and recording, or in any information storage or retrieval system, without permission in writing from the publishers.

Notice:
Product or corporate names may be trademarks or registered trademarks, and are used only for identification and explanation without intent to infringe.

English Version by permission of Tsinghua University Press.

British Library Cataloguing-in-Publication Data
A catalogue record for this book is available from the British Library

Library of Congress Cataloging-in-Publication Data
Names: Xu, Zhu, 1983– author. | Wu, Tong, 1954– author.
Title: Returning to scientific practice : a new reflection on philosophy of science / Zhu Xu, Tong Wu.
Other titles: Fǎu gui ke xuâe shâi jiáan. English
Description: Abingdon, Oxon ; New York, NY : Routledge, 2019. |
Series: China perspectives | Includes bibliographical references and index.
Identifiers: LCCN 2018042743 (print) | LCCN 2018049457 (ebook) |
ISBN 9781315727110 (ebook) | ISBN 9781138846975 (hardcover)
Subjects: LCSH: Science–Philosophy.
Classification: LCC Q175 (ebook) |
LCC Q175 .X813 2019 (print) | DDC 501–dc23
LC record available at https://lccn.loc.gov/2018042743

ISBN: 978-1-138-84697-5 (hbk)
ISBN: 978-0-367-72898-4 (pbk)

Typeset in Times New Roman
by Newgen Publishing UK

Contents

List of figures vii
List of tables ix
Postscript xi

Introduction: towards philosophy of scientific practice — 1

1 The origin of the concept of practice — 13
2 Scientific practice: significance, types, and scopes — 37
3 The nature of scientific practice — 61
4 The nature of knowledge: local knowledge — 82
5 Knowledge and power — 105
6 The contextual normativity of scientific practice — 127
7 Philosophy of scientific practice and naturalism (I) — 137
8 Philosophy of scientific practice and naturalism (II) — 147
9 Philosophy of scientific practice and relativism — 159
10 Partnering the philosophy of scientific practice: the philosophy of scientific experimentation — 172
11 New empiricism: a close relative of the philosophy of scientific practice — 194

12 The starting point of scientific research: opportunity, question, or observation?	208
13 A new solution for an old problem: the relationships of observation, experiment, and theory	224
14 New studies on replicability of scientific experiments	231
15 Local knowledge (I): traditional Chinese medicine (TCM)	243
16 Local knowledge (II): Chinese theory of *Fengshui*	264
17 Local knowledge (III): ethnobotany	281
18 Conclusion: scientific practice in ongoing and unlimited process	295
References	311
Index	321

Figures

2.1	Different preoccupations of the conversations	52
2.2	Geometrical representation of the complex number (Pickering, 1995, p. 122)	54
2.3	Addition of complex numbers in the geometrical representation (Pickering, 1995, p. 122)	55
2.4	Hamilton constructing quaternions	58
4.1	The scheme of interpretive anthropology	85
4.2	The double-foliot Japanese clock	100
4.3	The dials of the corrugated-panel pillar clock and the circle-graph clock	101
5.1	Arab scholars' estimation of how to upgrade personal status	118
10.1	Types of apparatus (Radder, 2003, p. 33)	185
10.2	Watson and Crick's model of DNA	187
10.3	Two versions of Michael Faraday's electromagnetic motor (Baird, see Radder, 2003, p. 48)	188
10.4	Barlow's Star, Peter Barlow's variant of Faraday's motor (Radder, 2003, p. 48)	188
11.1	Two views of science: left, unified hierarchical view; right, non-unified dappled view (From Cartwright, 1999, pp. 7–8)	195
11.2	The hard systems stances which Checkland against	196
11.3	The soft systems stances which Checkland support	196
11.4	Specific application of Coulomb's law with magnetic field (quoted from Cartwright, 1999, Figure 3-1b)	200
12.1	The relationship between opportunity, question, and observation	212
12.2	Laboratory conversation topics	213
12.3	The capital required by researchers' study	214
12.4	The structure chart of the hybrid tracking system of a magnetic field and image	221
14.1	Experimenter's regress	235
16.1	Optimal layout and sites of house, village, and city	272
16.2	The scientific explanation of a Fengshui layout	273
17.1	Folk Mongolian's named relationship with the Allium plant	284

17.2	The structure of a yurt	290
17.3	Yurts on Inner Mongolia's grass	290
18.1	The digitalization of watershed geomorphologic formation. (Taking Li Zixi, a tributary of the Jialing River, as an example. Quoted from the conclusion materials of Dan-xun Li's laboratory exploring lessons.)	298
18.2	The structure of the 3D digitization simulation platform for a river (quoted from Dan-xun Li, etc. materials of laboratory exploring lessons)	301
18.3	The evolution of scientific river research	302

Tables

1.1	Aristotle's three types of human act	14
1.2	The main differences between practical and theoretical hermeneutics by Rouse (1987, Chapter 3), concluding	32
5.1	How the Arab academic community gained its place and status	119
10.1	Types of experiment: all three with representing relations (Radder, 2003, p. 231)	178
11.1	Different expressions of Newton's second law	197
14.1	Types and ranges of reproducibility (Radder, 1992, p. 66)	237
14.2	Differences on the notion of replicability between SSK and the new experimentalism	241
17.1	'Large and small Hang'gai mountain': local knowledge in Mongolian folk songs	282
17.2	The relationship between Mongolian and academic names for Allium	285
17.3	The catalogue of national botany of Mongolian folk medicinal plants at Aru Corqin	287

Postscript

This book is a result of a collective study on philosophy of scientific practice (PSP), which began around 2002 and is still in an ongoing process. In 2002, our research group led by Prof. Wu Tong at Tsinghua University found that there was an apparently increasing interest in scientific practice, influenced by the historicistic turn in philosophy of science and the rising of sociology of scientific knowledge (SSK). The new challenge here is how to deal with those traditional problematics in philosophy of science by focusing on scientific practice. Prof. Wu and his research group believe that it is necessary for PSP to turn from the theory-dominant position to the practice-dominant position, in order to respond to those challenges. Accordingly, Prof. Wu leads an 'Contemporary Issues on Philosophy of Science and Technology' course, which organizes his doctoral students to study works of Joseph Rouse, Andrew Pickering, Ian Hacking, and Nancy Cartwright, and to clarify references to basic issues about main continental philosophers, such as Martin Heidegger and Michel Foucault, in order to understand PSP from its original connections with phenomenology and hermeneutics.

Undoubtedly, PSP is very different from the mainstream philosophy of science in analytic tradition. For one thing, it emphasizes that science should be at first thought of as a particular kind of practice intervening into the world, whereas the mainstream philosophy of science takes theoretical construction as the core of scientific enterprise. For another, it usually argues for an equality between scientific knowledge rising from Western culture and other local knowledge from non-Western community, and also argues that science would be nothing other than a kind of local knowledge, whereas for philosophers of science in analytic tradition, demarcation of science from non-science or especially pseudo-science has always taken an indispensable concern, since it is suggested that scientific knowledge should be the truth in a universal and non-local sense.

Therefore, PSP has also put forward some possibility to reinterpret epistemic status of local knowledge in Chinese tradition, which provides the most significant motivation for us to participate this study. In this book, we have selected three main cases to elaborate how PSP could do in that issue. Chinese medicine is a kind of local knowledge in medical practice for thousands of years to effectively protect Chinese people's health and to cure their diseases. However, it has been criticized as a pseudo-science after comparing with Western medicine based on modern science since the period of the Republic of China. Feng shui is another kind of local knowledge which deeply influences Chinese architecture and residence choice in thousands of years, and it is also thought of as pseudo-science for the same reasons as Chinese medicine. Ethnobotany, different from both aforementioned, is a discipline originated from Western academia, while merely taking a non-mainstream position in the development of botany or even biology. In view of PSP, however, ethnobotany could be irreplaceable to grasp the main significance of the local knowledge of botany in Mongolian living practices, which is also put as an example of a kind of local knowledge, thought belonging to traditional China, different from the view of Han people.

In a word, the focus of our collective study is not merely on theoretical construction of PSP, but also on the various applications of PSP, especially re-interpreting and demonstrating a variety of local knowledge from traditional China, which seems to be a genuine contribution for the international enterprise of philosophy of science, particularly made by Chinese scholars. It is our mission from which we have no reason to escape, but have hundreds of reasons to take without hesitation.

As the result of a collective study, this book is also written by different hands. Its original edition is drafted in Chinese; in accomplishing in English, it involved lots of translators and finally got edited by Xu Zhu, Associate Professor from East China Normal University in Shanghai. All the Chinese writers and English translators have made tremendous contributions for the accomplishment of this book. Most of them are very impressive scholars in the Chinese circle of philosophy of science and technology. Their names and affiliations are listed as follows:

Introduction: written by Prof. Wu Tong (Tsinghua University) and translated by Dr. Su Li (Xidian University);

Chapter 1: written by Prof. Wu Tong (Tsinghua University) and translated by Dong Xiaoju (doctoral candidate, Tsinghua University);

Chapter 2: written by Prof. Wu Tong (Tsinghua University), Dr. Zhang Chunfeng, and Dr. Wu Shuxian, and translated by Prof. Wang Na (Beihang University);

Chapter 3: written by Prof. Wu Tong (Tsinghua University) and translated by Dong Xiaoju (doctoral candidate, Tsinghua University);

Chapter 4: written by Prof. Wu Tong (Tsinghua University) and translated by Prof. Xu Zhu (East China Normal University) and Prof. Su Zhan (University of Chinese Academy of Sciences);

Chapter 5: written by Dr. Lv Zhenhe, Prof. Ren Yufeng (Inner Mongolia University), and Prof. Wu Tong (Tsinghua University), and translated by Dr. Cui Bo;

Postscript xiii

Chapter 6: written by Prof. Yu Jinlong (Beihang University) and translated by Prof. Xu Zhu (East China Normal University);
Chapter 7: written by Dr. Tian Xiaofei and translated by Prof. Xu Zhu (East China Normal University);
Chapter 8: written and translated by Prof. Xu Zhu (East China Normal University);
Chapter 9: written by Xiao Biao and Prof. Wu Tong (Tsinghua University), and translated by Dr. Li Fangfang;
Chapter 10: written by Prof. Wu Tong (Tsinghua University) and Zheng Jinlian, and translated by Zheng Jinlian;
Chapter 11: written by Prof. Wu Tong (Tsinghua University) and translated by Dr. Wang Dong (Beijing Technology and Business University);
Chapter 12: written by Prof. Wu Tong (Tsinghua University) and translated by Li Pei (doctoral candidate, Tsinghua University);
Chapter 13: written by Prof. Wu Tong (Tsinghua University) and translated by Dr. Su Li (Xidian University);
Chapter 14: written by Dr. He Huaqing and translated by Dr. Wang Dong (Beijing Technology and Business University);
Chapter 15: written by Prof. Wu Tong (Tsinghua University), Prof. Ma Xiaotong (China Academy of Chinese Medical Sciences), and Dr. Zhang Shuyan (Tianjin University);
Chapter 16: written by Dr. Li Jingjing (China University of Petroleum–Beijing) and Prof. Wu Tong (Tsinghua University), and translated by Dr. Li Jingjing (China University of Petroleum–Beijing);
Chapter 17: written by Prof. Wu Tong (Tsinghua University) and translated by Dr. Jiri Gele;
Chapter 18: written by Prof. Wu Tong (Tsinghua University) and translated by Jia Zhixiang (doctoral candidate, Tsinghua University).

Acknowledgements

We are grateful to China National Social Sciences Fund for the support (grant no. 05BZX029).

Introduction
Towards a philosophy of scientific practice

Since the 20th century, mainstream Anglo-American philosophy of science has encountered insurmountable difficulties. Both logical positivism and historicism underwent attacks internally and externally, on issues such as scientific explanation and realism/anti-realism debates. Various models of unified science failed. By the 1990s, it was indisputable that the philosophy of science had entered into a kind of 'Warring States' era. We are led to ask several questions: for example, where will the philosophy of science go? What went wrong? And how can we break through those obstacles in the way of developing philosophy of science?

Many schools of thought have proposed different programs filled not merely with conflicts, disagreements, and mutual criticism, but also with dialogue, mutual development, and synthesis. One of them, 'The Philosophy of Scientific Practice' (hereafter PSP), seeks a practicalist approach to philosophy of science by incorporating SSK, anthropology, phenomenology, and hermeneutics of science.

1 PSP approaches

Philosophy of science can be divided historically into two main schools: positivism and historicism. Positivism holds that theoretical reason should be separated from practical reason in science. Logical analysis of theoretical reason is the only way to understand scientific rationality, while practical reason is classified into ethics, sociology, psychology, and other disciplines. (In fact, many philosophers used to treat practical reason as such.) Many historicists inevitably slipped into scepticism since they neither accepted positivism nor successfully reconciled theoretical with practical reason. It is however undeniable that some historicist work – especially Kuhn's theory of scientific revolution, based upon his practical conception of the 'paradigm' – came to greatly inspire philosophers of science later.

PSP arose in the 1990s from philosophical naturalism. It regards scientific activity as something specified by human culture and social practice. And it attempts to carry out an in-depth research on the main features of scientific practice. The understanding on scientific rationality requires us to abandon the formerly accepted boundary between theoretical and practical reason, and to carry out various empirical researches besides philosophical speculations.

Currently, philosophy of scientific practice can be generally divided into three approaches: the hermeneutic, the new experimentalist, and the embodied cognition approach. In a narrow sense, PSP particularly refers to the hermeneutic approach, mainly represented by Joseph Rouse, which will be mostly the main research object in this book.

The embodied cognition approach stresses on the mechanisms and functions of cognitive practice in building knowledge. Some philosophers, such as P. Churchland, R. Giere and H. Simon, suggest that philosophy of science should be replaced with the cognitive science of science. They have provided some micro-mechanisms of practices by describing the inner mental mechanism of cognitive agents, which is criticized as cognitive individualism. A recent approach of cognitive science is most related to embodied cognition, which points out that the body of agent is also involved in cognition with special influences and functions. This approach mainly takes the human being as an agent extending mind into an objective world – the so-called notion of extended mind. That strongly supports the PSP from perspectives of individual cognition and interactions between human bodies and surroundings.

J. Rouse (1987, 1996, 2002) is a principal representative of the hermeneutic approach. He claims that M. Heidegger and T. Kuhn are his predecessors.

The new experimentalist approach is related with lots of philosophers of science including Ian Hacking (1983), Allan Franklin (1999), Peter Galison (1997), David Gooding (1990), Deborah G. Mayo (1996) and H. Radder. As a matter of fact, sociologists of scientific knowledge (SSK) and scholars undertaking science studies, such as Latour and Woolgar (1986), Michel Lynch (1993), Karin D. Knorr-Cetina (1981) who advocates laboratory study, Andrew Pickering (1992) who studies sciences as culture and practices, and some feminist philosophers of sciences like Evelyn Fox Keller (1985).

SSK's contributions to the study of scientific practice are gradually recognized. It takes practical reason as an important notion related with specific contexts, and locates practice in cultural settings, providing important incentive to the progress of the PSP. Feminism indicates that scientific knowledge may be related to regions, culture and gender practice, which provides justification to scientific practices.

A new hermeneutic theory of practice is proposed regarding functions of scientific reason in philosophy of science by taking scientific practices as a starting point. This theory re-examines classical issues in philosophy of science, such as explanation, confirmation, the relation of science and value, and theoretical changes of science. And it integrates these classical issues with philosophy of technology, sociology and psychology of science. Relevant fields include viewpoints and methods from functions of experiments, the relation between observation and opportunity, between knowledge and power, and perspectives from hermeneutic and phenomenology, in order to understand scientific practices more effectively.

2 The hermeneutic background of PSP

As is known, in the 20th century philosophy of science was dominated by the analytic tradition, which originated from Frege's philosophy and mathematical logic. However, there are still other sources. Phenomenology and hermeneutics are main sources of continental philosophy. The two traditions of philosophy often go apart and then against each other, as if the two fields have already been divided. Both of them argue that hermeneutics can only apply to humanities and social sciences, while scientific explanation can fit all natural sciences.

Therefore, though it has a long history, hermeneutics remains a theory thriving in humanities, and has marginal effects on philosophy of science over a rather long time. Not until now it begins to exert influences on philosophy of science from reflections through Dilthey. That influence is originated from Kuhn. According to Rouse, the influence of hermeneutics has surely close relation with the renaissance of pragmatism. Pragmatist philosophers such as Rorty, Habermas, Bernstein, and Hesse have widely discussed over important effects of hermeneutics on philosophy of science (Rouse, 1987, p. 41).

Kuhn and Heidegger are the two important figures directly affecting PSP. And they are related by their background of hermeneutics. Rouse believes that he himself subjects to the direct influence of Heidegger's practical hermeneutics, on which we will discuss in a special section. Now let us turn to Kuhn's hermeneutic ideology and its influence on PSP.

When Kuhn began to study the history of science, he spontaneously thought upon some questions relevant to hermeneutics when reading Aristotle's physics. The question is: why could a sage at that time made such a mistake that we regard as so simple? After repeatedly reading, Kuhn found suddenly one day that it was different kens and interpretations that led Aristotle to concern different issues and to solve different problems. Besides, it's absurd and inadvisable to judge an era as inferior to the later. This is certainly an issue of hermeneutics. Kuhn's study on history of science is undoubtedly affected by hermeneutics. On the reading of Aristotle's literature, Kuhn points out in the preface of *The Essential Tension*:

> Consciously or not, they (historians) are all agents of the hermeneutic method. In my case, however, the discovery of hermeneutics did more than make history seem consequential. The most immediate and decisive effect was instead on my view of science. That is the aspect of my encounter with Aristotle that has led to my recounting it here.
>
> (Kuhn, 1981, V)

Kuhn thought upon it in around 1947 and did not begin to use the term 'hermeneutics' until 1947 (Kuhn, 1981, III–VII). Kuhn firstly realize and apply hermeneutic thinking in studying philosophy of science. Needless to say, Rouse, the advocator of PSP, is strongly and directly influenced by Kuhn.

3 Comparison: traditional philosophy of science and PSP

Traditional philosophy of science regards theory as prior to experiment and observation in science, which are considered as meaningful only in theoretical context. The description of theory provides a scope for explaining observation in the construction and operation of experiment. Therefore, observation and experiment are the medium in transferring and applying research results, which indicates that observation/experiment are infiltrated by theoretical proposition. According to traditional philosophy of science, there are accidental factors generated by scientific knowledge, such as local research places, experiment construction, technical facilities, particular social network of researchers, and practical problems in research. It proposes that sciences are universal propositions, that theory is the ultimate outcome of research, and that the aim of science is to put forward better theories.

Traditional philosophy of science also considers science as a representational system, aiming to accurately describe the world independent from our representation. Truth is the correspondence between our representation and the real world. The importance of observation is to link the represented world with the world itself. Only in sensory experience does the world affect us by restricting the possibility of representation of the world.

Those are widely accepted view from traditional philosophy of science. Those basic ideas have not dramatically changed, though there are ideas on whether the representation has really connected the world with models, and on serious attacks from historicism to logical positivism. Due to the domination of those views, traditional philosophy of science is now in a predicament, and cannot break away from its limitation.

The overall feature of traditional philosophy of science, according to PSP, the new approach of philosophy of science, is very precisely and outstandingly called 'theory-dominated philosophy of science' (Hacking, 1983, p. 185; Rouse, 1987, p. 27), namely a view of science to ascribe priority to theory. On the contrary, the PSP is against traditional philosophy of science by proposing an opposite idea 'practice-dominated philosophy of science'.

First, PSP holds that the fundamental mistake of traditional philosophy of science rests on neglecting functions and significance of scientific practice, or practical features of scientific research, while merely regarding science as a sophisticated system of knowledge, or a logical net of propositions, especially employing physics as an example. In a word, traditional philosophy of science is theory-dominated.

Rouse, who has proposed PSP to integrate new experimentalism, pragmatism, and hermeneutics, points out in a more direct but profound way that the problem of traditional philosophy of science does not lie in its negligence of one side of science (experiment) and emphasis on the other side (theory), but in its distortions of scientific images and its viewpoints on scientific practices. Scientific practice has provided theory with most of the basic materials, but these skills and feats are rarely praised to the extent that they deserve (Rouse, 1987, IV). Rouse prefers to consider science as activity. He proposes a hermeneutic notion of practice to enrich the meaning of practice through his criticism on Dilthey's hermeneutics (the idea of 'hermeneutics is applicable only to humanities'), his interpretation on Heidegger's practical hermeneutics (while criticizing Heidegger's residual of the theory-dominated thought of hermeneutics), and his new discovery and exploration on Kuhn's notion of paradigm (differentiating Kuhn I and Kuhn II). For example, he takes science as a realm of practice rather than a net of propositions. In the first place, science involves operation and intervention in the world, namely the method acting on the world rather than representing the world (Rouse, 1987, pp. 26, 38, 129). Scientific practice occurs in practical background of skill, practice and tool, not in a systemized theoretical background (Rouse, 1987, pp. 95–6). In *Being and Time*, Heidegger regards daily practice as hermeneutics of human existence per se, which means that the daily practice itself embodies the interpretation and description of the world. Similarly, the intervention of scientists, technical experts, engineers as well as other agents of scientific practices with the world around them, is their interpretation and description of the intervened world, without any need of further representation (this indicates that tools, equipment, and laboratories also contain some kind of knowledge, especially the knowledge of objective matters referred by new experimentalists). When we employ tools, some significance is gained in this process. In other word, we are interpreting the world and ourselves in what we are doing or use to do, and in how we do. Therefore, PSP puts practice in a primary position, since practice shapes human being and the world.

Second, PSP has proposed profound criticism on some important viewpoints of the traditional philosophy of science.

The issue about the connection between the world and its representation is actually shared by realism and anti-realism. It is also one of the most controversial issues of philosophy. Transcendentalists and empiricists were both trapped in the past. The transcendentalists completely separate reason from experience and the world. The empiricists, though admitting the essential link between being and presence, interpret the presence mistakenly from experience in an excessively narrow sense. Philosophy of science involves some viewpoints from Heidegger's practical hermeneutics, and it combines the hermeneutic aspect with Kuhn's notion of paradigm to provide a careful and witty answer: the problem doesn't lie in how we reach the represented world *per se* from the linguistic representation. We have always participated in the world in practices. Therefore, the whole problem of access to the world (the resort to observation is a response to the question, for example) would be eliminated (Rouse, 1987, p. 143), since our practice is a part of world activity:

> Only by interfering with the world can we discover what the world is like. The world is not an unreachable thing opposite our theory and observation. It is something presented in our practice and something belonging to us it resists or accepts when we act on it. Scientific research has changed the world and the way by which the world is known together with other things we do. Instead of knowing the world in way of subject representation, we grasp and comprehend the possibility for us to discover our own self as actors. Turning to operation from representation and from that know to how know, do not deny the commonsense idea of revealing the surrounding world.
>
> (Rouse, 1987, p. 25)

This fundamentally changes many ideas. For instance, in PSP, knowledge, though recognized as representation by the tradition philosophy of science, is no more merely representational thing (such as text, thought or chart), but a component of the present and existing practical mode of interaction, namely a component of the method of interaction with the world. Scientific conception and theory can be understandable only as a component of the more extensive social practice and material practice.

The important category clearly presented to us is no longer the relation between the observable and the unobservable. On the contrary, the questions should be described as following: what can be used; what we must take into consideration during the use; and what is the goal we pursue. What answers these problems is the decision of action by the community of agents, rather than the physiological threshold of sense organ (Rouse, 1987, p. 143). Therefore, it is engineering approach rather than theorization that prove scientific realism of theoretical entity (Hacking, 1983, p. 274).

In this way, we can see that the PSP will definitely lead to a pluralistic conception of research and to generate other important difference from the traditional philosophy of science: Is knowledge local or universal? Undoubtedly, the traditional philosophy of science defines scientific knowledge as universal at the very beginning, and it always regards the process of abstract acquisition as a universalized process. Even though scientific knowledge is supposed to have some local traits, it only admits the locality of the origin and occurrence of knowledge, though knowledge is still universal in nature.

According to the accepted view, there is a process in which scientific knowledge ultimately turns into the universal from something local. This process is called 'delocalized' or 'decontextualized'. It includes three aspects: thematization of scientific object, delocalization of scientific object, and unindexed scientific target, no matter whether they regard knowledge as activity or system. PSP greatly varies from this viewpoint. It holds that knowledge is local at any time. The process recognized as universalization by traditional philosophy is actually a process in which local knowledge is standardized. When there is insufficient consistent description and explanation, scientific knowledge could be still in circumstances as ability of using particular exemplars, for instance, the current complexity research. Kuhn did so by extending the notion of paradigm. Rouse also had the same idea. Therefore, the skill and practice in the local, material, and social context are essentially important for all description, comprehension and interpretation.

According to PSP, scientific knowledge and activity must be local. This is demonstrated by all of things generated and needed by scientific knowledge: specific laboratory, research program, local community, and research skill. The so-called universalization of scientific knowledge is merely a transfer from local knowledge to another place. The so-called decontextualization should actually be the standardization (which is to say, enacting norms to let scientific research in each area comply with a certain local norm, in order to set a scientific norm as a standard) of local scientific knowledge. Actually, in virtue of the current context of combining science and technology, it is common to argue that technical standards of different countries construct the so-called international standard in technical competition, and that developing countries are technically and politically restricted by the technical standards of developed countries.

The new philosophy of scientific knowledge also has some features of postmodernity. As is known, traditional philosophy of science is identical with the ideological basis required by modernity. In other words, it is modernity that creates the traditional philosophy of science. According to Rouse, the traditional and often quoted conception of modernity involves the following features (Rouse, J., p. 49):

1. Secularization – religious practice and belief are not simply declined, but also that various realms of human life are separated from theology, including the privatization of religious and other fundamental matters of concern into individual beliefs and commitments.
2. 'Humanity' (the unity of abstractly differentiated *individuals*) is constituted as the *subject* of representation and knowledge, sources of all the values, and the possessor of rights morality and dignity.
3. The development of distinct domains of knowledge and practice (for example, law, the economy, science) in which autonomy is institutionally recognized and protected.
4. Rationalization – formal procedures of calculation developed more intensively, extended to more and more domains of human practice, and unified across domains.

5 The rapid growth of science and technology as the essentially modern human practices and the correlated understanding of nature as inert object of knowledge (science is perhaps the only human activity for which the appellation modern has almost no controversial force).
6 The expansion and concentration of productive resources (usually identified with the beginnings of a capitalist mode of production).
7 The global extension of European ('modern') culture (colonialism and postcolonial modernization/Westernization).
8 The self-referential narrative legitimation of modernity as the progressive realization of freedom and truth.

In a superficial view, the pursuit of unifying all theoretical and practical sense with a computational science (such as mathematics, formal logic and theoretical physics) seems to be pursuing unified science. However, it actually has a deep aim of promoting the 'unilateralism' and 'colonialism' in academic enterprise, an ideological and political (powerful) feature of development of science since early modern era. The rationalization based on this science is also taken as the formal feature of scientific method and economic behaviour. Since science and technology provide a typical example formed in secular field, it is considered as the basis and paradigm of economic expansion and westernization of politics and culture. The expansion of western culture not only resorts to modernism but also is regarded as a modernist explanation that has proven universal forms of human value and knowledge. From a more general perspective, all the modern legalized narratives resort to the value that defends secularization, rationalization, economic expansion, and allocates resources and authority to scientific research and technical development. As a matter of fact, this is also the most explicit feature of modern philosophy, as shown by positivism, for instance. Positivism strives to construct all kinds of propositions from atomic propositions, and pursues to the certainty of empirical verification. Their empiricist philosophy of science particularly stresses the universal aspect of individual experience (regarding the individual's knowledge as the knowledge of category), but neglects the difference between individual experience and the local characteristics of knowledge. The tradition of empiricism is taken as the main context during the 1930s and 1940s. It involved philosophers in considerably subtle and exquisite technical issues, and substantially influenced the discussion on science in many fields. To a great extent, the tradition of positivism seems to be not only a doctrine with influence and superior status in philosophy of science, but also the philosophy of science *per se*.

Positivism gradually declined during 1950s and 1960s, highlighting a series of new problems and themes in philosophy of science. Different approaches were emerging after positivism, such as rational criticism and historicism, which can be called as post-positivist philosophy of science. This kind of philosophy is usually linked with Kuhn and Feyerabend, who often emphasize the anti-modernism purport of argument, in particular the case of Feyerabend's efforts: science is no longer exceptional compared with other cultures; scientific practice is related with specific community and belief; while formal approach is disputed. This trend itself begins to reveal the movement inside the dialectics of modern and anti-modern narratives. However, Kuhn and Feyerabend's work also have obvious modernist purport based on empiricism. Nevertheless, the well-known post-empiricists, including Lakatos, Toulmin, Marry Hesse, and Larry Laudan, also have more remarkable modernist features of empiricism, which is indicated by Bruno Latour in his homonymous work: we've never been modern. In fact, many philosophers of science have never been post-modern.

Modernism approves the increasingly specialized autonomy of knowledge, and strives to unify social practices by a formal program of inferences. Once the autonomy achieves its appeal of separating from the outside, they would begin to promote unification in their own. Positivists affiliate scientific requirement of discipline autonomy to the unification of science. Most of post-positivist critics have abandoned formalism, and turned to empiricism for the autonomy of discipline and internal research program thereof. Even though they turn against formalism, most philosophers still have a strong preference for rational description of meta-science. Adhering to the requirement of traditional philosophy of science, they appeal for experience on one hand, but never change the way of appeal for standardization on the other. This concern for reason is closely connected with the rational pattern that has a notable significance in the modern narrative: this concern requires to differ the 'internal' history of science from 'external' social or cultural influence, and to define scientific change and development as an internal process changing as stipulated by a set of paradigms, according to the internal rule of science or historicism. As a matter of fact, the explanation of standardization about scientists' spiritual temperament in Merton's sociology of science consolidates this kind of explanation. Hence, the post-empiricist philosophy of science actually discards formalism instead of modernism. Certainly, the formalism herein is not the one stressing the appearance of form, but claiming that science can be ultimately expressed via a universal formal system, such as formal or mathematical logic. This is a kind of philosophy of science that commits to formalism at least in epistemology. Nowadays, the epistemological commitment for formalism seems to distract philosophy from science in itself so as to pay more attention to logic, and it has deviated from all the insights of scientific practice or success. With few correlations with scientific practices, positivism, as it were, has just provided a scientific model, but no science has actually ever been similar to it. However, philosophy of science in the sense of post-empiricism no longer concerns formalism, but just replaces formalism with the pattern of scientific progress in history of science, and thus it's still a paradigm of the modernistic view of progress. After separating from standardization, progress will be the process of 'palace coup' and nightmare.

4 Some unsolved problems within and beyond PSP

It is certain that many problems in PSP still need further research. The most important problem is that, as we mentioned, the hermeneutic approach is still in conflict with New experimentalism, which hinders the general understanding of PSP.

The first important conflict is in the fact that the hermeneutic approach is basically metaphysical, while New experimentalism is an approach mainly supported by physics rather than metaphysics. The hermeneutic approach conducts in-depth interpretation on scientific practice, scientific thought, and scientific operation, by the means of philosophical arguments in both Anglo-American and Continental way. In contrast, New experimentalism deals with scientific experiments by undertaking research on scientific experiments, instruments, and tools. The New experimentalism directly derives from philosophical reflections on scientific activity, and criticism on historicist philosophy of science. It is important to show how the two approaches integrate through argumentation and criticism.

The second conflict is about understanding the relation between theory and experiment. In Rouse's view, there seems to be no essential distinction made between experiment and theory, since the notion of experiment involves both laboratory and theoretical activities. Discursive practice itself contains speech acts, when we regard science as activity and practice rather than a net of statements and beliefs. As far as new experimentalism is concerned, if experiment cannot be differed from theory, the position of historicist philosophy of science would be returned, and then fail to elaborate the priority of experiment. Nowadays both the hermeneutics of scientific practice and new experimentalism must be based on practice, in order to renew the understanding of 'experience'. There will be a solution in a certain sense arguably.

Moreover, it is still necessary to investigate further into the ethical, practical, and epistemological sense, considered in experimental design, in which the connection among logical construction of mature science and cognitive competence of human subjects from the role of images, models, metaphor, and computer simulation.

New experimentalism and hermeneutics of practice should continue not only to criticize the 'theory-dominated' philosophy of science, but also to break through a 'practice-dominated' concept. If 'theory-dominated' is wrong, 'practice-dominated' can be seen as having merely realized an overturn, and as a highlight of another element inside the structure. How could 'practice-dominated' be a new proper basis of the philosophy of science? Is it another question of foundationalism? People can reasonably ask 'Why must science be "dominated" by something?' Can we have a view of science without 'dominated' thinking pattern? This question may lead to a new approach of holism or coherentism.

There is a third approach of cognitive science, such as embodied cognition which involves the main body intentionality of cognitive practice, and the conceptual content of practice. Although the cognitive science approach does not involve much conceptual relation with the notion of practice, it is indeed connected with each other in many ways. But the practical implication is not yet revealed. This is also a new direction for further study.

In PSP, there is an issue on the local nature of knowledge, which is totally different from the traditional view on the universality of knowledge. How can we confirm that locality is the nature of knowledge, not merely for knowledge in non-western areas? That is also a question hard to be demonstrated, for it needs both argumentation and explanation. Rouse's argument upon the nature of locality are unfolded mainly through the relation between knowledge and power, which is demonstrated by knowledge transferring in a process of standardization. This is surely an advisable route, but its argumentation appears to be external from the perspective of traditional view. Undoubtedly, the power referred to by Rouse is an element of scientific knowledge and practice, rather than something outside. However, there is actually other way of demonstration. Nancy Cartwright (1999), as a new experimentalist, argues that the reason for why the expansion of local knowledge looks like universal knowledge is because of the setting of the *ceteris paribus* conditions in the function of nomological machine. This kind of argumentation is a method for demonstrating inside knowledge.

Anyway, PSP is a new orientation of philosophy of science. It extends research to the realms ignored by traditional philosophy of science, and it explores new fields outside the traditional philosophy of science – science as a space of practice, which has provided new light for solving several problems in traditional philosophy of science by opening up a new route as practical understanding of science.

The research of the book is divided into two parts: one theoretical, the other applied. The theoretical part deals with the content of PSP *per se*, such as the nature of practice, standard of practice, the nature of local knowledge, the feature of naturalism, and its relation with relativism. Besides, the philosophy of New experimentalism and Neo-empiricism are closely related to PSP. The application part is to solve the issues concerning the relation among observation, experiment, and theory, including the origin of scientific research and local knowledge in the context of China, without distinguishing theory from practical application. In the theoretical part, we make necessary demonstration and defence for the further establishment of philosophy of science as well as the rationality of its viewpoints.

In the application part, I discuss my own knowledge and practice within the viewpoints of PSP for peers to test and criticize. I hope readers can gain some understanding of PSP, and support our efforts to develop and explain local knowledge with PSP.

1 The origin of the concept of practice

Compared to the traditional theory-dominant philosophy of science, PSP is at least one of the practice-dominant philosophy of science (in a strong sense), or a philosophy of science based on practice (in a weak sense). The notion of scientific practice plays an important role in PSP, a practice-dominant or practice-based view. It is possible to propose a question, such as 'what contribution PSP makes for the notion of practice?' or 'Which normative roles do the practice-dominant concepts have on PSP?' Compared to the scientific explanation in traditional philosophy of science, what advantages does the scientific explanation based on the concept of practice have? To understand these problems, we need to study practice itself historically. In fact, there is a long history of practical philosophy in the western. At least, from Aristotle, there are systematic interpretations of practice. In modern times, this tradition has evolved more intensely. For example, Marxism is a philosophy based on practice, which is nonnegligible for any research of practice. We can have a better understanding of scientific practice if we interpret this tradition.

1.1 The origin of the concept of practice

1.1.1 Aristotle's practice

In ancient Greek times, the meaning of the word 'practice' was the acts of all lives. And in the history of philosophy, the concept of practice usually traced back to Aristotle. Before Aristotle, there were several discussions about 'practice'. Many philosophers such as Plato and Hippocrates had used the concept of 'practice'. But there is no systematic theory. And Aristotle not only put forward the concept of practice, clearly, but also set up a set of philosophy system with practical characteristic. Aristotle used practice in various senses. Aristotle's 'practice' was put forward in ethical and political senses, which related to people's correct behaviours. His practice refers to the behaviours of goodness. Some researchers pointed out that in *Nicomachean ethics* Aristotle divided people's acts into three types: theoretical acts, poiesis, and praxis. Some scholar (Rotenstreich, 1977, p. 18) summarized as follows (Table 1.1):

Table 1.1 Aristotle's three types of human act

	Theoretical sphere	*Practical sphere*	*Poetic sphere*
1. activity	viewing	act	doing
2. type of knowledge	science	deliberation	skill
3. end attained	happiness	proper life	welfare

Different from 'poetic acts', the purpose of 'practice' is not outside of oneself, but in itself, and the purpose is just itself; but the purpose of 'poetic acts' is the result which is produced by it, and it does not constitute purpose in its own right. So Aristotle said that

> Making and acting are different; so that the reasoned state of capacity to act is different from the reasoned state of capacity to make. Hence too they are not included one in the other; for neither is acting making nor is making acting.
>
> (Aristotle, 2009, p. 105)

Aristotle also said that 'Intellect itself, however, moves nothing, but only the intellect which aims at an end, and is practical, for this rules the productive intellect as well' (Aristotle, 2009, p. 103). Therefore, in some degree, praxis is also associated with poiesis. According to W.D. Ross, a famous expert on Aristotle, there are five states of mind by virtue of which we reach truth: science, art, practical wisdom, intuitive reason, and theoretical wisdom. 1) Science is: (a) concerned with what is necessary and eternal, and (b) communicable by teaching. Science is: 'the disposition by virtue of which we demonstrate', 2) art is something that we use to deal with the contingent, and the disposition by which we make things by the aid of a true rule, 3) practical wisdom is the power of good deliberation, it is a true disposition towards action, by the aid of a rule, with regard to things good and bad for men, and 4) intuitive reason is that by which we grasp the ultimate premises from which science takes its start (It grasps the first principles by 'induction'.), 5) theoretical wisdom is the union of intuition and science, directed to the loftiest objects: it is as much superior to practical wisdom. According to Aristotle, there are two basic principles of praxis: one is that it is the purpose; another is that it is not the biological or production activities which maintain the material life of human beings, not the activities between human and nature, but is the ethical and political activities among people in a broad sense. Since practice is the rational activity of human, the practice of ethical and political implication should have the normativity and directivity of goodness. Therefore, Aristotle's notion of practice is normative in ethics.[1] Of course, for Aristotle, the praxis is far inferior to the theory and clearly differ from the latter. The tradition from Aristotle has obviously influenced the understanding of theory, praxis, and their relationship in later generations. It makes separation between theory and praxis, which makes theory superior to praxis. It neglects that praxis is the foundation of theory and theory is not the only purpose of human. Theory still contains the purpose and need to solve the problems. What we most concerned here is the relationship and distinction between theory and practice, made by Aristotle. According to Xu Changfu who has done an in-depth study on Aristotle, the virtue of theory is contemplation, while the virtue of practice is act. While theoretical knowledge can be acquired merely by consideration of universality, practical knowledge must be acquired by specific operation. The significance of theoretical science is to provide knowledge. Though practical science provides knowledge, the significance of practical science for people is to live well, rather than to acquire knowledge (Xu Changfu, 2004, p. 59). In the era of Aristotle, the purpose of the distinction between the practice and poiesis is to distinguish telos and activity as means. According to Aristotle poiesis is the means to achieve the end. In social sense, distinction between praxis and poiesis correspond to masters and slaves, employers and craftsmen.

What is the reason for such a division of Aristotle? I think Xu Changfu's explanation is well. In fact, people may live in two worlds at the same time, one is the world of physics, that is the field of nature; the other is the world of nomos, that is the field of autonomous living. Therefore, man has always had a dual nature: being-in-itself and being-for-itself. Moreover, its nature of being-for-itself also needs to be based on the nature of being-in-itself. Thus, people need a kind of activity to overcome the bondage of natural necessity, and to create conditions for autonomous living. Such an activity is not the contemplation of nature, but an action on outside nature through inside nature and its extension. In this way, the activity produced by natural necessity is first to seek the survival materials and expansion of the living area. Such an activity is the conscious activity, which is forced by the natural necessity. Or it is creating freedom, but it is still in the stage of trying to break out of its necessity. Thus, it is only a means of freedom. However, due to the openness of activity and the world, natural world is related to the theory and poiesis of human, and the human world is also related to the praxis and poiesis, and the area of telos related to the theory and praxis at the same time. Thus, this becomes the intrinsic motivation for Aristotle's three classifications of human activities (Xu Changfu, 2004, pp. 59–60). The conception of Praxis has always transformed in the era of modernity. The content of Aristotle's poiesis has now been taken into the understanding of Practice.

Gadamer made a further expansion of Aristotle's notion of praxis in the modern sense. For example, when Gadamer discussed Aristotle's philosophy of practice, he pointed out that people must be aware of the word 'Praxis' first. The concept of praxis should not be understood narrowly, for example, it cannot be understood as the practical application of scientific theory. Gadamer pointed out that the opposition of theory and praxis makes the praxis and practice of theory totally different. To be sure, the practice of theory belongs to praxis.[2] But it is not everything. Praxis means more. It is a whole, which includes our practice, activities, and the self-adjustment of humans in this world – which is to say, it also includes our political, political consultation, and legislative activities. Our practice – is our form of life

(Lebensform). 'Praxis' in this sense is the theme of Aristotle's philosophy of practice (Gadamer, Jin Huimin (trans.), 2005, p. 7). Gadamer's hermeneutics of practice has been the important resource of hermeneutics of scientific practice raised by Rouse. We can find it in *Knowledge and Power*.

1.1.2 Practice in Kant

After Aristotle, German philosophers, influenced by the Greek philosophy especially the philosophy of Aristotle, have attached great importance to practical problems, especially Kant. For example, he distinguished 'speculative reason' from 'practical reason'. In *Critique of Judgment*, he further distinguished two kinds of 'practice' concept, one of which is 'approaching the nature', the other is 'approaching freedom'. The latter related to the ethical and political activities.

There is a difference between the practice of concept of nature and the practice of concept of freedom in Kant. In the introduction in *Critique of Judgment*, Kant said that:

> If one divides philosophy, insofar as it contains principles of rational cognition of things by means of concepts (not merely, like logic, principles of the form of thinking in greenmail without distinction of objects), into theoretical and practical, as is customary, then one proceeds entirely correctly.... There are, however, only two sorts of concepts that allow an equal number of distinct principles of the possibility of their objects: namely the concepts of nature and the concept of freedom.... Thus, philosophy is justifiably divided into two parts, namely, the theoretical, as natural philosophy and the practical, as moral philosophy. Hitherto, however, a great misuse of these expressions for the division of the different principles, and with them also of philosophy, has prevailed: for that which is practical in accordance with the concepts of nature has been taken to be the same as that which is practical in accordance with the concepts of freedom; thus, under the same designations of theoretical and practical philosophy, a division has been made through which, in fact, nothing has been divided.
> (Kant, 2000, p. 59)

For Kant, there is an essential difference between practice which referred to the concept of nature and the concept of freedom. What's the difference? The notion referred to nature is technically practical, while the notion referred to freedom is morally practical. The technical practice belongs to the theoretical philosophy (as a doctrine of nature), because

> all of these contain only rules of skill, which are thus only technically practical, for producing an effect that is possible in accordance with natural concepts of causes and effects which, since they belong to theoretical philosophy, are subject to these precepts as mere corollaries of it

(of natural science), and thus cannot demand a place in a special philosophy which is called practical.

(Kant, 2000, p. 61)

On the contrary, the moral practice is 'founded entirely on the concept of freedom, ... which do not ... rest on sensible conditions', 'but on a suppressible principle, and require a second part of philosophy for themselves alone, alongside the theoretical part, under the name of practical philosophy (Kant, 2000, p. 61).

Therefore, Kant admitted that theoretical philosophy (including nature science) can be practiced, and its principles come from theoretical knowledge of nature (being the principles of technical practice). However, the practice in the strict sense belongs to the scope of practical reason; it is a practical activity under the guidance of the moral law which belongs to ontological domain. A persuasive idea is to take the notion referred to nature also as practice, which belongs to the field of either ontology or epistemology. It is the practical activities under the guidance of human cognition. They both have differences and connections.

I think the current views develop and transform the tradition from Aristotle to Kant, which have changed the people's views about the field and basic essence of practice. However, this development is at the expense of losing Aristotle's political aspect and Kant's moral practice. Reconsidering Aristotle and Kant is a supplement to the current views about fields and essence of practice, which makes the practice not only have the natural dimension, but also dimensions of social politics, morality and value. This makes meaning of practice much richer, which have the content of interaction with men and social norm as well as nature. This is the explanation resource which combining nature and society that the new scientific practice precisely requires.[3]

1.2 Marx's notion of practice

Practice is the basic characteristic of Marxist philosophy, which is identified by most of researchers on Marx. This kind of practice can be understood from two aspects. First, it takes practice as its foundation, and the practice is not only the basis of its epistemology, but ontology. Second, it concerns the practical issues of social reality, rather than the pure research on issues of thought. The fundamental task of Marxist philosophy is to change the world or discuss how to change the world with practice. In Marx's *Theses On Feuerbach*, he criticized the materialism, and pointed out that 'The chief defect of all previous materialism – that of Feuerbach included – is that things, reality, sensuousness, are conceived only in the form of the object, or of contemplation, but not as human sensuous activity, practice, not subjectively'. He insisted that 'it does not explain practice from the idea but explains the formation of ideas from material practice'.

In *Theses on Feuerbach*, there are famous words that 'Philosophers have hitherto only interpreted the world in various ways; the point, however, is to change it'. This shows that the difference between Marxist philosophy and the philosophy of the past is that Marxist philosophy is not only the interpretation of the world, but also to change the world resorting to practice. Here, the interpretation and practice are not opposite, but interactional and including with each other. In Marx, interpretation and transformation are not opposite. In the term of practical hermeneutic, the interpretation of the world and the practice of changing the world cannot be separated from each other.

From the beginning of 1980s, a major point of views in the study of Chinese Marxist philosophy, is to argue that the Marxist philosophy is a kind of practical materialism or practical ontology. Some scholars even describe Marxist philosophy as 'Practical Hermeneutics' (Wujin Yu, 2001). However, there are many controversies about it. Some people think that Marx's concept of practice refers to the 'sensuous activity' or 'objective activity'. Other people think that Marx's practice mainly refers to the material production activities of human beings. It is the objective material activities of human beings, which has the purpose of transforming the world consciously. And one of the most important activities is the human activity on nature. Some scholars criticized such understanding and pointed out that the definition of interpersonal actions (social and political) in Aristotle's practice is often ignored in the research on Marxism (Zhang Rulun, 2005, p. 160).

Some scholars believe that there has been a long tradition of practical philosophy in the western philosophy, and Aristotle is the founder of this tradition. And Marx's philosophy of practice is the inheritance and the creative transformation of this tradition. And various kinds of modern western philosophy of practice are the new states of such a tradition (Xu Changfu, 2004, p. 61). Just as previously noted, Aristotle divided human activities into three kinds: theory activity, praxis and poiesis. The related researches in the later ages can be regarded as the repair, split and transformation of these three activities. For example, by the way of investigating scientific and technological activities, Francis Bacon' work can be treated as a successful instance which combine Aristotle's theory activity with poiesis activity, and of course, this is also a reactionary of the meaning of Aristotle's praxis.

In this sense, Marx's concept of practice can also be regarded as a creative transformation of Aristotle's activity of trisection. In Marx, the most obvious difference is that praxis and poiesis is no longer distinguishable. Poiesis is the concrete form and expansion of praxis, and praxis is the essential content of poiesis. Another major innovation of Marx's concept of practice is to put production practice as its primary content, so as to form a systematic set of Marxism. For example, all other practical labours are placed on the basis of production practices; all the human society and its historical products, including material life, economic life, political life, and even the spiritual life (of course including cultural life) are regarded as the productive achievements of human practice activities. And all kinds of institutions, such as the economic system of capitalism, are treated naturally as the specific mechanism to develop the potential of human practices and the necessary link to achieving a certain ideal state. In this way, Communism is such a life form and the social formation that the potential of production of human practices can obtain complete liberation.

Of course, recently, some scholars have criticized this kind of interpretation of Marx, especially the interpretation of the concept of praxis in Marx. They have criticized the standpoint which regards the production practices and the daily practices of human beings as the primordial content of the concept of praxis, which is popular in the Marxism. They consider that Marx's concept of practice first refers to expanding relationships among people. Of course, Marx pays most attention to revolutionary practice, it is 'a question of revolutionizing the existing world, of practically coming to grips with and changing the things found in existence'.[4] Of course, revolution is only a part of the general practice of material life, one special part. As for Marx, there are only historical nature and natural history, thus there is no nature opposite to the human or their history. Therefore, it can be considered that the practice of material life and production is necessary and even the most important human practice. Marx starts from here and considers that other patterns of practice are consequently determined by it. In this way, Marx and Engels regard the material production and life as the basic form of practice. It can be treated as a transformation of Aristotle's form of practice, but also can be regarded as a deviation from Aristotle's notion of practice. If we put Marx into the era, this is the consequence of the development of the era, and it is also the consequence of such a reality that the development of science and technology has gradually become an important content of the practice.

There is also a characteristic of Marxist practice, which is great momentum. Marx's practice is similar to a social movement. The person in Marx's philosophy is a class, so Marxist practice is about the practice of this class, not concrete practice. Especially in Historical Materialism and Dialectical Materialism, the human being has become abstract. This characteristic also becomes one kind of shortcoming in the research, because research on practice can't be deeper. This is the failing of Marx's philosophy on practice, but also the need for supplement by other philosophy. Of course, Marx's mission is the liberation of the entire human society. He and his colleagues cannot just focus on the scientific practices. We cannot make excessive demands on Marx.

From the view of practical hermeneutics, Marx's views on practice and the role of the practice on hermeneutics have contributed to the Hermeneutics. Those views are:

First, practice is the basis of all understanding and interpretation activities (Wujin Yu, 2001, pp. 82–5). Wujin Yu thinks that this is Marx's first contribution to the Hermeneutics. According to Wujin Yu: a) Marx pointed out that all the understanding and interpretation of the activities originated in practice. Because all of basic conditions of understanding come from practice (interpreter's living, sound mind and the language); b) Marx pointed out that all understanding and interpretation activities are to point to practical activities from the content point of view. In ***The outline of Feuerbach***, Marx said: 'All social life is essentially practical. All mysteries which lead theory to mysticism find their rational solution in human practice and in the comprehension of this practice'; c) Marx pointed out that all the activities of understanding and interpretation are serving the people's survival practice activities. So, Marx clarified the ontological premise of all activities of comprehension and interpretation by introducing the concept of practice into Hermeneutics.

Second, the historical nature of practical activities is the basic feature of all understanding and interpretation activities (Wujin Yu, 2001, pp. 85–8). Wujin Yu studied the characteristics of Marx's historical materialism. And He pointed out that, in the sense of practical hermeneutics, Marx regarded any practical activity as a real person's activity in the given historical conditions. And the historical nature of this kind of practice inevitably lead to the historical activities of understanding and interpretation. For instance, a) morality, religion, metaphysics, and other forms of consciousness have been stripped of the appearance of independence after revelation of their practical basis; the history of concept is also the history of basic practices. b) The ruling class occupies the dominant position in the material practice, and their thoughts and ideas are bound to occupy the dominant position in the understanding and interpretation activities. It is a kind of hermeneutics of power. Marx was aware of the inherent relationship between power and interpretation activities far earlier than Nietzsche and Foucault. c) In modern times, the historical characteristics of alienated labour will inevitably have a profound impact on the understanding and interpretation of modern society.

Third, Ideological criticism is the correct way to entering the circle of hermeneutics (Wujin Yu, 2001, pp. 88–9). Marx understood historical activity of understanding and interpretation. He does not reject this historical character. So, how we enter the circle of hermeneutics has become an important issue. Marx believed that the ideology of a certain period constituted a general background of the activities of understanding and interpretation in this period. Therefore, when understander and interpreter put themselves in this context, they cannot find their preconception's problems if there is no awareness of this kind of background and without deep reflection and criticism of ideology.

Forth, deconstruct 'independent realm' of linguistic illusion, reveal the origin of language in the practices of human existence (Wujin Yu, 2001, p. 89). This shows that it is not enough to discuss the activities of understanding and interpretation in the linguistic turn. Language is closely related to the human practice, and greatly influenced by the practical situation.

Fifth, the introduction of Hermeneutics' methods (Wujin Yu, 2001, p. 90). According to Wujin Yu, Marx advocated two methods, the first method is reduction (of course, it is related to that researcher is now standing in the position of Phenomenology and Hermeneutics to review Marx). It is much like the phenomenological reduction because that Marx not only reduce it to the text, but also reduce the text to the real life which is the activities of survival practice itself. The second method is archaeology. History itself is the premise and foundation of activity of understanding and interpretation. Of course, I think that the explanation here has gone beyond Marx. When philosophers of science quote Wittgenstein, they are not to be faithful to Wittgenstein, but to start from Wittgenstein. Similarly, I think that for the application of Marx's practice, we can develop the new practical interpretation of the present age without distorting the original meaning of Marx's.

1.3 Scientific practice and Marx's practice: connection and difference

In fact, Marx's practice is also an indirect source of PSP in spite of the fact that Rouse repeatedly distinguished between PSP and practical materialism. According to Wujin Yu, Many of Marx's ideas in practical hermeneutic are in great agreement with the views of Rouse and Heidegger. Rouse's predecessor, Heidegger, was fully aware of the important effect of Marx on practical issues, whether or not Rouse has absorbed Marx's idea.

Of course, other researchers of scientific practice are not unaware of the impact of Marx. Confined to the length, we primarily discuss the influences that Marx's practical thoughts have on the sociology of science, SSK, and PSP.

1.3.1 From Marx to STS:[5] practical thoughts embodied in social research of science

Marx's concern for the scientific practices and its consequences are known to all. Of course, Marx's concern for science and its progress is the need for social transformation. Marx and Engels have always regarded scientific progress as the result of the biggest driving force – social economy. Science is also the locomotive of historical progress.

Obviously, according to Marx, science and mathematics are not eternal and universal 'transcendental objects' in Plato's kingdom, and are not found in some way. They are neither product of 'pure' mental activity nor indent of 'genius'. In any social forms, dominant cognitive style comes from practical activities, and is compatible with the dominant mode of production and social benefits. This is indeed the most accepted view of Marxism about the science in society (Jasanoff, 2004, p. 102). And these views are exactly same with the thought that social factors are implicated in science emphasized by sociology of science and Sociology of knowledge. For example, they think that all knowledge is mediated by social practice, culture and history. And SSK or post SSK think that social practice, culture and history are co-evolution with all knowledge. All of them are the most prominent ideological basis in the research of science as practice and culture and the SSK.

In fact, STS do not deny the impact of the Marxism tradition.

First of all, Marx's views of practice influenced the scientific sociology. Marx's insight about social origin of science and technology is always one of resources of scientific sociology. Marx asserted in 1847:

> The same men who establish their social relations in conformity with the material productivity, produce also principles, ideas, and categories, in conformity with their social relations. Thus, the ideas, these categories, are as little eternal as the relations they express. They are historical and transitory products.

Bernal summarized Marx's contribution to understanding the relationship between science and the social as:

> The value of Marxism is as a method and a guide to action, not as a creed and a cosmogony. The relevance of Marxism to science is that it removes it from its imagined position of complete detachment and shows it as a part, but a critically important part, of economic and social development. In doing so it can serve to separate off the metaphysical elements which throughout the whole course of its history have penetrated scientific thought. It is to Marxism that we owe the consciousness of the hitherto unanalysed driving force of scientific advance.

Bernal even pointed out that: 'Already we have in the practice of science the prototype for all human common action' completely in accordance with Marx.

The influence of Marx's practical and historical viewpoint on western scholars is also strengthened by the Soviet scholar's analysis of scientific history with the view of Marxism. In 1932, when the Soviet delegation attended the International Conference on science history, they submitted a paper which the influence of economy and society on science in the age of Newton is analysed from in the Marxist point of view. Hessen's paper 'The Social and Economic Roots of Newton's Principia' had the most extensive influence.[6]

> In England, interest in dialectical materialism dates effectively from the world congress on the history of science in 1931, which was attended by a strong Russian delegation, who showed what a wealth of new ideas and points of view for understanding the history, the social function, and the working of science could be an were being produced by the application to science of Marxist theory.

Bernal even thought that: 'Hessen's article on Newton, which it contains, was for England the starting point of a new evaluation of the history of science'. And Norman W. Storer, the editor of 'The Sociology of Science', also pointed out that it affected a large number of researchers: 'its influence was mostly visible not in Stalin's Soviet Union, where Hessen soon disappeared from view, but in England, where it appeared in the far more discriminating historical work of scientists on the political left, such as Joseph Needham, J.D. Bernal, Lancelot Hogben, and J.B.S. Haldane, and in the rebuttals by such historians as Charles Singer, G.N. Clark, and Herbert Butterfield. In the United States, Hessen's essay and Clark's criticism of it were both taken into account in Merton's monograph'.

Merton, a landmark in the sociology of science, was also deeply influenced (by Marx's view of science and practice via Hessen.) by Hessen. Thus, he was influenced by Marx's view of science and practice. Merton admitted that his 'Science and Technology and Society in Seventeenth Century England' was influenced by Hessen's ideas and methods. In the addendum of Chapter 10, Merton defended Hessen when he mentioned a Clark's paper which think of Hessen's paper 'over-simplifies the social and economic aspects of the science of this period'. Merton pointed out that 'preceding three chapters of the present study, despite certain differences of interpretation, are heavily indebted to Hessen's work'. Merton did mention Hessen in his papers and writings many times.[7] Especially in Chapter 7, he pointed out that he would follow closely the technical analysis of Professor B. Hessen to discuss some technical and scientific problems raised by certain economic developments. He said, 'Professor Hessen's procedure, if carefully checked, provides a very useful basis for determining empirically the relations between economic and scientific development. For example, in Chapter 9 (interactions of science and military technique), Merton let people compare the paper by Hessen, and thought that this chapter is heavily indebted to Hessen. And the empirical classification in the Chapter 10 was also inspired by the Hessen and learned from the study of the Hessen. Of course, Merton also criticized the tendency of classing science.[8]

Marxian also influenced SSK and social studies of science later. For example, Merton also pointed out that Marx and Engels's views of science and practice had an important influence on the sociology of knowledge. Merton argues that 'Marxism is the storm center of *Wissens-soziologie*'. Marxist thought is always the focus of the sociology of knowledge: 'Paradigm for the Sociology of Knowledge'. Merton discussed the Sociology of Knowledge from Marxism to Mannheim. Merton also pointed out that 'the most distinctive feature of the Marxist sociology of knowledge imputation of function is its ascription, not to the society as a whole, but to distinct strata within the society. This holds not only for ideological thinking but also for natural science'. Merton meant that science and technology are tools of the ruling class for control. If it was a questionable proposition or view in the Merton era, especially those scholars who advocated scientific autonomy, it is not only the content of their theory itself, but also what they have been trying to prove for Heidegger, Foucault and Rouse. At this point or in some sense, Marx is their pioneer.

For example, in famous 'Handbook of Science and technology studies' (Jasanoff, 1995), Marx and Marxist contribution are mentioned many times. In Chapter 5 'The Theory Landscape in Science Studies', written by Sal Restivo, there is a special section 'Marxist Science Studies' discussing Marxist ideas of social studies of science. Although its conclusion has the characteristics of simplicity, it also objectively shows that Marx's social research of science is an important part of sociology of science and social research of science. In addition, in the discussion of social constructivism, it believes that one of the classical sources of social constructivism can also be related to Marx (Jasanoff, 1995, p. 103).

1.3.2 Isomorphism of practical activity and concept of construction[9]

When we apply the social constructivist theory of scientific practice to the laboratory research or observe the laboratory with the view of Sociology, we find that there is no original, bare 'fact'.

> Constructionist studies have revealed the ordinary working of things that are black-boxed as "objective" facts and "given" entities, and they have uncovered the mundane processes behind systems that appear monolithic, awe inspiring, inevitable.... Constructionism was the answer laboratory studies gave to the micro-processes they observed in realtime episodes of scientific work.
>
> (Knorr-Cetina, 1995, p. 148)

The basic idea of 'construction is Constructionism' ... just asked that 'reality', or 'nature', be considered as entities continually re-transcribed from within scientific and other activities (ibid., p. 149). If we compare it with Marx's concept of activity in this sense, what are the differences and connections between them? Let us look at Marx's classic discourse on the concept of practical activity again:

> The main defect of all hitherto existing materialism, that of Feuerbach included, is that the Object, actuality, sensuousness, are conceived only in the form of the object, or of contemplation, but not as human sensuous activity, practice, not subjectively. Hence it happened that the active side, in opposition to materialism, was developed by idealism – but only abstractly, since, of course, idealism does not know real, sensuous activity as such.
>
> But what distinguishes the worst architect from the best of bees is this, that the architect raises his structure in imagination before he erects it in reality. At the end of every labour process, we get a result that already existed in the imagination of the laborer at its commencement. He not only effects a change of form in the material on which he works, but he also realizes a purpose of his own that gives the law to his modus operandi.

After careful analysis, we will find that the concept of construction in constructivism has some similarities with the concept of practical activities mentioned by Marx. According to Marx, practice is the activity of reforming the world whose character is subjective effects of objective. Subjective effects of objective is subjective initiative, but what is construction? According to Knorr-Cetina construction are object and its process was changed by the activity of the actor. There is no bare fact in the process which people deal with real. No matter whether Marx's concept of practice or Knorr-Cetina's notion of construction contains the same theme, Practical activity is not a passive adaptation to the environment. People or actors affect formation of facts and practice model people and the surrounding environment. This is constructionism, and in fact, it is the core of Marxist practical activity.

1.4 Practical research on SSK in early and late stages

People think that the rise of SSK in the philosophy of science just opened a new dimension that is to identify the scientific knowledge is related to the social dimension; at the same time, it opens another dimension: dimension of practice.

Early SSK put forward the famous strong program in theory, in order to showing the social construction of scientific knowledge. It not only demonstrates that the truth of scientific knowledge is related to social context, but also try to prove how science and scientific research is to evolve. The latter part of SSK has made a careful study of the scientific laboratory research from practice and experience. In Pickering's book *The Science of Practice and Culture*, article 'From science as knowledge to science as practice' makes a clear statement of the two characteristics of SSK;

> First, as its name proclaimed, SSK insisted that science was interestingly and constitutively social all the way into its technical core: scientific knowledge itself had to be understood as a social product. Second, SSK was determinedly empirical and naturalistic. Just how scientific knowledge was social was to be explored through studies of real science, past and present. The apriorism of normative philosophical stereotypes was to be set aside.
> (Pickering, 1992, p. 1)

We cannot the blame experience approach of SSK loses the normative of philosophy. Because SSK research itself is a kind of research approach of practical sociology. In addition, SSK provides abundant experiences to the philosophy of science, and this indeed gives the transcendental philosophy of science a blow, those a priori, beautiful norms, in the face of real scientific research cases and strong practical persuasion, revealed the false normalization. They are castles in the air, no real explanation. Real science requires us to put forward a better norm for science and research based on SSK. And the establishment of a new norm must first return to the basis of the experience and practice of scientific research.

In the later period, the scientific practice of SSK has a very rich content and many kinds of approaches, among which the more classical ones are:

1. B. Latour and S. Woolgar's *'Laboratory Life'*. They intervened laboratory life in an anthropological way, which examines the scientists' discursive and scientific practices in the laboratory of life. This shows that the practice of the scientist is not only considering the forces of nature, but also the role of social forces. Their work, 'laboratory life', is the result of this study. Later, Latour cooperated with Michel Callon, and put forward the 'Action Network' research approach based on knowledge dissemination. This study lays the foundation for the Paris school. Of course, Latour and Woolgar took laboratory as machine inscribed, its main role is inscribed. They believed that the main role of scientific research is to characterize. Behind it also implies the influence of representationalism.

2. Knorr-Cetina's 'Laboratory Practice and Cultural Studies'. Her study focused on the role of scientific instruments, equipment and other physical force. Based on practice, Knorr-Cetina constructed a laboratory culture, and discussed the scientific laboratory as a unit of cultural characteristics. She adopted participatory observation of laboratory study, and integration into the anthropology of science and technology. This has promoted research and reflection on the work of science, and also contributed to discursive practices such as the development of practical research tools. Knorr-Cetina made people realize that laboratory is also a theoretical notion in an emergent theory of the types of productive locales for which laboratories stand in science. In sociology in general, localizing concepts are often associated with the small scale and the weak. Laboratory studies shed light on the power of locales in modern institutions and raise questions about the status of 'the local' in modern society in general (Knorr-Cetina, Jasanoff, et al., 1995, p. 142).

3. Lynch, Livingston, and Garfinkel (1983; Lynch, 1985; Livingston, 1986) studied the scientific practice from the perspective of ethnomethodology. And almost at the same time, another anthropologist Traweek (1988) was using this method and symbolic text analysis method to study the scientific practice of particle physicists at the Stanford Linear Particle Accelerator Center. The ethnomethodological concern with the fine-grained analysis of daily practices, have reinforced the interest in detailed description of scientific work, such as the 'embodied', 'circumstantially contingent' and 'unwitting' character (Knorr-Cetina, Jasanoff, et al., 1995, p. 148, 156). Fujimura's research can also be regarded as practice research for the anthropological approach. Through the study of on the changes of cancer gene research in the twentieth century, in particular, she pointed out that a technology standardization and the concept of flexibility make different subject object can use them to allow objects to be integrated on the border.

In the UK, according to Pickering's view, Gilbert, Mulkay, and Woolgar' discourse practice analysis, created the research of SSK studying itself, also further developed the concept of discursive practices.

For us, the most important thing is what the practical concept or practice thought of Constructivism in SSK. Through Cetina's analysis, we can understand the practical concept and thought of this kind of SSK constructivism.

According to Knorr-Cetina, constructionism has the following basic characteristics: first, with regard to the material world of independence and structural problems, constructionism does not solve the problem, just transfer the problem from the stage of the philosophical argument to the stage of empirical research, simply said, that is turn to the practice of research. Second, the Epistemological Characteristics of constructionism are the empirical knowledge of science and the social world. Constructivists advocate the practice of seeking truth and practical knowledge in the study of the real process of knowledge production. The third characteristic of constructionism as it emerged from laboratory studies is the emphasis placed in the respective studies upon the phenomenon that knowledge is worked out, accomplished, and implemented through practical activities that transform material entities and potentially also features of the social world (Knorr-Cetina, Jasanoff, et al., 1995, pp. 148, 150).

Pickering has a deeper understanding of the SSK concept of scientific practice, he pointed out sharply, SSK, as its name suggests, put perspective on knowledge that is, however, typically underwritten by a particular vision of scientific practice that goes broadly. Practice is a kind of service means, and it is the creative extension of the conceptual net to fit new circumstances. With the help of Callon's 'actor network', some scholars of SSK followed the thought of Ludwig Wittgenstein and Thomas Kuhn, think that the extension of the net is accomplished through a process of modelling or analogy, that modelling is an opened process, can plausibly proceed in an indefinite number of different directions; nothing within the net fixes its future development. Of course, these ideas are of great significance, however, it brings a problem to SSK, open endless. There should be many kinds of Science or numerous possibilities for development. But why science today still reflects some kind of unity? So SSK turn to social benefits to end the debate and the development of divisive, think, best expansion results of the concept of network is embodied in accordance with the best interests of the scientific community. Pickering believes that this is SSK's description of the practice. And this description is very narrow and inadequate (Pickering, 1992, p. 4).

Pickering further analyses the positive meaning and negative meaning of practice in the sense of SSK. Thus, we learn more about the concept of scientific practice in SSK, including its meaning and the significance of scientific research. By interpreting Pickering's view, we can see, the positive meaning of the concept of scientific practice in SSK is: SSK through scientific practice illustrates the operation of the conceptual network and the relationship between operating results and social construction. Scientific practice is to play the supplementary description function in the description of the scientific operation. Compared with the traditional philosophy of science, scientific practice has been used in the description, it acquired the status and significance. The negative meaning of the concept of scientific practice in SSK is: if SSK seen as scientific practice and scientific culture actual picture or normative picture, rather than just a supporting explanation to understand knowledge, then this is definitely a problem. This concept is wrong, its error is that it is still too ideal. Too simple, does not help us to grasp the complex situation of science and technology development. In particular, the mainstream of SSK takes the practice process as a process of interest negotiation. Too much of the reduction, of course, this kind of reduction is not a natural dimension but a social dimension (see Pickering, 1992, pp. 4–5).

In the early and middle period of SSK, the discussion on scientific practice is more abstract. Later, SSK scholars began to care about the physical dimension of scientific practice. Of course, this physical dimension in the SSK usually expressed as social benefits and social relativity, latter SSK study realized that we cannot use the social dimension to replace the physical dimension, the physical dimension also includes the role of nature. After the start of the study, to SSK unexpected is that, this dimension once launched, immediately open up on multiple dimensions: first, the in-depth study of the discourse and the text of the laboratory not only opens the door to the study of discourse practice, but also enriches the cognition and content of the concept practice. Second, the interpretation of laboratory research activities opens up the dimension toward hermeneutics. Third, the research about the laboratory material force action associated with the neo-experimentalism's scientific practice approach. It is precisely because of those points that the PSP get a lot of resources from the SSK.

1.5 The thought of practice in the tradition of hermeneutics in the modern continental philosophy

In the philosophical reflection on science in mainland Europe, hermeneutics gradually goes into the philosophy of science, plays the role of philosophy of science, or becomes the Continental philosophy of science. This includes hermeneutic tradition. And in such hermeneutics tradition, which goes through by Husserl, Heidegger and Gadamer, and of course the later Wittgenstein, and other scholars on phenomenology and Hermeneutics (Heelan, 1983; Babich, 2002), the hermeneutic implication in practice is getting more and more stronger.[10] And this constitutes the direct source of hermeneutics for modern PSP. This section, mainly based on Rouse's comments on Heidegger's practical hermeneutics, will discuss Heidegger's practical hermeneutics, which has a direct impact on PSP.

In the most three important books, Rouse quoted a large number of Heidegger's phenomenological views of hermeneutics, and he pointed out that Heidegger and Kuhn are the direct precursor of his PSP. However, Rouse did not cite Heidegger's views wholly intact in his philosophy of science. In order to criticizing the theory-dominant views, Rouse attributed the practical hermeneutics to Heidegger on the one hand, to criticize the tradition of theory hermeneutics by virtue of Heidegger's practical hermeneutics; on the other hand, he also criticized the of theory-dominant views in Heidegger's hermeneutics, and implemented the tradition of practical hermeneutics more profound and more thorough, which in a certain sense develops the practical hermeneutics.

First of all, according to Rouse, he believes that Heidegger described two fundamentally different senses of 'hermeneutics' in *Being and Time*. In the first part of the book, Heidegger regarded human existence itself as hermeneutical. The ways in which we are being in the world embodies a certain interpretation of the world and ourselves, which can itself be elucidated by the interpretation of our everyday practice. Therefore, the attempt to disclose the significance of our practices and the practices themselves are both hermeneutics.... In the second part, Heidegger explored a deeper version of hermeneutics. Such a deeper interpretation intends to reflect an apparent discovery of our everyday interpretation, which is what we make of ourselves and the world, and which reflects an attempt to disguise the lack or 'uncanniness' of the grounding of these interpretations. In this latter sense, hermeneutics represents the unmasking of such disguise, and permitting an 'authentically resolute' existence, which faces up to the uncanniness of being-in-the-world rather than fleeing from it (Rouse, 1987, p. 58). According to Rouse, in his later works Heidegger still insisted on the first account of everyday activity as hermeneutical, but abandoned the second sense of hermeneutics as an unmasking of a hidden truth. Thus, Rouse points out that although Heidegger's hermeneutics is the practical hermeneutics in general, rather than a theoretical one, but there are still some theory-dominant senses when it refers to science.

Secondly, what are the major points of Heidegger's practical hermeneutics? According to Rouse, it mainly includes as follows:

1 Everyday practices have reflected the interpretations of the world, which is the best interpretations of the life world; such a recognition makes us to pay more attention to the relationship between everyday practices and scientific practices.
2 In the process we do, we have interpreted ourselves and the world, our everyday practices and our bearings as we engage in them have make our present state, and repeatedly remake us. Therefore, the most important things are the reconstruction of the practices of doing and its significance.
3 Our practices and the interpretations they embody hang together. Any particular activity acquires its significance and its intelligibility by the coherence of practices, roles, and the equipment to which it belongs.
4 Our possibilities emerge from the way of being in the world, not from the fundamental beliefs.
5 Understanding is always local, existential knowledge; it thus is not a conceptualization of the world, but a performative grasp of how to cope with it.

The first sense of hermeneutics roots on a prior understanding of 'what do our everyday practices mean'. It even regards everyday practices as an interpretation itself. Why our everyday practices embody an interpretation of the world? What does it mean? In Heidegger's view, to begin with, we take account of things around us in various ways. We use equipment with acquiring an orientation, a focus, a significance, a function. Both what things are and how they are show up in the ways we deal with what surrounds us. Thus, we interpret ourselves and the world in what we do and how we do it; our everyday practices and our bearing as we engage in them make us what we are and repeatedly remake us (Rouse, 1987, p. 59). Second, according to Heidegger, space and time are organized by the way we live and we do within them. In sum, our everyday ways of engaging the world exhibit a style, which is itself an interpretation of what it is to be and how things are in the world (Rouse, 1987, p. 59). Third, why can we say that our practices and the interpretations they embody hang together? According to Heidegger, Rouse believes that we should consider this question through the dependence our practices have on the equipment and the dependence among equipment. Heidegger pointed out that:

> Taken strictly, there 'is' no such thing as an equipment.... Equipment always is in terms of its belonging to other equipment. These 'things' never show themselves proximally as they are for themselves, so as to add up to a sum of realia and fill up a room. What encounter as closest to us (though not thematically) is the room. What we encounter it not as something 'between four walls' in a geometrical spatial sense, but as equipment for residing. Out of this the 'arrangement' emerges, and it is in this that any 'individual' item of equipment shows itself.
>
> (Rouse, 1987, p. 60)[11]

Rouse thinks that Heidegger put the contextuality involved in practices in the first place; Heidegger is trying to articulate a sense of contextuality as determinative of what things are without treating contexts as things themselves. Rouse points out that the contexts of roles, practices, equipment, and goals within which our activities function both guide and make sense of what we do (Rouse, 1987, p. 60). This elucidates that in fact contextuality is more important than the things themselves in the hermeneutical sense. Only in the contextuality, we own the interpretation and we could interpret. Once the relationships between things and context can be articulated clearly, and the significance of the existence and evolution of things that in context can be revealed, it itself is the interpretation. In this case, the practical interpretations are not ineffable, but can be effable through the practice and its description, through the reveal of significance of practice, and through the significance of practice to the existence.

In order to making a connection between Heidegger's practical hermeneutics and the interpretation of scientific practice, norms of practice, Rouse tried to find the points from Heidegger which would be beneficial to him. For example, do practices have common norms or not? The solution to this question is actually the key to the interpretation of scientific practice that makes it normative and naturalistic. MacIntyre has attempted to articulate this contextuality of action in terms of shared 'schemata' for interpretation. Rouse find that Heidegger, however, would hasten to add that these schemata are not normally explicit prescriptions or rules. Most importantly, in *Being and Time,* Rouse find that among Heidegger's discussions about 'interpretation' on the issues such as the relationships between 'skills and practices' and 'beliefs, dispositions and rules', and the various possibilities when we face the future practices, there are various differences between practical hermeneutics and theoretical hermeneutics. One of the most important is that between theoretical hermeneutics and practical hermeneutics, there is a subtle but intrinsic difference: the former generally believes that we need some consistent theory presuppositions or general theory beliefs when the possibilities of the world emerge up. Quine just thinks so (Rouse therefore also puts Quine's theory as theoretical hermeneutics (Rouse, 1987, pp. 62–3)). However, practical hermeneutics believes that our possibilities emerge not from fundamental beliefs, but from a way of being in the world (Rouse, 1987, p. 62). Such a way of being in world is contextual and local. This interpretation offers a very good demonstration for the argument that the norms of practices come from practice itself and the situations of practice (practice itself is hermeneutics), and they evolve through practices and its contexts (for practical hermeneutics can be naturalistic). It also offers the fundamental preparation for the demonstration about the norms of scientific practices, which opens a door for understanding the locality.

The following is the main differences from practical hermeneutics to theoretical hermeneutics by Rouse concluding (Table 1.2).

Although Rouse's accounts of science absorbed the perspectives of Kuhn and New Empiricism, and the core part of Heidegger's hermeneutics, Rouse did not accept some Heidegger's views on scientific research. He said, 'It is ironic that despite my reliance upon central features of Heidegger's hermeneutics, I must begin by challenging the specific account of scientific investigation Heidegger provides in *Being and Time* …' (Rouse, 1987, p. 73). What does such challenge aim at? It aims at the theory-dominant views in Heidegger's early thought, the views of isolation of practices and theories, and the views which treat practices as subsidiary means of theories.

First, Rouse opposes that Heidegger isolated relationship between practices and theories by distinguishing readiness-to-hand from present-at-hand.

Heidegger puts forward the concept of 'readiness-to-hand' which is closely related to human being's practices and its context in the phenomenological and hermeneutical senses. And he insists that such state does not exhaust the ways things can be. He puts forward 'present-at-hand' in order to distinguishing from 'readiness-to-hand'. Such a state means that decontextualization of things can exist independent of the behavioural responses of persons within a

Table 1.2 The main differences between practical and theoretical hermeneutics by Rouse (1987, Chapter 3), concluding

	Practical hermeneutics	Theoretical hermeneutics
The context within which interpretation takes place	local context configured by equipment, persons, and physical setting	web of belief, universal
The location of interpretation	a configuration of presences and absences	a background of representations
Presuppositions	a form of life	theories
The form would take	a skilful knowing our way about in the world	theoretical knowledge of the world
Interpreters	embodied, situated within the world	a position over against the world we survey and represent
The disclose of interpretation	what it is to be, show it	what is the case, say it
The objects interpretation focuses on	what to be close	what happened

configuration of practices and functional equipment (Rouse, 1987, p. 74). Of course, Heidegger's view is distinct from the traditional view of science, as he also believes that science is the configuration of practices and only theoretical understanding is limited to science. That is, only in science we decontextualize things from the configuration of everyday functioning. Although scientific understanding is limited to sociality, but it disengages from locality and practical involvement in the ways of 'present-at-hand'. Heidegger believes he has found 'the ontological genesis of the theoretical attitude' like science (Rouse, 1987, pp. 74–5). But Rouse has derived the residue of the theory-dominant view of science. Do not say how readiness-to-hand change into present-at-hand, even if such momentary disengagement happened, how could practices of functional contextuality get replaced by 'mathematical projection of Nature' of a priori theoretical conception in a sudden? Practice is important, but here it is treated as a means that can provide a guarantee for theory. And a theory is a new way of facing world, which is different from practice; it comes from practice but disengages from the latter.

Second, in *Knowledge and Power*, Rouse's criticism about Heidegger's views and the theory-dominant tradition of these views mainly focuses on the following four aspects: 1) the misunderstanding of the practices of theoretical representation themselves; 2) the misunderstanding of the role and significance of experiments in science; 3) the misunderstanding of 'localization and contextualization'; 4) ignoring the social and functional context still dominates the acceptance of the most abstract theory. Next we are discussing these points.

1. Rouse thinks that Heidegger misunderstands the practices of theoretical representation themselves. He almost did not regard scientific research as an activity of doing. When he treats science as the decontextualized mathematical projection of Nature, Heidegger concerns about the conditions of theoretical cognition. He does not consider further how theoretical cognition develops. And in the activity, the problem is how to overcome this kind of activity characteristics. Heidegger's discussion on science is focused upon the interpretation of finished results and upon the relation between such interpretation and what takes place in everyday circumspective concern. The issue he takes up in not how we can do research in science but how we come to comprehend general theories (Rouse, 1987, p. 80). This is also the characteristic of philosophy of science with theory-dominant tradition. In fact, as long as we stand in the position of theory-dominant tradition, the scientific view of theory-dominance is very different from the scientific view of practice-dominance, the theory is opposite to the practice and is the ultimate goal of science, and practice is only a means to serve theory. However, as long as we can change the position and the perspective, and treat science as a kind of practical activity, then theories is not opposite to practices, but a special kind of various scientific activities. Because what we can see is not only the so-called final results of representational meaning of theoretical proposition, but a variety of activities engaged in the establishment of the theory. Such as thinking about how we can explain the practical problems encountered, and make it modelling, and whether the model we made corresponds to the problems we need to solve. Even the calculation of the proposition is a kind of historical, narrative and interactive activities related to the activities of various existing theories.

2. Rouse thinks that Heidegger misunderstands the role and significance of the experiments in science. In *Being and Time*, Heidegger implicitly assigns a secondary role to experiment within the sciences. Rouse thinks that Heidegger's discussion on science then proceeds to examine the genesis of such decontextualized theorizing while ignoring experimentation or laboratory work. Rouse believes such omission is significant and it suggests that though experimentation is of practical importance in the actual development of scientific knowledge, it can be largely neglected when we consider the philosophical or ontological significance of the sciences. When we understand theorizing, we seem to have understood what is essential about science (Rouse, 1987, p. 96). Rouse suggests three plausible accounts of this omission. First, Heidegger is only concerned to explicate how we understand theoretical assertions, and the problem Heidegger addresses

is not specific to the research process and activities, thus he mixes up someone not involved in science but tried to understand its claims and someone actively engaged in research. Second, he also accepts a prevalent view among philosophers that experimentation does not generate content apart from that provided by theory (Rouse, 1987, p. 96), therefore, there is no philosophical significance to care about experimentation. Third, the investigation of experimentation is inferior to the investigation of theory, as observation is theory laden and experimentation has no life of its own (Rouse, 1987, pp. 97–8). In the weaker sense, the investigation of theory is considered to be a consideration of experimentation. Theories are indeed important, as they can tell us where to look, what to look for, and what to make of what we see (Rouse, 1987, p. 100). However, the scientific view of theory-dominance at most regards the importance of observation and experimentation as a touchstone enabling us to test theory and to refine or correct it. In fact, the importance of experimentation is far from being this. Experimentation in science often takes its own course, exploring areas not yet theoretically articulated. Experimentation does more than fill in the gaps left by prior theorizing or check its adequacy, it also opens new domains for investigation and provides a practical grasp of that domain as a resource (Rouse, 1987, pp. 100–1).

3 Rouse's criticism of Heidegger's 'delocalization and decontextualization' is that in Heidegger's thought, he really thinks there is a kind of knowledge that can be universal and decontextual. Therefore, in such position, Heidegger's practical hermeneutics is not the real practical hermeneutics. Of course, because Heidegger has a huge influence, if we want to criticize him, we should choose an appropriate point of view, to do an appropriate conversion, otherwise it is difficult to convince others. Rouse indicates that Heidegger's theoretical projection of such decontextualization and delocalization could be better construed as a process of standardization of the local knowledge of science. Thus, the local scientific knowledge and practices can be reliably transferable to the new research contexts. Such transformation appears to be universal, or as Heidegger has said 'decontextualization'. And the project of local knowledge in order to rise it to universal knowledge and make it disengage from particular practical situations and interests is actually a consolidation of power, which enables power to be extended into new networks (Rouse, 1987, p. 79). For example, it is a case that the expansion of Western knowledge in modern times. Just think that if the East has always been strong, then whether the Eastern knowledge will become today's standardized knowledge? Since Eastern knowledge can be called local knowledge, is the Western knowledge not the local knowledge?

From Rouse's point of view, Heidegger agrees to defend science with social situation, but he does not insist on consistently. From the view that science is a practical activity, scientific research is an involved, practical activity rooted in the skilful grasp of specially constructed local situations, typically laboratories (Rouse, 1987, pp. 119–20). In fact, such situations are situated in society, or at least involve social situations. This is because the intelligibility of the activities of scientists and their results rests upon the anticipated response of others, and the discourse practice and the material practice among scientists associate with each other. According to Heidegger, science is an activity which belongs to *Dasein*, an interpretation from a notion of behavioural self-adjudication.

In science, the standards of rationality are not individual but social. They are embodied both in the formal institutions and informal interpersonal network (such as actor network). Scientists are disciplined constantly by the norms established by both the scientific community and the interaction between the community and the society. What's more, through repeatedly ask their predecessors and co-workers what they can rely upon, what they must make account of, what they can ignore or must do, scientists grasp the norms of scientific practice, or disciplined by these norms; at the same time, the formal or informal evaluations of their co-workers also adjust the rational or legal standards. In some cases, new scientific practices conflict with existing norms, or do not match them well, or there may be no standards of practice. Accordingly, these standards evolve, improve, or are even abandoned in practice. These practical standards of rational acceptability are a clear example of the behavioural self-adjudication which Heidegger describes as Dasein. And once the standards established, they will own social stability or social authority. As long as the actual processes of scientific research are not being ignored, we can draw a conclusion that scientific claims are thus established within a rhetorical space rather than a logical space. The standards of scientific research are practical, and the defences of these standards are situated within localized social networks (Rouse, 1987, pp. 120–2).

Furthermore, practical activities in social networks satisfy the existing norms while also modifying them. Normativity is such modified, established and consolidated by the actors and their practices of networks and situations. And in a certain extent it also changes with the change of situations. Therefore, the establishment of normativity is a naturalistic process, so it should comply with naturalism that we described and elucidated on such process. And it should be normative because this is the explanation for such problems as how the norms establish, how it run and how it may collapse, it should also have normativity.

In the next chapter, we will discuss the significance of the scientific practice in hermeneutical senses, how it criticizes the traditional philosophy of science, the kinds and the scope of scientific practices. The clarification of these questions is crucial for understanding of following issues: the objects of scientific practice, the kinds and the meaning of the concept of scientific practice in the PSP, and so on.

Notes

1 We believe that Rouse valued the norm of practices which have the significance of ethics, political science, and political philosophy. Or Rouse valued that practices at least have the meaning of normative direction. After proving that science participates in power, practical norm can be formed naturally through ethical dimension and political dimension. So the PSP with naturalism have the inherent norm. And it will not make his philosophy to be descriptive which could completely deconstruct philosophy and be criticized more strongly.
2 In fact, SSK and social and cultural studies on science also regarded mental activity and research activity of conceptual representation to be conceptual practices. Keep in mind that the emphasis here is on the 'activity', rather than representation in the form of text itself.
3 This is the reason why Rouse's PSP contains the dimension of political philosophy of science, with this dimension, the implication of ethical, moral and political disciplines come alongside naturally.
4 Yu Wujin contrasts with the German Edition and changes this paragraph into: 'to make the existing world revolutionary, to oppose and change the things we encounter in the practical way' (Wujin Yu, 2001, p. 48).
5 The STS here is both Science, Technology and Society and Science & Technology Studies. Social studies or cultural studies of Science are also the STS of this meaning.
6 About the life and the influence of Hessen, there are two papers to discuss this. In addition, Tang Wenpei, Peking University doctoral student, is investigating Hessen's papers and his influence. At the time of writing, her paper had not been completed. And when this book is modified, her doctoral dissertation will have been completed.
7 In *The Sociology of Science*, Merton mentioned Hessen and his influence, and admitted that he and Western scholars were influenced by Hessen five times.
8 For example, Merton pointed out that Hessen and Bukharin all thought that 'only "proletarian science" has valid insight into certain aspects of social reality'. Merton said that it's worth discussing.
9 Xu Zhu has made a great contribution to this section.
10 Rouse calls his PSP a 'hermeneutics of scientific practice' explicitly. In *Knowledge and Power*, Rouse also mentioned the phrase 'practical hermeneutics' several times. And by way of making a distinction between Heidegger's practical hermeneutics and theoretical hermeneutics, and criticizing the insufficiencies of Heidegger's practical hermeneutics, Rouse puts forward and demonstrates the main standpoints of his practical hermeneutics (Rouse, 1987, pp. 48–50, 58–68).
11 In the book *Being and Time* which translated by Chen Jiaying and Wang Qingjie (Beijing: SDX Joint Publishing, Revised Version, 2006, pp. 80–1), there is a slight difference from this section. Here, we discuss Rouse's quotation from Heidegger, so we adopt the English translations in the book *Knowledge and Power*.

2 Scientific practice
Significance, types, and scopes

2.1 The concept and significance of practice in PSP

In order to criticize the theory-dominated view of scientific practice, a notion of scientific practice is proposed in PSP. We will discuss the significance of scientific practice and several hermeneutical conceptions of scientific practice in this chapter.

2.1.1 The conception of scientific practice: as a critique to theory dominance

The standard view of science usually includes the following general positions. First, practice is opposite to theory. Overall philosophy has strengthened the view that theoretical property is the most important characteristic of scientific knowledge. Second, theories are superior to practice. Experiments are governed and guided by theories and designed for theories. Third, the standard view of science has almost completely ignored the role of laboratory practice. The role of laboratories is only accepted in the sense of genetics; Meanwhile the role of laboratories in scientific research is neglected primarily. According to Ian Hacking, philosophy of science has become so speculative that it has denied the existence of pre-theoretical observation and experiments.

In order to criticize the improper view of science, Joseph Rouse, an advocate to PSP, and Ian Hacking, a pioneer of the new experimentalism, takes scientific practice as a weapon in constructing PSP.

Let us begin with Rouse's view on the critique to the standard view of science and conception of scientific practice. Rouse and other researchers propose the following points:

The most widely known attempt to distinguish between the image of science one gets from regarding research and the one that result from looking at finished theories was developed by Thomas Kuhn. (Rouse, J., 1987, p. 80). Kuhn proposed in *The Structure of Scientific Revolutions*, that

> The image of science by which we are now possessed has previously been drawn, even by scientists themselves, mainly from the study of finished

scientific achievements as these are recorded in the classics and, more recently, in the textbooks from which each new scientific generation learns to practice its trade. Inevitably, however, the aim of such books is persuasive and pedagogic; a concept of science drawn from them is no more likely to fit the enterprise that produced them than an image of a national culture drawn from a tourist brochure or a language text.

(Kuhn, T., 1970, p. 1)

Rouse thinks of Kuhn's paradigm as not a uniform idea, but a set of models to solve problems including instruments, examples and beliefs, which are a primary normative description to practice.

The research of scientific theories is also a distinct scientific activity. Rouse proposes that theoretical thinking and constructing as activities are also practices. In *Knowledge and Power: Toward a Political Philosophy of Science*, Rouse firstly pays attention to taking the concept of scientific practice as an alternative to the representationalist conception of science. The former conception of science has provided support to the legitimating project. The conception of science as practice emphasizes that practice cannot be distinguished from theories. Theorization is also a kind of scientific practice like other aspects of scientific activities. As a practical activity, it reconstructs the world through description.

How do we think about 'Theory as Practice', which is the part of the second headline of section 2 in chapter 4 of Rouse's *Knowledge and Power*, and 'Research as Action'?

At first, when we don't distinguish theories from experimental activities, but take both of them as scientific research activities, we will consider that there are quite a bit and basic practical activities in theoretical research. Theoretical activities cannot be distinguished from practical research activities. For example, for the accepted theories at present, we are supposed to confirm the constitution of their main issues, to specify some potential incompatibility within the theories, to build new experiments to testify the predictions from theories, and to assess some content that has not been clarified. All of these are regarded as a theoretical task. But these tasks are also subject to local context. At the same time, scientific research has characteristics of opportunism. Problem cannot be independent from its local context. If we get rid of the context in which the problem is rising, there will not be any opportunity. In practice, we cannot distinguish theoretical research from experimental one. Those people who believe in traditional view of science usually consider that it is a theory that determines the 'problem' that we can distinguish. Nevertheless this view is unacceptable, since not all theoretically identifiable problems can be taken as opportunities for new research. For instance, why do we choose those problems rather than others? Why do we need to study a problem in this way rather than in that way? All of these depend on the subjects to study them. If no one would like to study these problems, they would not even arise in the study of which we are currently engaged.

Scientific research is the practice and activity in which we engage. Science is not only the way that we observe and know the world as a spectator, but also a way that we deal with and intervene in the world. We are not living in a network of scientific concepts, but living in the world of practice constructed by scientific research. Many philosophers of science ignore this. For example, even if he proposed a hermeneutics of practice, Heidegger also thought of science as a decontextualized, mathematical projection of the natural world (Rouse, 1987, p. 80). Most of them simply take science as a tool to know the world, and then it seems that science can be distinguished with technology. According to Rouse, from the view of PSP, when we cannot make a difference between theories and experimental activities any longer, we do not make a difference between technical and research activities. If we take science as active practice, we should pay heed to the following: how does the theoretical knowing develop? How should the knowing difficulties be overcome? Here what people concerned is not how we can understand the general theories, but how scientists conduct their researches.

There is no unified theoretical background in scientific research. Rouse criticized M. Heidegger, W.V.O. Quine, and M. Hesse. He believes that a unified theoretical background appears in Heidegger's 'mathematical projection of the natural world', Quine's 'total theory' or 'the total system of science', and Hesse's model on the state of the learning machine at any given time. It is the ingrained prejudices of traditional philosophy of science that scientific theories must be able to form a unified and coherent network of theoretical system. In fact, in the process of doing science, there is not a stable understanding background. Science is taken as a dynamic process which is always different, unfinished and open-ended. Even if it helps to ensure a background, it is changing constantly; Secondly, theories and experience are not always fit to each other. They will meet only on some key points, but they are not totally coherent. 'Instead of being seen as a network of interconnected sentences or conceptual scheme, theories are taken to be extendable models' (Rouse, 1987, p. 83).

Kuhn argues that theories are not primarily sentential systems whose representational content is firstly learned and only then exerted on specific situations, perhaps with the help of additional bridge principles. He believes that the content of theories is embedded in standard and exemplary solutions to model problems. Learning theory is not for anything else, but learning to understand these problem solutions in such a way made in model cases can be extended and transformed to deal with a range of more or less similar cases. Rouse also points out that, from this perspective, theories are tools by which one learns how to use, rather than sentences of which implications one comes to know. Instead of being systems of sentences of which applications are deductively derivable, theories constitute a loosely connected set of models extendable by analogy. One increases grasp of the theory by moving from one particular case to another, rather than from theoretical generalization to a specific application (Rouse, 1987, p. 83).

Rouse indicates that Nancy Cartwright absorbs and extends Kuhn's view, and she does not think of theories as the unified whole but as extendable models. Cartwright follows Kuhn in insisting that the content of a theory is contained in model treatments of specific problems: 'On the simulacrum account, models are essential to theories. Without them, there is just abstract mathematical structure, formulae with holes in them, bearing no relation to reality' (Cartwright, 1983, p. 159). According to Cartwright's view, the explanatory power requires the theoretical models we use to be rather remote from any actual empirical situation. And they fit only highly fictionalized objects (Cartwright, 1983, p. 136). She said,

> we live in a dappled world, a world rich in different things, with the different natures, behaving in different ways. The laws that describe this world are a patchwork, not a pyramid. They do not take after the austere, elegant and abstract structure of a system of axioms and theorems.
>
> (Cartwright, 1999, p. 1)

It is thus, many philosophers of science, such as Kuhn, Cartwright, as well as Rouse that challenges the traditional view of theory as a unified, systematic deductive structure. In this respect Ian Hacking goes on even further. In discussing the kinds of theoretical principles in science, Hacking tells us that theories may be sweeping and suggestive speculations that only point us in a general direction, or they may be elegant formal mathematical representations or physical models that give us a qualitative understanding of causal interactions and mathematical representations which are more or less *ad hoc*. One does different things with theories of these various types. For example, Hacking points out there are six levels of theoretical ideas in this case by researching Faraday's electromagnetic experiment and theoretical study, according to the research on the scientific discovery of Maxwell and Lorentz (Hacking, 1983, p. 212):

1 Motivated by faith in the unity of science, Faraday speculates that there *must* be a few connection between electromagnetism and light.
2 There is a Faraday's analogy with Brewster's discovery: something electromagnetic may affect polarizing properties.
3 Airy provides an *ad hoc* mathematical representation.
4 Kelvin gives a physical model, requiring a mechanical picture of rotating molecules in glass.
5 Maxwell uses symmetry arguments to provide a formal analysis within the new electromagnetic theory.
6 Lorentz provides a physical explanation within the electron theory.

So which theoretical thought will you accept in the end? Or do you adopt six kinds of theoretical thought together? In either way, there is no unified theoretical interpretation. Therefore, scientists are looking for interpretation in different theoretical views in fact.

Of course, someone would say there may be a variety of theoretical speculations in scientific discoveries. Once a scientific discovery is accepted successfully, there is a unified theoretical interpretation. However, the theory is no longer a posterior after scientific discovery, not hindsight. Even theories are only for justifying. There may be several explanations on the same question. As Cartwright said, a unified explanation is extremely rare in science.

2.1.2 Practical character of scientific research activities

Scientific research activities are not intended to make the theory better but to explore the unknown world. Scientific research involves the intervening and opportunistic characteristics. According to the traditional view of science, scientific research is considered as activity while scientific knowledge is a theoretical system. Scientific research should be understood to solve the conflict and embarrassing oversight of the theoretical system. That is to say the primary goal of scientific research is to solve theoretical problems. In fact, the foremost goal of scientific research is not theoretical but to solve the problems at hand. Most of these problems come from the research practice at hand. Scientific research cannot escape from a need for practice.

Thus, what guides the research? According to traditional view of science, scientific research must be based on theories. But Hacking argues that scientific research may not be guided by theories. According to Karin D. Knorr-Cetina's view, scientists carry out research in accordance with all the resource at hand, and their strategies of research and knowledge production are basically opportunistic (Knorr-Cetina, 2001, pp. 63–71). According to Latour and Woolgar, 'Nonetheless, informal communication is the rule. Formal communication is the exception, as an *a posteriori* rationalization of the real process' (Latour & Woolgar, 1986, p. 252). Rouse indicates that the assessment of research opportunities constitutes what Heidegger would call 'circumspective concern'. It is a practical assessment of what makes sense, on the assumption of the resources available and the aims and standards in a given scientific practice. Much of what one considers in such circumspection is the current state of knowledge, including what results are (and are not) reliable bases for further work, what tools and techniques are sufficiently precise and illuminating, and whether perspective achievements would constitute a significant advance or not (Rouse, 1987, p. 88).[1]

Much of experimental studies are mainly restricted and guided by existing resources. In the traditional view of science, there is a conflict between theoretical predictions and empirical results. The predicted consequences require to modify the theory. The collision is boring, unless it heralds a new breakthrough. Scientists' experimental efforts seem to aim at resolving the conflict between experiment and theory systematically, with a strong destination. While in the view of PSP, the cause of the conflict between theories and experiments is not a substantive defect of theories, but the particular situation of experiments and descriptive construction. In fact, there are a lot of experiments that scientists just want to use the existing resources: equipment, technology, trained experimenter, and relative scientific achievements (Knorr-Cetina, 2001, pp. 53–68).

In various methodological principles on scientific research, there is a principle often named as 'feasibility principle', which means that scientific research needs to correspond to the capabilities and based on objective and subjective criteria.[2] In the view of PSP, there are constraints for scientific research from the local context. They must consider what skills and methods are suitable for their research programs. Evaluation on research opportunities certainly includes intervention and skilful know-how about the objects and practice, including the assessment of the researchers on the theoretical level, practical skills, and the experimental situation of which researchers can make use, etc.

2.1.3 Case studies on scientific practice

In a real case, the aim of research was restricted by the approaches and conditions so that the aim changed before and after, which means realistic conditions restricted the research. A new experimental method is produced in the process of experiment surprisingly.

The case is described as follows: there is a variety of cutting-edge science, technology and engineering research in the Department of Precision Instrument, Tsinghua University in Beijing. I met a PhD student who was major in optical engineering research in 2005. His doctoral dissertation was on the measurement problem of flying insects. The purpose of this research was to make early preparation for micro aircraft research, namely the bionic research of micro aircraft. The researcher originally intended to gain systematic data on insect flight, and on that basis, he wanted to come up with a new theory.

However, this pre-set goal was not reached, but only built a new preliminary measurement system of insect flight (of course, which is quite amazing). The main reason was that we only had an intuitive experience, without any theory about free flight of insects, such as bees flying. The research opportunity of insects' space flight depended on many factors, in which the most important bottleneck factor was whether we had better observation instrument to tracking flight. The researcher finished a hybrid system consists of two subsystems: the magnetic field sensing coil measuring system and image tracking system. By using this system, the researcher tracked and shot the bumblebee's in-flight position (hovering, flying straight and turning) and got the relevant data. At the same time the researcher also successfully measured the wings' fan angle and torsion angle by bonding a miniature sensing coil on the bumblebee's wings (Cai Zhijian, D., 2005).

This case suggests, first, the experiment in the case has not been inspired by some drone's flight theory. On the contrary, the initial objective of the research was just to construct a theory. The experiment was carried out in the dark. Second, though there is the conception on the experiment, the feasibility of the experiment decided whether the experiment completes or not. Third, in the process of the experiment, what we can discover is unpredictable in advance, and the discovery is opportunistic. Obviously, this case supports the views of PSP very well.

I think the method of bonding the coil will become one of the preliminary standardized research methods on insect flight through technical practice, though it was originally an opportunistic effort. Moreover, this case also provides us an example that theories are prior to experiments. Experimentation has its own life. (We will detail this case further in Chapter 12, in order to illustrate the opportunistic characteristics of scientific research.)

2.2 The main types of scientific practice I: scientific practice, experimental practice, and laboratory practice

According to Rouse, if all research is scientific practice, how can we distinguish the different types of scientific practice? This is really a problem, which needs us to give a detailed argument. In my opinion, scientific practices can be distinguished as in different types with no essential difference in all types of scientific practices.

In Rouse's view, we can specify four types of practice. They are: discursive, scientific, experimental, and laboratory. Discursive practice occurs more frequently in Rouse's account, especially in *How Scientific Practices Matter*. We will study it in the next section. Here we focus upon three main practices: general scientific, experimental (the one of scientific practice whose feature is experiment and activity), and laboratory (this kind of practice proceeds in local laboratory, and laboratory take a key role in it. In other words, it is the scientific practice that occurs in a particular laboratory or by means of specific laboratory.)

2.2.1 General scientific practice is not abstract

Here we use 'general' to indicate all scientific research as activities are scientific practice. It is a practice-dominated view. Here, the practice is no longer a position opposite to the theory, but a more fundamental concept. We discuss 'experimental practice' here, and the meaning of scientific practice in the next section. It should be noted that all of the scientific research in the sense of activities is scientific practice, which does not deny the existence of theoretical activity. As an activity, the process of raising and resolving theoretical problems also belongs to scientific practice.

'Experimental practice' has highlighted the significance in scientific practice. It reflects that scientific knowledge is local, contextual, and anti-modernistic.

In Rouse's view, scientific research is a 'circumspective activity', by taking place against a practical background of skills, practice, and equipment (including theoretical models) rather than a systematic background of theory (Rouse, 1987, pp. 95–6).

44 Significance, types, and scopes

Compared with theories, scientific experiments are not on a secondary position. Although experiments and theories are mutual penetration, experiments are more fundamental in the sense of activity. However, why many philosophers of science including Heidegger neglected experimental and laboratory study? The reason is that the researchers who are affected by the theory-dominant view of science believe that experiments can be largely neglected when we consider the philosophical or ontological significance of science. It seems to me that we can understand what is essential about science if we have understood *theorizing*. What philosophers of science concerned is how we understand theoretical propositions. The other reason is that experimentation has no cognitive content apart from that provided by theory. For example, Karl Popper believed that 'The theoretician puts certain definite question to the experimenter, and the latter ...' (Popper, 1999, p. 99). In contrast of this view, Thomas Kuhn argued that scientific paradigms are the fundamental locus of professional commitment is science, and he points out that he takes paradigms to be 'accepted examples of actual scientific practice ..., which include law, theory, applications, and instrumentation together' (Kuhn, 1970, p. 10). Kuhn was still analysing some important roles of standard experimental procedures, instruments in the scientific discovery. He pointed out that

> in short, consciously or not, the decision to employ a particular piece of apparatus and to be used in a particular way carries an assumption that only certain sorts of circumstances will arise. There are instrumental as well as theoretical expectations, and they have often played a decisive role in scientific development.
>
> (Kuhn, 1970, p. 59)

'Neither oxygen nor X-rays emerged without a further process of experimentation and assimilation (Kuhn, 1970, p. 57). These examples stress the importance of experimentation and apparatus in paradigm.

'Practice is the sole criterion for testing truth', which indicates a view of Instrumentalism and subordination status about practice.[3] However, even behind such a Theory of Truth, it also implies a theory-dominated position. This position only thinks of practices as tools of testing theories. This is an Instrumentalist view of theory of truth, which is problematic and unfair to experimental practice, since the roles of experimental practice are far more than that.

2.2.2 One of the role of experimental practice is constructing phenomena

In discussing the role of experiments in 'creation of phenomena', Hacking goes further than Kuhn. Hacking indicates that 'one role of experiments is so neglected that we lack a name for it. I call it the creation of phenomena'. Scientists often create the phenomena which then become the centrepieces of theory (Hacking, 1983, p. 220). Hacking took for example the effect discovered by E.H. Hall to illustrate that the discovery of the Hall effect was not for testing the theory but an exploration guiding by optional comments. 'Maxwell had said there might be some sort of island in those uncharted waters', then Hall went to look for it (Hacking, 1983, pp. 224–5). In a theory-dominated view, phenomenon is waiting for us to discover. In contrast, in the view of PSP, Hall effect does not exist without the specific instruments. Isolation is the key factor to the existence of Hall Effect, since nature is comprehensive with no pure isolation.

Experiments have their own lives. We just consider the practice as a tool and standards for testing theory, which obscures the other roles of practice. As Hacking has emphasized, experimentation in science often takes its own course in exploring the areas not yet theoretically articulated. It is often guided not by explicit theoretical hypotheses, but only by a suggestion of something worth examining and a sense of how to proceed with the exploration (Hacking, 1983, pp. 149–66; Rouse, 1987, p. 100).

Therefore, 'the principal reason for this degree of independence is that useful experimental techniques, instruments and procedures are sufficiently difficult to develop and refine that they often play a leading role in directing experimental work' (Rouse, 1987, p. 100). Maybe we are used to being guided by the view of theory–dominance. So we will inevitably ask: what is the role of theory neglected in the view of practice-dominance? Is experimentation useful if we do not understand what experimentation reveals and why it is important? I believe it is a wrong question, since we limit our understanding to the range of the theoretical representation. Experimentation has its own life. From a research perspective, experimenters adjust and grasp the apparatus not in accordance with the subject of the research, but 'a practical grasp of the workings of her apparatus and its possibilities and limitations. This 'feel' is for the instruments, more a practical craft knowledge than a theoretical representation' (Rouse, 1987, p. 100). If there is no such skilled craft knowledge, we cannot operate the apparatus, obtain a practical observation, and make an accurate judgment to the range of experimental results. We cannot do scientific research without the use of apparatus, or the grasp on skilful know-how. It is particularly evident in the computer field. When we deal with some problems by using computers, we do not need to grasp theoretical knowledge in computer science, but merely have all kind of skilful know-how of using computers. The skilful knowing-how even determines the research situation to some extent. For example, Holland, known as the father of genetic algorithms, cannot create the field of artificial life in complexity science research without a lot of skilled craft knowledge and training.[4] Thus, experimentation does more than fill in the gap left by prior theorizing or check the adequacy. It opens new domains for investigation, refines them to make them suitable for theoretical reflection, and provides a practical grasp of that domain as a resource. The creation of new phenomena, and practical understanding of scientists in the instrumental context of laboratory with which this creation occurs, cannot be easily subordinated to a theory-dominant picture about how science develops (Rouse, 1987, pp. 100–1).

2.2.3 Laboratories and their apparatus are local, contextual, and practical in scientific research

In a traditional view of science, laboratories only play a role in producing scientific knowledge, which will disappear as a local context later. Rouse reevaluates the status of the laboratory in scientific research in his book *Knowledge and Power*. He indicates that the laboratory is a key place, a local context in scientific research which plays three important roles: isolating, controlling, and tracking the microscopic world. Knorr-Cetina argues that laboratory science is even the most important cultural practice of science. We believe that these views in laboratory studies can be also stressed.

The significance of the laboratory is constructing the phenomenal microworld. The relationship between laboratories and scientists is just as the one between research foundation and scientific research. A laboratory is a locus for construction of the phenomenal micro-world, where scientists can show their talents completely. Scientists construct a precise, artificial, and simplistic 'world' in a laboratory, while avoiding the complex nature of phenomena. It is not denied to take a positive role of experimentation constructing microworld in our traditional textbooks. For example, experimentation can simplify complex phenomena, strengthen the experimental conditions, and so on.[5] Indeed, the experimental subject is strictly controlled in laboratories, and the interaction is assumed to be 'isolated' or 'isolate' in accordance with the control conditions, so that each experiment can be easily manipulated, tracked, and to make a clear enough result. Therefore, laboratory practice is essential and indispensable for scientific research.

The first role of laboratory practice: isolation and highlighting. Clearly, laboratories have become a specific artificial world. Any possible external influence is isolated, and a number of factors needed in constructing phenomena are highlighted in laboratories. Our chemicals in laboratories are processed and preserved carefully; the bunch of particles constructed by us are also adjusted and emitted by specific instruments; our mice used in experiments are also specifically artificial objects. For example, Onco – Mouse™ is used in the cancer research is standardized by DuPont company (Fujimura, 1996, pp. 7–9). Therefore, we are able to know exactly what exists in a laboratory, a constructed microscopic world and even how much dust grains are permitted by experimental conditions, in order to isolate the external causal influence strictly from the experiments. We introduce a control factor into the experiment every time in order to highlight the effect of certain factor, which is the construction in some sense. There is not one single effect of one factor in nature. Thus, the construction of the phenomenon is not in accordance with nature, but with artificial microscopic world, which is the practical construction of laboratories.

The second role of laboratory practice: manipulation and intervention. It is also apparent that the aim to construct an isolated micro-world is not only to show the phenomena of artificial micro-world, but also to manipulate it in a particular way. Therefore, it seems to break through the traditional view of science once again. Scientists introduce an artificial micro-world intentionally, where they do not just observe the events, but let the events happen in laboratories. Therefore, what scientists studied is a constructed and isolated system, in which they are not spectators, but actors, participants and practitioners. The way of their doing science is not to 'see', but to 'do'.

The third role of laboratory practice: tracking the micro-world. Rouse indicates that the tracking is not just 'observation' in the empiricist sense, not a perceptual awareness of the experimental result, but something related to the initial construction and the whole experimental process. Tracking is not just monitoring the results of the experimentation, rather it is monitoring the proper function of experimentation. This is circumspective concern and activity in Heidegger's words. Tracking is related to experimenters' practical skills, such as the 'perception' to experimental design, and the experience and attention to the experimental operation, etc. There are also components of constructing the micro-world in laboratories. By tracking, a variety of micro-world events can become visible phenomena in laboratories, in which phenomena become records, and classified, coded, archived and handled in laboratories. Latour and Woolgar imagine a picture as following in their book *Laboratory Life: The Construction of Scientific Facts*: entering a laboratory at night, an observer opens one of the large refrigerators peels off the labels, throws them away and returns the naked samples to the refrigerator. What would happen? The laboratory was in an extreme confusion. It would take up to five, ten and even fifteen years to replace the labels (Rouse, 1987, p. 104). This suggests that scientists need to construct a place to making things clear and visible. That is rightly the significance of laboratories in scientific research.

To a large extent, there are circumstances of experimental design subjecting to theoretical need. Scientists produce similar experimental constructs for theoretical models in laboratory micro-world. Like theoretical models, the experimental constructs are also isolated system. They do not emerge in the more complicated real world. These conditions in which experimental design goes beyond the theoretical model and experimental construction does not guarantee pre-existing theories. Therefore, more precisely, the experimental design and theoretical models are often combined

The fourth role of laboratory practice: constructing artificial research environment and the knowledge of artificial nature. A laboratory is almost an artificial research context, which is an artificial world constructed according to research need and under the support of the conditions above-mentioned. Even the natural objects have been transformed into artificial-natural objects. For example, are Chinese goldfish natural objects or artifacts? They are improved natural objects, thus no longer pure natural. Since there are human factors involved, Chinese goldfish has become a mixture of 'technical facilities – natural objects', or called as 'artificial-natural objects'. The research subjects in laboratories are

48 *Significance, types, and scopes*

almost artificial-natural objects. We can take nature transformed into artificial objects by constructing apparatus, instruments and laboratory facilities, and make research subjects in nature to be strictly controlled by man-made condition. These phenomena produced in the laboratory come back to nature by expanding laboratory conditions to controlled nature. Or in a weakened sense, nature has been domesticated in the laboratory. A drosophila colony is a domesticated version of an orchard replete with a breeding population of fruit flies that display variation by selection. A tokamak is a domesticated version of a star. A powerful magnetic field confines hydrogen atoms in a small volume, fusion to helium being ignited by an external energy source (Harré, R., in Radder, 2003, p. 27). Domestication means that certain properties of natural objects are lost, while some others highlighted, and some new properties produced in humanist condition. At the same time, this domestication also indicates that artificial-natural objects are still a part of nature.

2.3 Three main types of practice (II): thought experiments, discursive practice, and conceptual practice

2.3.1 Thought experiments

Concerning to thought experiments, some people have employed it in the history of science, in which Galileo and Einstein are the pioneers. Whether the thought experiments are species of scientific experiment has also long been questionable (Kuhn, 1981, pp. 237–63; Shu Weiguang, 1982, pp. 1–10). In particular, whether or not is a thought experiment in physics a kind of scientific experiment? If it is, what is its characteristic? What is the difference between thought and general experiments?

Thought experiment is often defined as follows: it is the ideal process shaped by mind; it is a thinking process of logical reasoning and an important approach of theoretical studies. Thought experiments are frequently discussed in the circle of Dialectics of Nature in China, which indicates that we have a position of not placing thought experiments into scientific practice.

Kuhn criticized the wrong views on taking thought experiments merely as conceptual tools of scientists (Kuhn, 1981, pp. 239, 250). He indicates that

> They arose, that is, not from his [scientists'] mental equipment alone, but from difficulties discovered in the attempt to fit that equipment to previously unassimilated experience. Nature rather than logic alone was responsible for the apparent confusion ... the scientist learns about the world as well as about his concepts. Historically their role is very close to the double one played by actual laboratory experiments and observation. First, thought experiments can disclose nature's failure to conform to a previously held set of expectation. In addition, they can suggest particular ways in which both expectation and theory must henceforth be revised.
>
> (Kuhn, 1977, p. 261)

Significance, types, and scopes 49

One of the differences between thought and laboratory experiments is that thought experiments can only rely on the information at hand, especially pre-existing experience. It is not entirely feasible that we discover the same capacity of nature relying on pre-existing experience. This needs the thinking from excellent thinker. Therefore, constructing a thought experiment puts a high requirement to the experimenter. If material factors have an important role in a material experiment, as one type of scientific practice, the mental factors have a more important role in a thought experiment. Certainly, if we take thought experiments as one type of activities, they are also species of scientific practice.

According to Kuhn (1977, p. 263), 'thought experiment is one of the essential analytic tools which are deployed during crisis and which then help to promote basic conceptual reform'. Thought experiments can highlight abnormal situation prominently, promote to doubt the accepted results, and abandon the old conceptions in order to prepare for new conceptual tools.

Of course, Kuhn sometimes still stands in a position of 'theory-dominated' in traditional philosophy of science. For example, he does not admit experiments as prior to theories (Kuhn, 1981, p. 200). In Kuhn's view, the significance and roles of thought experiments in the sense of discovery are not the objects of studying philosophy of science, since many philosopher of science distinguish the context of discovery from the context of justification. However, Kuhn has paid attention to the relationship between the concept of experiments and real world very earlier. He emphasizes that thought experiments may reveal the conflict between concepts and real world, and provide the knowledge of the real world, which is not an easy task at that time.

The above analysis on the roles and significance of thought experiments implies that such experiments are not pure conceptual tools but scientific practice, including the material factors of nature. Chinese academics in particular, influenced deeply by Marxism, do not need to analyse further whether or not thought experiments belong to scientific practice; but it is still a very important question in history, which relates to our understanding of practice in Marxist tradition.

In 1980s, China has just gone through the 'Cultural Revolution', and many theoretical and practical aspects in society need to bring order out of chaos. The situation is similar in philosophy of science. Shu Weiguang, a professor in Jilin University, at that time argued that the 'practice is an inherent nature of thought experiments'. He presented two propositions (Shu Weiguang, 1982, p. 2):

Proposition: A scientific experiment is a form of practice;
Proposition: A thought experiment is a form of scientific experiments.

He argued that we should not only admit that 'it relates to the Marxist philosophy that scientists and philosophers study thought experiments', but also to 'answer the further question: how do the thought experiments with

50 *Significance, types, and scopes*

concepts or words represent the nature of practice? This is a fairly difficult question' (Shu Weiguang, 1982, p. 2).

Shu Weiguang indicates that, generally speaking, the relationship between thought experiments and practice or the nature of thought experiments includes the following aspects (Shu Weiguang, 1982, p. 3):

1 Thought experiments strive for contacting with nature (or reality) beyond the scope of the concept or logic.
2 There is important inter-continuity between thought experiments and material experiments.
3 Thought experiments may serve as the guide under certain condition.
4 A thought experiment can take the place of a real experiment to work to a great extent (although cannot completely replace);
5 Thought and real experiments subject to a same family in history and tradition.

Shu Weiguang points out the reason why thought experiment is one kind of scientific experiment with a pragmatic nature. And he further claims the pragmatic nature of thought experiments. His arguments could be summarized as follows:

1 Thought experiments require conditions, not only conceptual, logical, or linguistic, but also physical plausibility. The envisaged condition is coherent to the physical condition. 'The Physical conditions of plausibility of thought experiments should strive to be or approximately to be the conditions actual in nature, which is the reason why it can provide the new knowledge on nature' (Shu Weiguang, 1982, p. 4). Cartwright names this set as 'opportunity set' or 'nomological machine' (Cartwright, 1999, pp. 49–58).
2 A thought experiment has perceptual and experimental characteristics, although it relies upon logical reasoning and includes theoretical thinking. The operating process of a thought experiment is perceptual, but the analysis of the process is logical reasoning. A thought experiment is a unity of 'thought' and 'experimentation', which is a tool not only for discovery, but also for justification and critique. If a thought experiment does not have practical nature, it cannot provide true and justified knowledge.
3 The pragmatic nature of a thought experiment is also a historical fact. History of science has proved that it is hard to distinguish a thought experiment from real experiments, neither upon experimenters nor experimental subjects. For example, Galileo's argumentation on free falling objects with different weights is considered as a thought experiment, but it could also be a laboratory experiment with ideal conditions.

According to Rouse's view, as long as we think of science as activities, the focus we exploring is not only a system of propositional statement or a result of theories, but also a process of scientist research, which is one of the content or process of scientific practice.

2.3.2 Discursive practice

There are two perspectives of observation to discursive practices of science. First, it is the scientists' own practical activities based on material practice, and it is scientists' understanding about the meaning of their own activities. It is also the scientists' explanatory activities to their colleagues. Second, it is sociologists' activity of recording and reinterpreting scientists' discursive practices. Therefore, discursive practices are usually to two orders. The first order is discursive practice by scientists, while the second is sociologists' interpretation of scientists' discursive practice.

Latour and Woolgar give a full description to discursive practices in the book *Laboratory Life: The Construction of Scientific Facts*. They argue that there are four different types of exchanges in discursive practices in laboratory according to their observation.

The first type involves 'known facts'. Scientists often discuss the fact just being established. The discussion often relates to the statements such as 'Eh, has someone already done that', 'Is there a paper on that method', 'When you try this buffer what happens', and so on (Latour & Woolgar, 1986, p. 160). Scientists often want to know how long scientists have known this fact. The main role of exchanging views is spreading information, which is helpful to rediscovering those practices related to the current issues of concern (Latour & Woolgar, 1986, p. 161).

The second type originates in the course of some practical activities. For example, such statements are common in a context of completing certain essay. 'How many rats should I use for the control?' 'Where did you put the samples?' 'Give me the pipette', and 'It is now ten minutes since the injection' (Latour & Woolgar, 1986, p. 161). These exchanges concern a correct way of actions and the evaluation on the reliability of a specific method. That is the Heidegger's conception of 'practical circumspection'.

The third type of exchange focuses primarily on theoretical matters. This one does not concern the past state of knowledge, the relative efficacy of different techniques, or specific scientists and papers, but concerns mostly the future prospects of a new theory. And discussions are also closely related to other topics such as the future development of discipline and of laboratories, which perhaps are the assessments of the research projects (see Latour & Woolgar, 1986, pp. 162–3).

The fourth type of conversational exchange characterizes discussion by participants from other researchers. This exchange often involves the personnel evaluation of other researchers working in the lab, including his ability, achievement, and integrity (see Latour & Woolgar, 1986, pp. 163–4).

After discussing the four types of discursive practice, Latour and Woolgar also claim that the theme is often transformed in discursive practice. And

52 *Significance, types, and scopes*

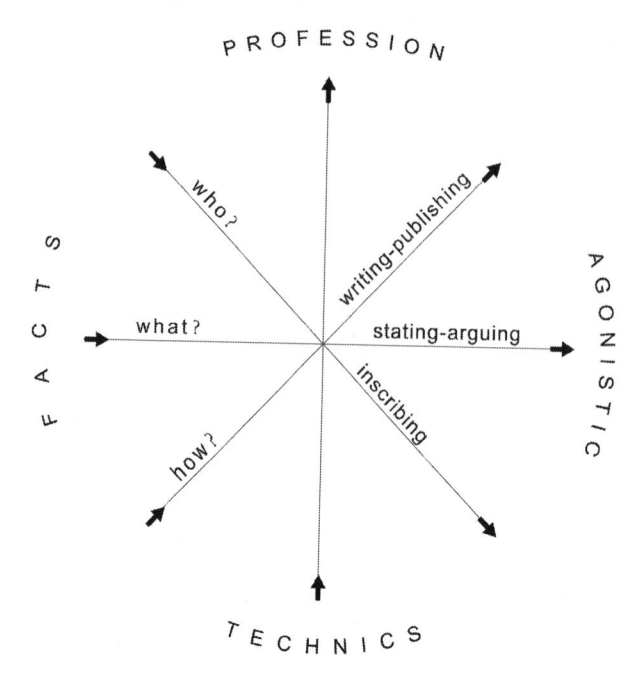

Figure 2.1 Different preoccupations of the conversations.

the abundance of discursive practice is also neglected in studying scientific practice. First, conversational materials exhibit quite clearly how a myriad of different types of interests and preoccupations are intermeshed in scientists' discussions (Figure 2.1). Second, we have presented evidence to indicate the extreme difficulty of identifying purely descriptive, technical, or theoretical discussion. Third, we have suggested that the mysterious thought process employed by scientists in their setting is not strikingly different from those techniques employed to muddle through in daily life (Latour & Woolgar, 1986, pp. 166–1)

2.3.3 Conceptual practice[6]

David Bloor, one of founders of Strong Program (Edinburgh School) wrote an article, 'Wittgenstein and Mannheim on the Sociology of Mathematics', published in *Studies in History and Philosophy of Science*, vol. 2, in 1973. In this article, he discussed mathematical and logical problems which seem irrelevant to social practice. He discovered mathematical formulas as social construction by means of Wittgenstein's discussion on the paradox of rule-following and implied the convention of rules and possible solutions to this in practice. He suggested that the process of construction is not the process of pure conceptual thinking, but rather related to social practice (Bloor, 1991, pp. 173–91).

On the same direction, Pickering and Stephanides give us a case of conceptual practice, which is about Hamilton thinking of mathematical concepts (Pickering, 1992, pp. 139–67).

First, Pickering and Stephanides indicate the studies on practice made by historians, philosophers, sociologists, and other scholars in the past still focus on experimentation itself, by exploring the relationship between laboratories and social world outside, such as Hacking's research on laboratories, and Gooding's research on experimental practice. However, we have known little about the conceptual practice of theories, so our next target is to concern 'how can the workings of the mind lead the mind itself into problem' (Pickering, 1995, p. 114). Pickering and Stephanides argue that we cannot claim to have an analysis of scientific practice until we can suggest answers to the questions as follows.

What is conceptual practice? According to Pickering and Stephanides, conceptual practice is a process of modelling, which can be decomposed into three primitive elements: *bridging, transcription, and filling* (Pickering, 1995, p. 116). In our opinion, conceptual practice focuses on the activities of concept, rather than the concepts as propositions or statements themselves; its main point is not what has been represented, but the process of thinking about how the concept becomes clear, how to bridge, transcribe and fill among different concepts.

Second, Pickering and Stephanides argue that, since conceptual practice is a process of modelling, general practice is also a process of modelling, based on the concept of practice, particularly in relation to theory development. Therefore, it is necessary to study conceptual practice. How do we get the idea by using symbols marked on paper? How can the workings of the mind lead into problem? Such problems need to be resolved in studying conceptual practice. Pickering's concerns are the dialectic features of conceptual practice, and any resistance or confusion in conceptual practice. That is also the central point of the practical process of scientific knowledge, since we are not able to transform the practical activities, structure and results into representation until we solve the problem of conceptual practice. If the practice of intervention in the world cannot be linked to the practice of representing the world, science and knowledge cannot achieve yet normative understanding and interpretation.

Pickering and Stephanides' study on conceptual practice takes as an example the process of Hamilton, an Irish mathematician, who is constructing quaternion system. The reason for selecting the case in history of mathematics is that it is free from material limits of experimental practice in science. Meanwhile, Hamilton's theory of quaternion has important historical significance in history of mathematics, which marks an important turning point in the development of mathematics. Moreover, Hamilton also left a lot of notes guiding him to discover a theory of quaternion.

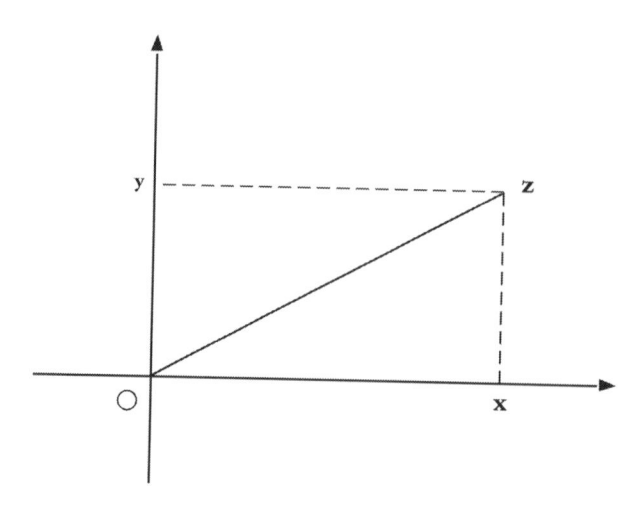

Figure 2.2 Geometrical representation of the complex number (Pickering, 1995, p. 122).

Now, let us review the practical process of constructing quaternion as a concept.

The early 19th century was a time of crisis in the foundations of algebra, suffering the problem of how the 'absurd' quantities – negative numbers and their square roots – should be understood. Various moves were made in the debate upon the absurd quantities, which led to an association between algebra and geometry, and the association in question consisted in establishing a *one-to-one correspondence* between the elements and operations of complex algebra and a particular geometry system.

We know that the standard algebraic notation for a complex number is $x + iy$, where x and y are real numbers and $i^2 = -1$. Positive real numbers can be thought of as representing measurable quantities or magnitudes. But what i and -1 might stand for? The geometrical response to such questions was to think of x and y not as quantities or magnitudes, but as coordinates of the endpoint of a line segment terminating at the origin in some 'complex' two-dimensional plane. Thus, the x-axis illustrates the real component of a given complex number, and the y-axis the imaginary part, the part multiplied by *i* in the algebraic expression (Figure 2.2) (Pickering, 1995, pp. 121–3).

In algebraic notation, addition of two complex numbers was defined as

$(a + ib) + (c + id) = (a + c) + i(b + d)$, and the corresponding rule for line segments was that the x-coordinates of the segments to be summed, and likewise for the y-coordinate (Figure 2.3).

Accordingly, the rules for multiplication of two complex numbers are more complicated, and we do not discuss here. All in all, for 19th-century mathematics, it solved the foundational problems centred on the absurd numbers to establish a one-to-one correspondence between complex algebra and a

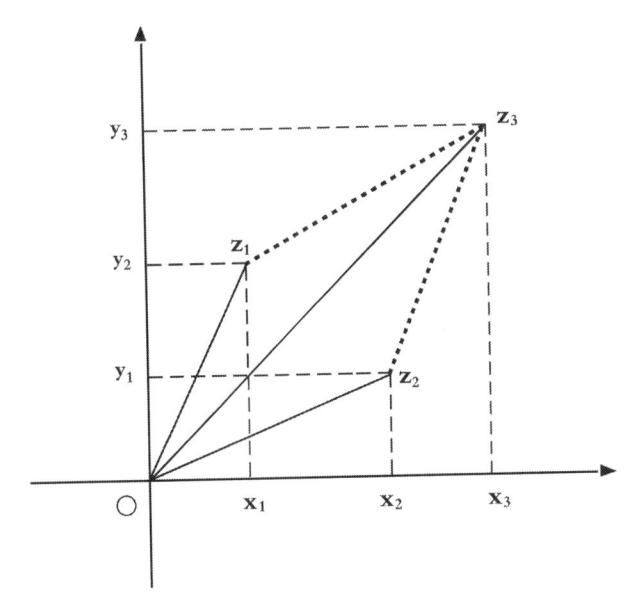

Figure 2.3 Addition of complex numbers in the geometrical representation (Pickering, 1995, p. 122).

particular geometry system. Furthermore, it is easier to expansion of mathematical operation. Since the geometrical representation of complex numbers theory can expand in a two-dimensional space. Is it possible for complex-number theory to correspond to three-dimensional space? Hamilton's job was to accomplish this task, linking algebra and geometry.

Hamilton's thought practice began with thought of segment in a two-dimensional plane, which is a natural process. He tried to extend it to a three-dimensional plane by moving from the usual $x + iy$ notation to $x + iy + jz$. He took the square of triplet as his initial object of study:

$$t^2 = (x + iy + jz)^2 = x^2 - y^2 - z^2 + 2ixy + 2jxz + 2ijyz.$$

(Pickering, 1995, p. 129)

But he needs to deal with $2ijyz$, since it does not correspond to the inferential rule. In Pickering's opinion, Hamilton encountered a '*resistance*', while Hamilton tried to assume either $ij = 0$ or $ij = -ji$ to accommodate this '*resistance*'. For the geometrical representation of the square of triplet, he assumed a new necessary condition that the square of triplet defined in three-dimensional space by itself and x-axis. Therefore, he constructed a new 'accommodation' for the '*resistance*' and began to try thinking the multiplication of two triplets:

56 Significance, types, and scopes

$$(a + ib + jc)(x + iy + jz) = ax-by-cz + i(ay + bx) + j(az + cx) + ij(bz-cy).$$

(Pickering, 1995, p. 133)

However, Hamilton encountered here a new *'resistance'*, defined as bz-cy. He proposed a new quantity k, a *'new imaginary'*, which succeeded in balancing the moduli of the left- and right-hand sides of the equation. More precisely, Hamilton actually defined a new bridgehead leading from two-place representations of complex algebra to a four-place system – through what Hamilton called *quaternions*. By turning to four-dimensional space, Hamilton finally adapted successfully to the resistance. He encountered and obtained a general rule for quaternion multiplication when extended into three-dimensional space.

We have reviewed the process of Hamilton constructing a theory of quaternion briefly. Now we analyse how Pickering and Stephanides elaborate the mangle of conceptual practice.

Pickering and Stephanides proposed 'the mangle of practice' based on Latour's 'Actor-network theory'. 'Mangle' is a process of both human agency and material agency participating. In this process, scientists are the actor capturing material agency by instruments. Human agency and material agency are intertwining and interacting in the process, and the interaction of human agency and material agency is performed by the process of resistance and accommodation. Resistance is a failure to intentional capture the material agency; accommodation is positive human strategies to adapting resistance. This positive response includes modifying the goals and motivations, questioning machines or physical form, as well as adjusting framework of human actions and social relationship around human actions. Thus the whole process is producing resistance, conducting accommodation against the resistance, producing new resistance, and conducting new accommodation against the new resistance. The process is open-ended, in which scientific culture has been extended and reconstructed.

For Pickering,

> Science is operational, in which the roles of both human agency and material agency are prominent. Scientists are the actors who capture the material agency by machines. Mangle integrates various agencies, so that human agency and material agency appear at the same time; mangle also makes human agency and material agency intertwine and inter-defined.
> (Xing Dongmei, preface of Chinese translation of *The Mangle of Practice*, 2004)

The dialectic of resistance and accommodation which is both goal-directed and goal-modified are the general characteristics of scientific practice.

Pickering and Stephanides think of conceptual practice as similar to general scientific practice. It is a process of modelling following the dialectic of resistance and accommodation. Well, how do they analyse the mangle of practice in conceptual practice?

Significance, types, and scopes 57

First, Pickering and Stephanides argue that the case of Hamilton constructing quaternion can be divided into three primitive stages: bridging, transcription, and filling. For the stages of bridging and filling can be understood as free moves in the process of modelling; transcription is forced moves surrounded by forces. Accordingly, we need to understand several concepts.

Modelling actually refers to creative extension to present cultural elements, the most prominent feature of which is open-ended, since it always directs to an association of variety of intertwined elements.

The bridging stage is actually a process of trying and expanding from known thought to the goal of seeking. Pickering stresses another concept – a bridgehead, which provides space for the transcription process, and indicates a specific target of the modelling process. Meanwhile, the bridgehead is tentative and modifiable, while this trial or modification does not necessarily guarantee any success. Obviously, this is a process of free thought. Nobody can restrict your thinking activities. As for Hamilton's case, the bridgehead should be the one-to-one correspondence between complex algebra and geometry system establishing in two-dimension plane in the process of constructing a theory of quaternion.

The stage of transcription is equivalent to copy a basic operating system or to a system according to bridgehead. The dialectic of resistance and accommodation in Pickering's concern is mainly reflected in this stage. We can think of resistance as the difficulty in fracture of thinking according to the bridgehead. For example, in the case of quaternion, the square of triplet ($2ijyz$), multiplication of two triplets ($bz-cy$), and geometric representation are all resistances encountered by Hamilton. His continuity of thinking was interrupted. Thus, it is very easy to understand his needs for the help of inspiration. It is on the way to the Royal College to attend a meeting that he got the revelation of quaternions.

The stage of filling is also necessary. In response to resistance scientists have to seek new theory constantly, sometimes even to change the bridgehead. The stage of filling, different from transcription, is free moves. Actors are free to choose a variety of means of settlement, while transcription is marking tentative choices and judgments in an indefinitely extended space of culture. In the case of quaternion, Hamilton assumed $ij = 0$ or $ji = -ij$, and proposed a new quantity k, which are all the process of filling.

It is worthwhile to notice that Pickering and Stephanides emphasize also that the stage of bridging, transcription, and filling cannot be analysed completely in the process of scientific practice, for it is a complex process including variety of intertwined factors.

As a result, we may be able to outline the dialectic of resistance and accommodation in conceptual practice. See Figure 2.4.

In Figure 2.4, bold characters represent the bridgehead of Hamilton constructing quaternion. Italic characters represent two primary resistances he encountered. Thus, we can find the mangle in conceptual practice. It also contains the dialectic of resistance and accommodation. Based on that, Pickering and Stephanides claim that there is the same nature in conceptual practice and scientific practice.

58 Significance, types, and scopes

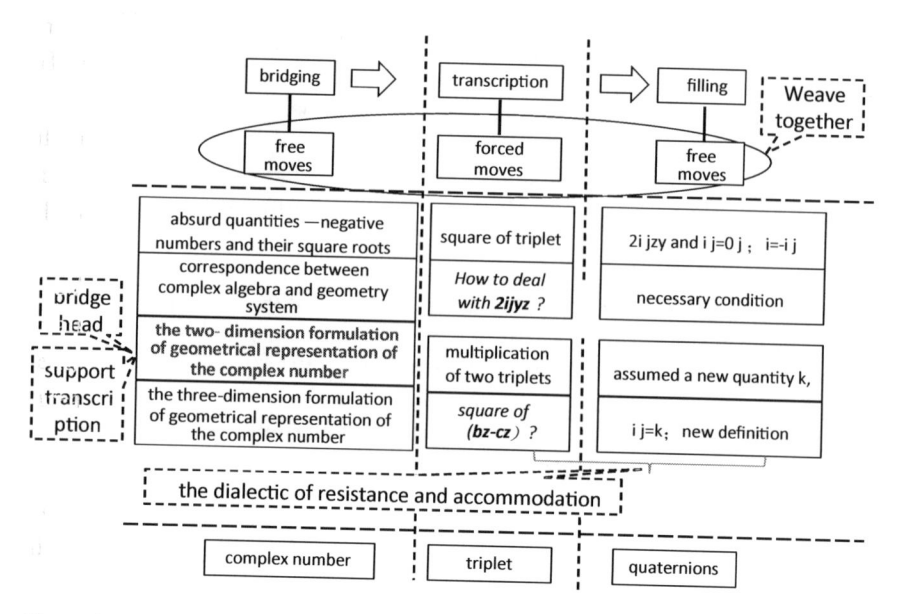

Figure 2.4 Hamilton constructing quaternions.

Then, the important significance of conceptual practice could be clarified as follows:

Firstly, Pickering and Stephanides think that the framework developed in experiments and sociological analysis can be applied also to conceptual practice. They show a picture of thinking of activities running their own as dividing into three sections: bridging, transcription, and filling. Hamilton's own thinking activities have become a problem in Pickering and Stephanides' studies. The result illustrates that the conceptual practice as thinking activity is a process in which free and forced moves are intertwined. It is also the dialectics of resistance and accommodation. Hamilton, as a practical subject, will yield to certain strength, whereas any scientist can find the resistance only by constantly trying in infinite possibilities.

Secondly, the study on conceptual practice is helpful to understand the debate between objectivity and relativity of scientific knowledge. Pickering tries to go beyond the old lines of argumentation, since in the old arguments objectivity and relativity are always described as totally opposite to each other. Pickering's view based on scientific practice is beyond both two views, while directing toward a true but also historical knowledge. In his opinion, knowledge is objective, relative, as well as historical. Hamilton's work allows us to know that objectivity is not absolute, but it also retains some elements of objectivity.

Thirdly, as an extension of culture, scientific practice is a process of open-ended modelling. This modelling process occurs in fields of multiply culture, and the generation of conceptual practice is also revolving around the connection of multiple cultural elements. Transcription in conceptual practices represents a complex discipline force, which causes the open and complex features of conceptual practice. Therefore, Pickering advocates integrating scientific practice and scientific culture together, and considers scientific culture as a cyborg. This huge combination is a joint between conceptual force based on conceptual operating and material force based on material operating, in which conceptual elements and other elements of scientific culture are co-evolution. 'Mangle' (by Pickering's words) puts scientific culture into a dynamic, historical, and practical background, while SSK theory is not almost involved in the development of social factors on their own.

Finally, Pickering's study on scientific practice is a reflection of PSP. People ignored the practical features of thinking activities when they study in general scientific practice, experimentation studies, and laboratory works. However, Pickering concerns and puts conceptual practice into scientific practice.

We have discussed the role of scientific practice in criticizing the traditional view of science and interpreted the meaning and effect of experimental practice and laboratory practice, which are the core concepts of scientific practice. According to SSK and PSP, a laboratory is not just the main place where experimental practice is carried out, but also something with a special role and significance. It highlights the importance of practice in scientific research and criticizes traditional view of science which neglected the role of practice. The discussion has focussed on the denotation of the concept of practice in a typological sense. We will discuss the nature of practice further in the next chapter.

Notes

1 About opportunistic nature of research, we have a special chapter for discussion.
2 Methodological principles on how to doing science are discussed in almost all the textbooks of *Dialectics of Nature* in China. Specifically see Huang Shunji, *Introduction to Dialectics of Nature*, Beijing: Higher Education Press, 2004, p. 112; Zeng Guoping, *Contemporary Guide on Dialectics of Nature*, Beijing: Tsinghua University Press, 2006, pp. 154–5; Xu Weimin, *Nature, Science-Technology, Society and Dialectics*, Hangzhou: Zhejiang University Press, 2002, p. 208.
3 Although it played an important role in China's reform and opening up, that the guideline of 'Practice is the sole criterion for testing truth', to be an academic view, it is characterized as aiming at representing theory of truth and realistic political power.
4 On the importance of practical know-how in research on genetic algorithms, see also: Holland: 'Hidden Order – creating complex adaptive' translated by Zhou Xiaomu, Shanghai: Shanghai Science and Technology Education Press, 2000;

Holland: 'emergence', Chen Yu, M., Shanghai: Shanghai Science and Technology Press, 2001; and Wardrop: 'Complex – was born in the order and Chaos edge science,' Chen Ling, Beijing: Joint Publishing, 1997.
5 See, the Ministry of Education and the Social Science Research Division Group with ideological and political work, EASTERN editor, 2004, 'An Introduction to Dialectics of Nature', Beijing: Higher Education Press, p. 116.
6 This section is written by Zhang Chunfeng, Wu Shuxian and modified, added by Wu Tong.

3 The nature of scientific practice

Pierre Bourdieu (1990, p. 80) suggests that it is not easy to discuss practice. In fact, it is not enough to merely discuss the concept of practice (including its contents, types and domains). That is because the discussion at most leads us to understand the types of scientific practice, but not the nature of scientific practice. The latter needs to be explained in naturalistic and normative ways which grasp problems such as: the origin of scientific practice, the mechanism forming norms, the natural evolution of norms and their stability, and so on. Rouse did not accomplish this task in his book *Knowledge and Power*, which focused on and criticized the traditional philosophy of science. However, in *Engaging Science*, Rouse pays more attention to the nature of scientific practice; and furthermore, in the book *How Scientific Practice Matters*, he puts forward that the nature of practice is the key to how scientific practice matters.

The nature of scientific practice involves another extremely important issue, namely the normativity of PSP itself. Does scientific practice as a core concept possess normativity or not? If it does, where does the normativity come from? From inside or outside? How does the normativity develop? Is the norm a priori or empirical? Answering these questions will impact directly on understanding and grasping the nature of scientific practice.

3.1 The nature of scientific practice or various features

Rouse devotes a whole chapter to discuss the nature of scientific practice, in Chapter 5 of *Engaging Science*, to characterize the normativity of scientific practice, to differentiate the concept of practice in PSP from other concepts of practice in other philosophies, as well as to respond to some criticisms and refutations.

First, it should be noted that this kind of interpretation is based on the hermeneutic tradition of continental philosophy; Thus, the hermeneutical features of the concept of scientific practice are presented not only in the style but also in the elaborations of explanation. Especially, this kind of interpretation develops through criticisms and refutations that Rouse and other scholars conduct, the latter comes from a variety of different representative philosophical stands. Rouse often combines different points of view, endorses a certain view against another one, or takes a reasonable part of his view in order to criticize some unreasonable ones. Our task is to clarify the nature of the concept of scientific practice. And we should first introduce and elucidate Rouse's account on the nature of scientific practice as precisely as possible. Then we might be able to grasp the nature and meaning of scientific practice in a hermeneutic approach.

Rouse claims that his account for scientific practice could be summarized in ten propositions:

1. Practices are composed of temporally extended events or processes;
2. Practices are identifiable as patterns of ongoing engagement with the world, but these patterns exist only through their repetition or continuation;
3. These patterns are sustained only through the establishment and enforcement of 'norms';
4. Practices are therefore sustained only against resistance and difference and always engage relations of power;
5. The constitutive role of resistance and difference is a further reason why the identity of a practice is never entirely fixed by its history and thus why its constitutive pattern cannot be conclusively fixed by a rule (practices are open to continual reinterpretation and semantic drift);
6. Practices matter (there is always something at issue and at stake in practices and in the conflicts over their ongoing reproduction and reinterpretation);
7. Agency and the agents (not necessarily limited to individual human beings) who participate in practices are both partially constituted by how that participation actually develops, and in this sense, 'practice' is a more basic category than 'subject' or 'agent';
8. Practices are not just patterns of action, but the meaningful configurations of the world within which actions can take place intelligibly, and thus practices incorporate the objects that they are enacted with and on and the settings in which they are enacted;
9. Practices are always simultaneously material and discursive;
10. Practices are spatiotemporally open, that is, they do not demarcate and cannot be confined within spatially or temporally *bounded* regions of the world (Rouse, 1996, pp. 134–5).

3.2 The account and interpretation of the hermeneutic nature of scientific practice

Rouse describes ten characteristics of scientific practice in the hermeneutics way. At the first sight, we might say that Rouse's account of scientific practice is very unique. When we talk about practice, we might firstly conceptualize it in a Marxist context, in which practice is a kind of activity in order to change the world, and to actualize the agent's intention in objectivity. Here, it first refers to the subject of practice, the active man, and then refers to the object of practice, the material world. For Rouse, practices are composed of temporally extended events and processes. What is the uniqueness of such a definition or description of practice? And what is the content and meaning of its ten characteristics? These questions need to be explained, elucidated, and commented concretely. In order to answer those questions, we need to integrate the ten characteristics for the purpose of interpreting important aspects thoroughly.

3.2.1 The constitution or becoming of practice

(1) Initially, Rouse points out that practices are composed of temporally extended events and processes. Why does Rouse treat practice as processes or (time related) events, or put the temporality of practice first? There is no doubt that scientific practice is the practice of human beings as agents. However, Rouse does not emphasize the subject of practice just because the extended temporality of practice is a more significant implication. What's the meaning of the extended temporality of practice?

First, the meaning of term 'practice' is doing something; doing is trying and repeating actions, which contains a repeating or other continuous forms of behaviour. In practice, some of the events and processes are repeated; human beings, as the agents and participants of the practice, engage in such a repetitive and ongoing process of dealing with the outside world. Because the influences come from the outside world, a certain approach will be gradually remained and formed in the perceptual experience of human beings during such a process; or some distinguishable or repetitive patterns begin to stand out. Unnecessary actions will be abated, and the error will be realized and removed in the repetitive process. Obviously, those unnecessary actions are uneconomical since they need more physical power. By simplified and repetitive behaviours, events and processes in perception will be fixed in perception. Meanwhile, their impressions will be more profound, and their stimulations on the agents will be more stronger. Since practices involve a variety of activities, Rouse describes practice as event or process, which must be some describable actions. That identification of practice broadens the notion of practice, since this kind of practice can keep the possibility of unidentified, in which practice can combine the apparatus of actions with actions itself and produce the normative patterns and cognition of practice.

Second, emphasizing the extended temporality of practice involves the meaning of distinguishing cognitive practice from knowledge obtained by practice. The temporality of practice reflects its differences from knowledge achievements. Meanwhile, by the temporality of practice, knowledge itself reveals its time or historical and local dimensions, which opposes to a static account of knowledge. Compared to the traditional view of scientific knowledge, the view of scientific practice always holds that scientific knowledge is necessarily located within practices. Such a location necessarily requires knowledge to be local and temporally extended (can be modified, accumulated and

64 *Scientific practice*

replaced). The traditional account of scientific knowledge precisely opposes to this, just as Rouse argues that in the traditional account of scientific knowledge,

> Knowledge has not usually been thought to be temporally extended in the way that practices clearly are: on most accounts of knowledge, whenever knowledge exists, that knowledge is fully present.... The knowledge itself is a state rather than a process. The reasons or reliances that warrant it as knowledge may be temporally extended, but knowledge itself is something full present.
>
> (Rouse, 1996, pp. 135–6)

When Rouse discusses the temporality of scientific practice as the ongoing narrative reconstruction pf science, by absorbing knowledge into practice, it not only makes knowledge itself temporal, but also criticizes the static account of knowledge which once had a ruling role. Knowledge therefore becomes a historical thing, neither fully present nor no-time present. Pickering agrees on this account, who pointed out that, '... I seek a real-time understanding of practice' (Pickering, 1995, p. 3), and that '... the goals of scientific practice emerge in the real time of practice' (Pickering, 1995, pp. 19–20). Some important physicists try to introduce temporality into physical sciences in a naturalistic sense, like what Prigogine did.

Bourdieu, a French sociologist, also emphasizes the importance of temporality in practice. He points out that

> Practice unfolds in time and it has all the correlative properties, such as irreversibility, that synchronization destroys. Its temporal structure, that is, its rhythm, its tempo, and above all its directionality, is constitutive of its meaning. As with music, any manipulation of this structure, even a simple change in tempo, either acceleration or slowing down, subjects it to a destructuration that is irreducible to a simple change in an axis of reference. In short, because it is entirely immersed in the current of time, practice is inseparable from temporality, not only because it is played out in time, but also because it plays strategically with time and especially with tempo.
>
> (Bourdieu, 1990, p. 81)

This passage is an excellent comment on Rouse's action of putting the temporality of practice in the first place.

(2) There are agents involved in practice, of course. However, the category of agents are less important or fundamental than that of the circumstance of practices; it is not the most important element of practice. Bourdieu called the circumstance of practices as practical field. There seems to be some problems in such an elaboration. Someone may ask how practice exists without agents. Nevertheless, practice is dynamic rather than static, since there are agents participating in it, taking their roles. Therefore, everything is in the process of practice. There is no doubt that practices need agents and practices are agents'

practices. But in practice, agents are not a prior condition or mechanism that determine practice; instead, agents are determined by the processes of practice, by the fact of how they participate and how practices develop after their participation. In his seventh thesis, Rouse pointed it out that 'agency and agents who participate in practices are both partially constituted by how that participation actually develops, and in this sense, "practice" is a more basic category than "subject" or "agent"' (Rouse, 1996, p. 135). Here, Rouse's account of practice is a continuation of the thought of Aristotle. Aristotle said that 'For the things we have to learn before we can do them, we learn by doing them, e.g. men become builders by building and lyre-players by playing the lyre' (Aristotle, 2009, p. 23). Moreover, the agents of practices are not necessarily limited to individual human beings. For Rouse, we can see that practice is a more important and basic category than subjects. In Rouse's perspective, to say the former is more basic means that it focuses on the extension of processes and events in the temporality; While to focus on the latter would pay attention to the behaviours of subjects, which is more like a psychological issue than an issue of PSP.

Rouse thinks that the agents' affects on practice require much more attention and extended discussion. Just as Rouse points that various humanist interpretations of practice focus on what seems to be the obvious and crucial difference between practices and other recurring patterns within the world, namely, that practices are always done by someone or others. Moreover, something is crucially at stake in this point: agents, who engage in practices, may have moral and epistemic obligations to one another which differ significantly from their appropriate relations to nonagents (Rouse, 1996, p. 143). Indeed, practice always involves doings and doers, along with what they are done with and done on; practice is always associated with these things. That is very crucial, since it indicates that practices are not just what agents do but also what have been done by agents are intelligible; Therefore, practices are located within the complicated relation. Hence, we can conclude that the agents who engage in practices thus belong to the practice, rather than the reverse. Of course, the same agents who engage in one practice also typically participate in many other practices, so there are problems to define practice through the agents. In this sense, agents do exist outside the practice in which they participate but not being a part of any of those practices. Thus the agents do not have an identity which could be fully separated from the practice in which they participate. Thus, the identity of agents is partially constituted by those practices, and/or transformed by them. Therefore, the agents 'belong' to practices in both a weaker and a stronger sense. On the one hand, an adequate description of a practice necessarily includes a characterization of the agents who (might) participate in it. In a stronger sense, an agent belongs to a practice to the extent that an adequate understanding of the agent (and the agency of the agent) includes an account of its participation in that practice (Rouse, 1996, pp. 143–4).

Rouse points out that the significance of this point turns on the basis for distinguishing agents from nonagents. Rouse takes the view that for Dreyfus and Winch there seems to be an objective difference between those events that belong to the domain of the natural sciences and those which cannot be adequately understood unless taken as skilful comportment or rule-governed action. R. Brandom denies an objective difference between translatable practices and explainable actions, for he regards the same events to be understandable in either way. He insists that the difference between agents and nonagents is instead a social and practical difference with comparable moral and epistemic consequences. Just like Brandom, Rouse denies any objective difference between agents and nonagents. For Rouse, these differences are constituted within practices and intelligible only in terms of discriminations that get their identity and significance from practices. Nevertheless, community judgments are not fully authoritative in the way Brandom proposes (Rouse, 1996, p. 144).

The difficulty once again is that the identity and boundaries of communities may not be specified clearly or unambiguously. Who can be counted as an agent? What kind of individuals constitutes the community? Is it supposed to be determined by the responses of members in that community? In fact, we just need to make a sufficient agreement within the community about what counts as an appropriate performance of each of the practices. Yet an agreement that is sufficient to enable practices to proceed without infinite regress may nevertheless still involve resistance and dissent. The community may have either a tight structure or a loose connection. Not everyone participates equally in determining what can be counted as community responses, and the dominant interpretations are nevertheless not equally authoritative everywhere. Thus, the boundaries of the community of agents have often been contested and ambiguous: differences of gender, race or nationality, social status or class, and so forth, which have frequently marked out these contested regions. The ambiguity results from the interconnectedness of the boundaries of practices. Just as Fujimura described, only the integration of boundary objects could push forward the development of practices. If the difference between agents and nonagents is established within practices, the difference is determined neither subjectively nor socially. What matters most is that the difference is interconnecting with the world, which is determined by the ongoing practical interaction.

According to Rouse, the viewpoint of the agent is very important for understanding scientific practices. There is something wrong with the social constructivist accounts of scientific practice. In practice, we cannot say that social practice is most important. Although social constructivists hold that the authority of communities is most important on norms of correct or incorrect performance, which Brandom insists in an unconstrained way, they cannot be accounted for either as objectively determined (whether by the real nature of things or rational necessities of inquiry) or as socially determined. The normative configurations of the world, which emerge from practice, must be understood as prior to any distinctions between agents and their environment, the social and the natural, or human and nonhuman (Rouse, 1996, p. 146).

Besides, Rouse points out that the boundary between agents and nonagents is controversial, which cannot be resolved either objectively or socially. It is especially important for understanding the political and cultural significance of science and technology. The controversies about the boundaries are all scientific and technological practices. The boundaries that can be regarded as the final purpose are frequently at issue. Obviously, the agency and moral dignity of ecosystems, endangered species, or the biological environment have been fundamental questions for environmentalist politics and science. Many feminist theorists have found that gender politics are not just about relations among men and women but are focused precisely on how to understand agency, body, rationality, and the boundaries between nature and culture, which theorists regard as central to feminism. Questions about the moral standing of human research subjects, laboratory animals, foetal tissue, or genomes have been addressed directly at research practices themselves (Rouse, 1996, p. 137). But the social constructivist notions of practice are not appropriate to dimensions of contemporary scientific and political understanding, since it is the correct dimension to exclude the dimensions of nature, practices, and culture, which settle the dimension of social interests.

(3) Practices involve resistance and difference, which resist the patterns of practice. And the resistance and difference sometimes have constructive effects on the form of practices and its norms. Practice can not continue if it has no pattern. Since resistance and difference are obstacles to the forming of patterns of practices, practices always need to conflict with resistance and difference. The so-called resistance and difference in practice are led by problems and opportunities, as well as the opposite of existing practice led by lack of resources and conditions. If practices could overcome such reverses, then the formed patterns could be strengthened by practice, otherwise the existing patterns may be destroyed, dissipated or disrupted. The existing patterns may include potential instability, or no longer adapt to new situations or economics. In fact, such a breaking of patterns may help to form a new one. Therefore, the resistance and difference in practices can also play a constructive role. Moreover, this indicates that the patterns of practices could not come into being for only one time and be fixed historically. The patterns of practices are evolutive, changing with time and situations. It is this point that opposes to *a priori* normativism while insisting that practices could be normative; Their norms are empirical, a posteriori, historical and evolutive, and determined by the circumstances of practices. We name this as 'norms of circumstances' (Yu Jinlong, 2007; Yu Jinlong, Wu Tong, 2007). If we observe various kinds of scientific experiments, we can find out that any stereotype of experiment is always repeated over and over again, as well as the proposal and stabilization of any scientific concept, and the establishment of any theory is also the same. It implies that the establishment of patterns of practices and their outcomes are always an evolutive process.

Therefore, Rouse insists it is central to the notion of practice that practices are understood as enforced by agents. It implies that practices can exist only against a background of resistance. The resistance of practices encompasses both the resistance of things and the nonconformity of agents. And such a resistance also contains maladjustment to the existing patterns of practices. New practitioners must be socialized into scientific practice. Based on the deflection of their predecessors, they must be deflected, suppressed, or contained as the ways opposed to their original intention. In addition, the environment must be rearranged in ways beneficial to their ongoing reenactment. Consequently, practices always include and permeate power relations. That is another issue about practice which we shall discuss next.

(4) Power is another factor in practice. There are often some misinterpretations about such power. For one thing, what is the role of power in practices? Is power internal or external? For another, is it a violation of norms if we attribute the establishment of norms of practice to the engagement of power to some extent? There are important and close correlations between power relations and the practices which embody scientific knowledge. However, these correlations are always misunderstood or ignored, not only because of the embodiment of scientific knowledge, but also because of many standard interpretations of power within sociological and political theory, which are not sufficient to interpreting the power relations within the local practices of scientific knowledge. According to Rouse's discussion about the relation between knowledge and power, that kind of power is internal – microscopic, in a sense of Foucault, being a disciplinary field.

First, those relations of power which constitute the processes of practice could not simply be characterized as struggles between the defenders and opponents of particular practices. Such a description is inappropriate since it mistakenly suggests that power relations have been fixed on agents, rather than on the ongoing reenactments. Taking practices to be fixed in such way underestimates the importance of their normativity as well as their lack of temporal closure. Instead, just as Rouse says, the real characteristics and significance of practice is often at issue, and there are always conflicts and differences during its ongoing reenactments. As a result, practices are radically open: whether a subsequent action counts as a continuation, transformation, deviation, or opposition to a practice is never fixed by its past instances (Rouse, 1996, pp. 140–1). It implies that power relations in practices are not the one in the ideological struggles, but the relations internal to the processes of practices; the generations and solvents of conflicts and differences during such processes are themselves the generation, transformation and extension of power relations.

Second, the power relations reflect resistances and differences in practices. In other words, they reflect the contradictions and conflicts in a particular practice. In fact, power relations are those conflicts and struggles between the existing norms and the new resistances and differences. These resistances and differences are subtle and local. When new norms come into being from such resistances, new power, as disciplinarity will take hold. This is obviously correlated with the openness of practices. Therefore, it is important and fundamental to put the openness and temporal extension of practices in the first place. Conversely, temporal extension of practices, namely the conflicts of practices and their interpretations, is the competition during the history process which could also constitute rival interpretations.

However, such conflicting interpretations of practices need not to be articulated explicitly. We can also use common participations to explain a particular practice. Such common participations would involve the participations of power relations in a disciplinary sense. People continue practices since they participate in the reinterpretation of the practice. Hence, practices are open to semantic drift as well as principled reinterpretation. In the ongoing reproduction of practices, the continuity of practices depends importantly on the extent to which such continuity is enforced by normative response. Practice is policed thoroughly and rigorously to discourage 'deviation': the less deviation it is, the more susceptible to drift. However, such policing gets no guarantee from the community in a practice. The normative responses in policing are themselves parts of the same practices.

Here, we can give an example about the reinterpretation and the changes of power.

There are four views about the interpretation of meaning of the text: the meaning of the text author, the meaning of the independent text, the meaning of the text-reader, and the meaning of the readers.

The view of the author is that the meaning of the text is endowed by the author. That emphasizes the effects which the author has on the text. Therefore, if we intend to understand the text, we need to understand the author and the context by his creation.

The meaning of the independent text is the meaning of the text is determined by the text itself. Once a text is produced it has its own independent meaning which is not determined by author's intention. For example, the theory of self-organizing holds the view that things themselves have the capacity to develop orderly, as long as the conditions are appropriate.

The meaning of the text-reader means that the meaning of the text is endowed by both text itself and the readers. When a text is accomplished, it has its own historical construction of meaning. When readers encounter the text, the meaning of text is no longer independent since it has absorbed the understandings of the readers. The meaning of the text is a result of integration both of the history and of the current horizons.

The meaning of the readers means that the meaning of the text is endowed entirely by the readers. The meaning of the text drifts and depends on the people who will read it.

There are conflicts among those views if they are fixed. However, those four views could be fused partly in the practice of textual evolution.

The evolution of 'text-author-reader' presented above will lead to power relations and translations, which occurred in the process of meaning production and interpretation. Thus, during the first stage of the evolution process of text, that is, the intention of author comes as first before the text is produced. When the work has been accomplished, the intention of author takes a back seat and the meaning of the text itself becomes more important. When the text is read by readers, there are interactions between text and readers, and the meaning of the text is produced both by readers and the text. Therefore, the context gradually fades out, and the readers have become prominent increasingly, and the meaning constructed by the readers plays a more and more significant role in the generation and defence of the text. It seems that every process of generation and reading of the text involves some dominant ideologies to present the dominance relation and its evolutions.

The interpretation of the generation of meaning reflects the power relations among text, author and readers in the process of the construction of text meaning. It also presents the process of rewriting practice norms.

3.2.2 The relation between practice and patterns of practice

Moreover, because of the repetition of events and processes, practice is not just an action in the process and circumstance constructed by agents, but constitutes as various patterns of events and processes. In fact, before Rouse there have already been various humanitist accounts of practice and its norms. The most obvious difference among those accounts is the identifications of what governs the continuation of patterns. For Winch, it is rules; for Sellars or Brandom, it is community norms; for Dreyfus, it is the purposiveness of skilled activity. It seems that these kinds of view look similar. However, Rouse points out that there is a significant distinction, which is whether what governs the continuation of such patters is identifiable apart from its actual instantiation, or whether the patterns exist only through their continuing reenactment (Rouse, 1996, pp. 137–8). Winch here differs from Sellars, Brandom, Dreyfus, or Rouse. Winch thinks that a rule that governs practice could be articulated explicitly even in the absence of any actual activity in conformity to the rule. But for Brandom, no matter which event instantiates the practice, the communities that actually engage in a practice are the final judge for patterns of practice. Just as Brandom argues, 'The respect of similarity shared by correct gestures and distinguishing them from incorrect ones is just a response which the community whose practice the gesture is does or would make' (Brandom, 1979, p. 188). What Brandom means is that there is no question of conformity or nonconformity to a practice when it is absent of the right kind of community. However, there are some problems in Brandom's account. For Brandom, what constitutes the community is those people who share the same practices. In other words, the constitution of community need to be explained by practice, but practice need to be explained by community. So there is a hermeneutic circle. Moreover, in that circle, there needs to be no objective criterion of correctness apart from the holistic regularities manifested in the ongoing activities and responses of community. Hence, in the absence of a sustained tradition of practice, no pattern would fix its identity. In Brandom's account, there is a significant point that the practice is open and evolving with the community.

Dreyfus denies the patterns which make up practices as identifiable apart from their actual instantiation in various circumstances. First, skills do not exist outside the agents. Second, the skilful simulation manifest in practices is a flexible responsiveness to situations. Its purposiveness is not confined to predetermine formulas or already explicit purposes, but a response to the situation. Underlying Brandom's and Dreyfus's arguments is a common foundation, which is also a common theme: practices are open to be temporally extended as flexible patterns of response. At that very point, Rouse agrees with them. Or maybe Rouse draws on both Brandom and Dreyfus to form his own view that practices are temporally extended. Practices can be extended in novel ways, and therefore cannot be confined to rules specifiable in advance. That is the fundamental explanation about the constitution of norms in practical circumstance. Here, Rouse's account of practice draws on both Brandom and Dreyfus: practices require skilled comportment and social norms. Such norms are constituted according to and flexible to the circumstances (Rouse, 1996, p. 136).

Practices can, of course, die out and the norms may be destroyed, when the pattern of activity is no longer reenacted, or when the circumstances within which those activities would be intelligible no longer obtain. Skills can be damaged, communities based on some practices can dissolve, and social norms about whether practices are suitable can be also abandoned. However, even if that happens, the present and the future of practices can be constituted gradually by further actions, events, and processes without specified norms. Norms do die out without practice. However, when practices activate, the norms can be reactivated, and even long-abandoned practices can be reactivated. That claim does not contradict the earlier claim (that is the patterns composing practices do not exist apart from their actual instantiation). Thus, practices always include a horizontal future (Rouse, 1996, p. 136).

According to Rouse, a practice 'includes' its future in a very strong sense. That is because the nature of practice at present depends to some extent on how its future develops. This identification of present practice with the a future is heightened in the case of scientific practices, which are oriented toward the disclosure of what is presently unknown. In Rouse's points of view we can know that the present practices often include many dimensions open to the future which we do not know at present. But as time goes by, the perspectives and problems in cognition included in the practices and their circumstances will unfold and present themselves gradually, thus something of the future will appear. That is not mysticism, but illustrated from scientific discoveries. Here, Rouse takes Hans-Jorg Rheinberger, a biologist, an example, since Rheinberger put this point in discussing the beginning of modern cellbiology and virology:

> In the spontaneous recurrence of the scientist the new becomes something already present, albeit hidden, as the research goal from the beginning: a vanishing point, a teleological focus. Without the avian sarcoma virus of 1950 [Peyton] Rous' sarcoma agent would have remained something different. But: The virus pf 1950 must been seen as the condition of

possibility for looking at Rous' agent as that which it had not been: the future virus. The new is not the new at the beginning of its emergence.

(Rouse, 1996, p. 137)

This case illustrates that practices, especially objects, activities and their norms of scientific practices, are open to the future and becomes more and more clear. Thus, it explains that practices are composed of temporal events and processes. It also elucidates metaphorically but unambiguously that the norms of practices are constituted gradually and more and more clearer according to the explicitness of practical circumstances.

3.2.3 Patterns of practice and their norms

In Rouse's perspective, the patterns of actions and events which constitute practices are in a crucial sense normative. By comparing the differences among Winch, Bradom and Dreyfus's views, Rouse puts forward his own ideas about the relationship between practices and patterns or norms of practice.

Rouse believes that a pattern can constitute practice, but that is not due to its correctness. In other words, not only the correct but also the incorrect patterns can constitute practices, though the incorrect norms of practices may lead to wrong actions and results. Scientific practices can not only generate the correct norms, but also the incorrect ones. The important question is why it is also normative.

Winch offers the first account. He argues that an instance of correct practice is to conform to the rules constitutive of a practice.

For Brandom, instances of a practice are determined entirely by the response of the community engaging in the practice. A performance is correct only if the community responds to it as correct. Thus, practices always encompass some events as differential responses to the correct or incorrect continuations of the appropriate patterns, and the practice is sustained by the effectiveness of these differential responses in creating and sustaining patterns of mostly correct continuations. Moreover, such differential responses to correct and incorrect practice will often be counted as normative responses only on the grounds of further differential responses to them. The result is not an infinite regress but an interconnected network of practices identifiable only by interpretive participation and translation (Rouse, 1996, p. 138).

Dreyfus offers a third account. For him, success and failure in attaining the proximate end of the practice are both constitutive for correct performance. For example, any activity that succeeds in driving nails is a correct instance of hammering. However, the process of driving nails is always accompanied by practice of incorrectness and their modifications. Here, we must recognize that social norms are necessary to practices, since 'without social norms there would be no way to set up the complicated referential totality involving nails, boards, houses, etc., in which driving nails has its place, and so no role for [hammering]'. Furthermore, skilful practice is constituted more fundamentally by proximately successful performance. The most important thing is that the norms which make sense of the ends of skilful activity can only constitute practice like hammering but not 'the natural kinds and causal laws that make nails and pounding possible' (Rouse, 1996, p. 138). According to all three views, the 'norms' policing and thereby maintaining practices need not be explicitly articulated principles, nor do they necessarily function as justification for the practice they sustain. Norms and practices are going on all together.

Rouse's view draws on both Brandom and Dreyfus's accounts, though he rejects Winch's view since it rules out the possibility of changing and developing practice. Violation of a rule that previously encompassed merely correct instances of practices may nevertheless mark the continuation of a significant pattern. But in the following aspects, Rouse thinks that Brandom and Dreyfus also make some mistake, for both of them propose a final judge for the norms of practice. Brandom concludes that how the relevant community responds to the practices is constitutive of practical norms. And Dreyfus appeals to the implicit telos of skilful activity. Moreover, the finality of these criteria stems from their significance. They understand it as a free human subject: a recognized member of a community or a skilful body. Brandom and Dreyfus only claim that human being can turn into subjects through participating practices, but what it is to be a subject is not situated historically and culturally. Therefore, in any case, neither of them could propose the criteria which can have alleged constitutive finality, since for them neither belongs to a community, nor what would count as a significant end for assessing the success of skilled practice can be fixed in advance.

Rouse believes that the patterns constitutive of practices require ongoing adjustments of agential relations to those they work with, to what they work on, and to whom and what they work for; None of these relations automatically takes priority in sustaining an ongoing pattern of practice (Rouse, 1996, p. 139).

3.2.4 The most important issues in practice

Rouse realizes that there are difficulties in specifying a practice by its constitutive patterns, since practices are defined not only by specific activities that compose them but also by what is 'at issue' in the practices, and therefore by what is at stake in their success and continuation (Rouse, 1996, p. 142). In his sixth thesis, Rouse indicates that practices matter. Why is it important? To what extent does it constitute the nature of science? In Rouse's perspective, practice does matter in a crucial sense, in which one cannot understood practice unless she has already grasped what is at issue in the practice. In other words, Rouse pays much attention to the key points in practices, and stresses our recognitions of those points. The issues and stakes in practices are not only prevented from reinterpretations and drifts, but also constructing the patterns of ongoing practices which involve the conflicts upon those issues.

What issues do practices focus on? What stakes do practices argue for? Or maybe we can ask questions in a reverse aspect. What issues do practices not focus on? What stakes do practices not argue for? There is no common rule to answer these questions, while the answers must be searched through some specific practice. The point here is that, according to PSP, the most important thing in practices is what issues do agents mostly concern. The issues of norms by which people follow retreat to the back. The most crucial thing is the questions on what is at stakes or what is at issue. What we can know at most might be that there are something temporal and with uncertain content which impacts the norms of practices.

3.2.5 Practice involves not just patterns of action, but also meaningful configurations of the world

Various humanist accounts can be understood as attempts to reinterpret practice from the view of intentionality. Practice could be interpreted as rule-governed behaviour, responsive social recognition, and skilful comportment. Those accounts are all different alternatives to mental representation of the world. However, the humanist interpretations of practice still display an important continuity with the Cartesian tradition. The Cartesian dualism still has influenced these accounts. In this sense, we can still classify the humanist accounts of practice into 'inside' and 'outside', no matter whether the 'inside' of practice is a constitutive rule, pattern of social recognition, or skilful engagement.

In order to understand the difference between PSP and humanist accounts, it will be helpful to take a survey of the argumentative lines on intentionality. Appeals to practice as naturalistic accounts treat meaning as fully explicable in causal processes of information flow or as naturally selected responses to an environment. Non-naturalistic accounts typically characterize meaning as constructed from, rather than discovered in, the natural world. Non-naturalism locates the meaning in the 'intrinsic' intentionality of the mental. However, the humanist accounts of practices often locate intentionality in the irreducible interaction of subjects with surroundings. In the most prominent versions, this interactive construction of meaning is accomplished either by interpretation (Davidson), social recognition (Brandom), or skilful comportment (Dreyfus' interpretation of Heidegger). Rouse thinks that each approach captures something right, but all these three go wrong because they still let a vestigially subjective construction of meaning opposed to its naturalistic emergence.

Brandom proposes his idea about traditional distinctions between subjects and objects. He indicates that the distinctions should be like this: 'Social practices constitute a thing-kind, individuated by communal responses, whose instances are whatever some community takes them to be. Objective kinds are those whose instances are what they are regardless of what any particular community takes them to be' (Brandom, 1979, p. 188). In this distinction, 'subjects' are not individual minds but reconstructed as social patterns of recognitional response. In effect, Brandom distinguishes the 'social world' from the 'natural world'. In the former community responses are authoritative, whereas in the 'natural world' the determinations of objects 'are what they are regardless of what any particular community takes them to be' (Brandom,

1979, p. 188). Here, Brandom tries to stake out a middle ground between naturalists, for whom the meaningfulness of the 'social world' is naturally explicable, and social constructivists, who insist on the ontological priority of the social world through which community responses determine what is (that is, what can count as) a natural determination. By identifying these positions, Rouse criticizes all of three positions in general but to draw on useful ideas from them. For example, Rouse points out that social constructivists are right to recognize that 'natural' objects do not 'have' identities or determinate properties and relations out of practices, but they are mistaken (along with Brandom) in taking community responses to be authoritative within the field. As we have already suggested, the identity of the relevant community member may be at issue in articulations of practice, while the results could not yet be confirmed. Part of the error is in understanding the significance of the materiality or 'embodiment' of social practices, as well as overestimating the clarity and unity of 'communities' and their responses to new performances. In effect, if we can understand practices as the field within which both the determinations of objects and the responses of agents emerge as some intelligible things, such a notion can be still non-naturalistic, i.e. the normativity of practices can not be reduced to causal relations. It even eliminates the vestigial sense that meaning is bestowed by the activity of subjects (Rouse, 1996, pp. 148–9). The subjects can be elucidated in two distinct ways through the practices which are meaningful configurations of the world. First, who can count as agents, along with what it is to be an agent, is historically situated, which emerges from practices rather than any other way. Second, the ongoing patterns that compose practices are patterns of the world in which the activities are situated. The world in which practices engage gains significance from the ongoing practices. It is the practice that configures the subjects, not in a reverse direction. The world obtains its configuration and significance in practices, which is a process of producing meaning, rather than of imposing something meaningful onto a meaningless world.

Rouse points out that the contribution of Dreyfus's account to the significance of practices is twofold. It emphasizes the bodily aspects of practices, and then it recognizes that the identity of practices depends on a right kind of situation or environment as well as on a right kind of activity. For Dreyfus, the paradigm of intentional direction is neither linguistic nor perceptual representation, but skilled bodily activity. Such activities are gradually emerging repertoires of bodily possibilities. Dreyfus emphasizes that practices often draw on the material contours of bodies of agents to accomplish skilful activities. According to Dreyfus's first account, Brandom's account of social practices fails, since it is an oddly return to a functionalism in philosophy of mind. It seems indifferent to how social practices are materially realized, but merely cares about theories of language as an ideal structure.

For Rouse, the second contribution of Dreyfus's account is more crucial. To be sure, we know that skills belong to things in some aspects. For example, drawing deftly with a pen responds to it differently from accurately throwing.

Today, people make a laser pointer in the form of a pen the main function of which is to give presentations. Here, the function of instruction is different from writing. Thus, the relevant 'properties' of things depend on their particular purposes. What's more, skilful use is always responsive to their particular capacities. Someone who knows how to write readily adapts to the distinct characteristics of ballpoint, writing brush and fountain pens. Furthermore, we may ask what will happen if one engage in skilful activity without the right sort of equipment in the right surroundings. If there is no substitute, he will not be able to engage in skilful activities. For instance, if a table-tennis player uses another player's bat, even though the two bats are similar, he will not be able to play as well. Therefore, in the absence of an appropriate object, skilfulness decreases. If he is attempting deliberately to use other's bat to hit a similar strike, it requires a physically different movement. In such a practice, it requires a constitution of significantly different activities, since one's bodily comportment responds to the particular resistances actually offered by the object. Just as when someone makes use of another's bat, he will use a slightly different force and angle to hit the ball. Therefore, Rouse indicates that in any skilled activities as a mode of intentional direction, the relational complex of the agent situated in a particular scene is important prior to both actions and purposes of agents, and also prior to any independently definable properties of surrounding things. Things become manifest according to their significance for possible action, while the activities and possibilities are configured by what the surroundings of the agent afford (Rouse, 1996, p. 150).

Thus, in a much more abstract and profound sense, Rouse characterizes practices as situated patterns of activity (Rouse, 1996, p. 150). Here, Rouse puts forward his own account of practices. According to Rouse, the situation involves a sense of priority, which is the relational complex of embodied agents in meaningfully configured settings for possible actions. We should also extend that priority of the situation to the agents and objects and their incorporated interactions. That is one reason why we call normativity of practices as normativity of situation. Traditional representationalist believes that the agents 'found' the patterns of practices depending on an 'original' intentionality of skilled comportment. In effect, this case can only acquire their significance as skilful practice within larger configurations of practice. Admittedly, the dependence of the supposedly 'imaginary' practices on appropriate material settings is more complicated, since most interesting practices involve multiple relations of objects in the world, which not only can function as equipment, proximate objects, obstacles, but also can function as surrounding environment and much more. Practices also involve complex social relations and norms in more obvious ways. It just reminds us that even though we can overcome the complexities in practices with ease, all analysis of the situations are still complicated in representations.

Admittedly, scientific practices convey knowledge, not just practice. It leads people to obtain skilfulness and capacities in the local situations through practices. It also promotes people to master experience and knowledge which practices highlight. It includes not only manipulations and skilled performances, but also acquaintances and discriminations with these performances. We can not identify these skilled comportments as all kinds of practices, just as what practices in laboratory look like. Not even we can disentangle practices abstractly from social norms and linguistic discriminations that are supposedly 'founded' on individual coping skill. Moreover, scientific practices, even these especially clear cases of activities, can not be sensibly understood apart from their local material settings. That is the central point of Rouse's extended discussion in *Knowledge and Power* (Chapter 4) on the construction and manipulation of controlled micro-worlds, within the institutional settings of laboratories, clinics, or field sites.

Therefore, it is appropriate and important to recognize practices as embodied and situated. However, if we account for intentionality as founded on skilled comportment, such accounts will be somehow inappropriate. We are unable to account for the normative determination of skilfulness by construing skilled comportment as a basis for intentionality without referring to social responses in putative skills and achievements.

The most important thing remains that practices should be understood not as a substitutional notion for human subjectivity, but as a larger relational complex in which subjects (agents) and subjectivity (agency) are constituted (Rouse, 1996, p. 152). The central claim is that the meaningfulness of agents' situation is not bestowed by either individual comportment or social norms in abstraction from their material realization; Instead, it emerges from the ongoing interactions of agents with their surroundings. The situation in which agents locate themselves is already meaningful, not because meaning is grounded in natural causality, but because agents are always responding to specific configurations of possibilities for action, which emerges from past practices. Thus, the temporality of practices becomes an alternative to both naturalistic and subjectivist accounts of intentionality (Rouse, 1996, p. 153).

3.2.6 *Practice is always simultaneously material and discursive*

In practices, it is often intelligible to understand materiality as the content of practices. Yet it is difficult to understand the discursive engagement of agents as the content of practices. To be sure, it is an extension of scientific practices that discursive practices have become one part of scientific practices. First, we must notice that we treat discourses as practices rather than as meaningful representations of states. This is a criticism on the general representationalist account. Rouse indicates that we should understand language use in terms of discursive practices rather than as meaningful representations of states of affairs. This standpoint opposes the representationalist accounts of language in scientific practice and knowledge. Second, Rouse points out that we can not diminish the discursive dimensions of practices. If it happens, the extension of practices will scale down, and the practices in scientific representations will be lost, which will lead the analysis of practices not to be involved in texts and discourses.

Recently, there is a tendency in philosophy, sociology and historiography of science to reduce the linguistic dimensions of scientific practice to something like 'rhetorical force' or 'literary technology'. In effect, it is inadequate to understand scientific practice in these terms, since scientific practice (including metaphors and models as well as supposedly literal discourse) is crucially rich, inventive, and important, which includes both discursive and material dimensions. It may locate scientific practice in a representationalist framework in terms of 'rhetoric' or 'literary technology', which will lead to the one-dimension misinterpretation of scientific practice. An important motivation for more recent emphasis on the materiality of social practice within science studies has clearly been dissatisfaction with the internalist tradition in philosophy and history of science, which presumes a representationalist view on scientific language. The internalist and representationalist traditions have realized some problems which focus on the dimension of discursive practices and activities. They intend to locate the discursive dimension of scientific practice in a more prominent position than the material dimension, which turns out to be problematic. Since the material and discursive dimensions of practices are closely linked, it is impossible to discriminate them abstractly. In different situations, the material and discursive dimensions are different. One can emphasize the materiality of scientific practice consonant with the work of the 'practice industry', while still justifying the discursive richness of scientific practices. The crucial move is away from conceptions of signification as an ideal (just as Frege's account), more precisely, as expressions of meaning or content that could be identically expressed both in discourses and materialities. For example, new experimentalists search for the knowledge of things which expressed in instruments, tools and equipments: model knowledge which provides material representations, working knowledge which presents phenomenons, and measuring knowledge that integrate both model and working knowledge (Radder, 2003, p. 54). And Baird also classifies the knowledge of things into three kinds: model knowledge – it instructs, manifests and explains things indirectly; working knowledge – the skilful knowledge that contained in materiality; and measuring knowledge – the knowledge that use tools to represent the measure of objects. The world is not in one 'way' apart from the signifying practices through which representation, articulation, and elucidation may be alternative to those meaningful practices. Otherwise the practices may look like out of function. In fact, the world not only links agents to the surroundings through practices, but also applies these practices to the agents and their surroundings, since these practices make us have various and ongoing connections to the surrounding world, and only through these practices can the world count as meaningful in some particular way.

Not all material things are meaningful, of course – only some of them. On one hand, we could use language and characters to represent these meaningful things. On the other hand, meaningful objects are not all representational.

For example, the 'Tao' in the Chinese sage Lao Tzu's expression is not representational but still experienced by us. Now, it is possible to represent some important things, due to the existence of representation, with the signification located in a specifiable part of the world, namely, within the things that function as signs. However, the signification exhibits only by appealing to more wide-ranging practices that open up to the world as a whole. Only by signifying, metaphorizing, and experiencing the surroundings as configured by ongoing practices, can signs count as more than marks and noises. Consider Wittgenstein's discussion of 'expecting' for a moment: it is in language that an expectation and its fulfilment make contact (Wittgenstein, 1953, par. 445). Wittgenstein notes that the various behaviours and surroundings that count as 'expecting' do not have some unifying properties (or even family resemblances), and these surroundings are prior to using the term 'expect' when talk about practices. Therefore, it is the surroundings that lead to expectation, not the expectation leads to the configured significance of surroundings. These practices themselves enable the world to display itself as a significant configuration. To that extent, the linguistic practice introduces this possible discrimination into the world. However, the term of 'expect' is not always meaningful apart from the ongoing practices in such situations. The significant configuration of the world happens simultaneously. Thus, in such a process the practices are carried out all the time. When we adequately understand those characteristics of signs, we can see that they are thoroughly material. They are referring to as the same thing by actual repetitions under different circumstances, not through any ideal replicability. Therefore, their significance is located within the world without any limited boundaries. Just in this sense, Rouse indicates that the openness of signifying practices is spatial as well as temporal. They open up to the world as a whole even as themselves situated within it (Rouse, 1996, p. 155).

3.2.7 Practice is open spatiotemporally

How can we understand practices as both local and open simultaneously? For Rouse, practices are materially located within the world without any identity with temporal or spatial regions, which is very important for understanding scientific knowledge from the view of PSP (Rouse, 1996, p. 155). As an naturalist, the eliminativism claims that it is not necessary to talk about epistemic terms such as truth, justification, reasons, and so on. For example, the neurophysiological eliminativism considers brain processes as an organism or central nervous system, and sociological eliminativism locates the considerations of social mechanisms within specific networks replaced by elements. The eliminativists intend to eliminate all talks about knowledge since knowledge can be equated (in principle) with an fully specifiable domain of causal processes, no matter whether those processes are located 'in the head' or in the interactions of social agents. However, the claim on knowledge as spatiotemporally bounded conflicts with the spatiotemporal openness of practices.

If knowledge is located in practices, it can not be spatiotemporally bounded. Therefore, Rouse indicates on purpose that it may be useful to understand the spatiotemporal openness of practice as an alternative to naturalistic eliminativism, since such an eliminativism replaces discussion of knowledge and justification with discussion of the causal structure of bounded region in a spatial or temporal way.

Rouse attributes this conflict to his earlier discussion on the normative characteristics of practices. In the beginning, Rouse suggests repeatedly that practices are iterable, and their iterations are always subject both to deliberate reinterpretation and to semantic drift. In its history the knowledge-producing practice was adequately explained by an eliminativist account. For example, by interpreting the cognitive process as causal interaction between the central nervous systems of individual organisms and the physical information in their environment, or as the causal product of complex networks of social exchange, the cognitive practices could be explained adequately. However, that does not guarantee that the subsequent historical development of the cognitive practices would continue to be so explainable. Rouse indicates that one way of interpreting this point relies on the normative characteristics of practices: any such confinement of epistemic practices within spatiotemporal bounds would require that such confinement be enforced, and that the relevant practices need not themselves be confinable within the same spatio-temporal boundaries (Rouse, 1996, p. 156).

The claim of the openness of practices is not just for the eliminativism, but also suggests a more general lesson. In the ways that fundamentally affect any particular practices, the significance of practices are always interconnected with the ongoing patterns of development, rather than isolated from one another. Just as Heidegger's 'Dasein', the practices of being-in-the-world deal not only with the world around us, but also with the world which we have already perceived. A particular practice always associates with others, and in communication with others including the subjects, themes, and the contexts of practices. Thus, local worlds are connected with each other. By learning, communicating, negotiating, and associating between knowledge and power, local practices constitute some standard networks of knowledge. It illustrates that a typical practice needs other practices to enforce its norms, to put forward its necessary equipment and resources, to train its agents, and to confer significance on it or to undercut its previous significance. Furthermore, since such practices are usually considered to be intelligible, these various kinds of practices and agents could together help to configure the world around the agents.

Therefore, in order to understand how scientific knowledge is situated within practices, we need to consider how practices are connected to one another, for knowledge will be established only through these interconnections. Just as Rouse says, 'scientific knowing is not located in some privileged type of practice, whether it be experimental manipulation, theoretical modelling, or reasoning from evidence, but in the ways these practices and others become intelligible together' (Rouse, 1996, p. 156). That is the circumstance of process in which practices give out the norms of practices which restrict and conflict with practices in the future. Therefore, the circumstance of process also includes the process of resolving the conflicting.

The discussions about practices do not explain the locality of knowledge completely. In addition to the locality of practice itself required by the PSP, the production and defence of knowledge should be also local. And it also need some cases and facts to support its arguments. Moreover, as for PSP, naturalism is one of its fundamental characteristics, which may conflict with the normative essence of practices. How could the account of PSP be both naturalistic and normative? Such problems may involve the naturalistic nature of PSP, the discourse about norms of scientific practices, and the relationship between knowledge and power. Those are not only important issues internal to PSP, but also very significant to our contemporary understanding of science in general. In the next chapter, we will discuss these important issues.

4 The nature of knowledge
Local knowledge

In this chapter we will compare the notion of local knowledge in anthropology with the one in PSP, in characterizing the relationship between both notions. Then we could clarify the role and significance of the notion of local knowledge for PSP, which demonstrates the 'locality' as a basic feature of all knowledge recognized according to a view of western philosophy. Therefore, the conclusion on the conception of knowledge would be seriously different from the traditional view in western epistemology, in which the universality is taken as the basic ideal of knowledge.

Recently, 'local knowledge' is a very popular issue. First, for instance, it is the central notion in Geertz's interpretive anthropology. Though it originally arises from anthropology, the notion of local knowledge recently appears in many papers on PSP. It is really confusing whether these two notions refer to a same thing. Therefore, it is necessary for philosophers of science to compare Geertz's conception with its exercise in PSP. To figure out the distinction, and to explain the notion of local knowledge in PSP, are both in a profound significance.

In addition, the idea of local knowledge has already been a strong challenge to the traditional view of knowledge. Is the nature of knowledge universal or local? In what sense we think of knowledge as local or universal? People continue to raise these issues, indicating that these fundamental issues concerning the nature of knowledge are very confusing. It is time to make certain responses to those important issues.

4.1 The conceptual background of 'local knowledge'

The first background comes from the history of anthropology, in which there has been a 'methodological debate between universalism and historical particularism' (Ye, S.X., 2001, p. 121). The doctrine of universalism is to find a common structure or universal laws of human culture. This tendency in natural sciences is more obvious. It appears very plausible for science to be dominated by universalism. However, we could doubt whether there is any universal law for natural phenomena. Universalism distinguishes scientific knowledge from local knowledge. However, local knowledge can become more universal through the process of decontextualization. In both sociology and anthropology, historical particularism is somehow reasonable, since it stresses the particularity of different cultures, and advocates field and case studies to interpret different cultural phenomena. It does not explain them in a universalist way, but only to provide particular and empirical explanations. In 1960s, sociologists and anthropologists mostly believe in structuralism and universalism. What they emphasize is social structure and evolution (i.e., system can also be taken as one of the editions of universalism); therefore, Human disappears. In order to refute universalism which ignores diversities of culture, symbolic and hermeneutical anthropology is supposed to argue for local knowledge. For instance, Geertz is one of the most famous representatives as hermeneutical anthropologists.

The second background is the global context of modernization and globalization, in virtue of which science is flourished in western world since 17th century. Globalization and modernization cleans up local or national characters in almost every issues, while making culture increasingly converged. This philosophical approach also provides a unified notion of 'modernity' in narrative framework.

'Modernity' is the name of a narrative strategy, which depicts the historical construction of the modern world. Some characteristics of modernity are cited below from J. Rouse (1996, p. 49):

1 Secularization – not simply the decline of religious practice and belief but also the separation of various realms of human life from theology and the privatization of religion and other fundamental matters of concern into individual beliefs and commitments;
2 The constitution of 'humanity' (the unity of abstractly differentiated *individuals*) as the *subject* of representation and knowledge, source of all values, and possessor of rights and moral dignity;
3 The development of distinct domains of knowledge and practices (for example, law, the economy, science) whose autonomy is institutionally recognized and protected;
4 Rationalization – formal procedures of calculation developed more intensively, extended to more and more domains of human practice, and unified across domains;
5 The rapid growth of science and technology as the quintessentially modern human practice and correlated understanding of nature as inert objects of knowledge (science is perhaps the only human activity for which the appellation *modern* has almost no controversial force);
6 The expansion and concentration of production resources (usually identified with the beginnings of capitalism as a mode of production);
7 The extension of European ('modern') culture across the globe (colonialism and postcolonial modernization/Westernization);
8 The self-referential narrative legitimation of modernity itself as the progressive realization of freedom and truth.

Therefore, modernity also naturally become hostile to the process and thought of locality. Many modern philosophers have pointed out the problems and shortcomings upon that issue, who starts to study knowledge arising in non-western nations, in order to realize the importance and significance of cultural diversity. 'Local knowledge' is emerged in such cultural background.

Thirdly, post-modern thought is rising in both academic and cultural contexts. Local knowledge in PSP imposes direct impacts on the academic background, for instance, the practice of SSK. Scientific knowledge has been regarded undoubtedly as a kind of universal knowledge. Although the notion of local knowledge has been rising in anthropology, it cannot be accomplished merely in anthropology, which at most makes people realize that, along with the modernization and decolonization, there is also a process of natural science to conquer other local knowledge in non-wester peoples. Thus, it is necessary to inquire a universal scientific knowledge. The post-SSK laboratory studies greatly promote the study of scientific practice. Scientific practice has been thought of as an active feature of scientists in the world, in which PSP started to (as Heidegger puts) become a specific 'by-hands' analysis or a powerful tool of interpretation, while promoting cultural studies as well as anthropology, ethnography, which directly contributes the notion of scientific practice and local knowledge.

4.2 Geertz's notion of local knowledge

To propose a notion of local knowledge is one of the most influential academic contribution from Geertz. Geertz's notion of local knowledge, in anthropology, is about the knowledge in folk model. According to Wang Hailong, the Chinese translator of Geertz's *Local Knowledge*, it belongs to the 'post-colonial/post-modern discourse', rather than the epistemology of interpretive anthropology (Figure 4.1). According to the translator's interpretation, 'local knowledge' requires a kind of ontological status of knowledge that arise naturally from local culture.

Geertz's notion of local knowledge is not constructed totally by himself. It should be derived from the investigations of knowledge, concerning to the 'ethno-', especially the 'ethno-sciences'. When they study knowledge of nature in non-Western culture, anthropologists added the prefix words 'ethno-', in order to indicate that knowledge is related to local races and other ethnic concerns, and that the knowledge is inseparable to its region; therefore the Western, which is not included in the same region, would not have the same knowledge. In fact, such investigations are carried out very early. For example, one of 'ethno-sciences', known as 'ethnobotany' in the Western tradition, can be traced back to the colonization and exploration in ancient Greek, Roman, and Islamic regions (Minnis, 2000, p. 6). The naming of 'ethnobotany' has been also more than 100 years, which is originated in the United States. In 1896, J. Harshberger, a botanist in University of Pennsylvania, makes a

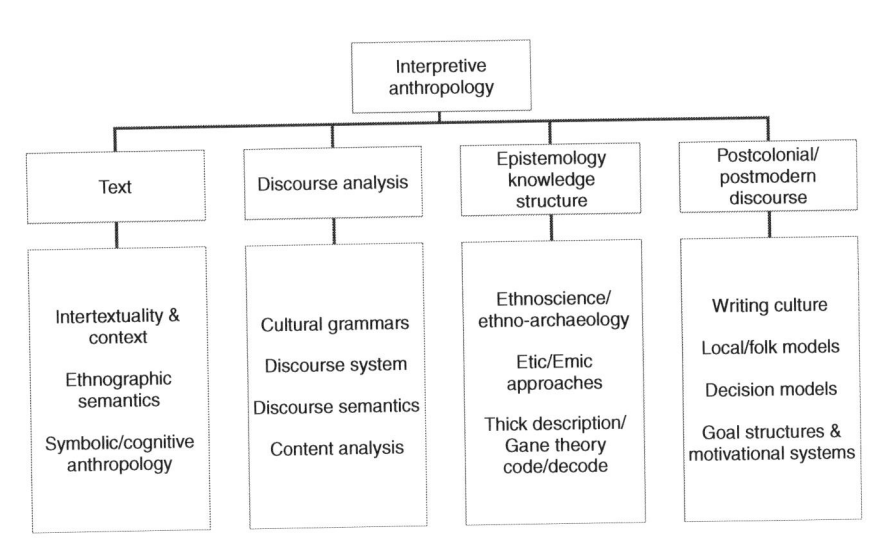

```
                    ┌─────────────────┐
                    │  Interpretive   │
                    │  anthropology   │
                    └─────────────────┘
         ┌──────────────┬──────┴──────┬──────────────┐
    ┌─────────┐  ┌─────────────┐ ┌────────────┐ ┌──────────────┐
    │  Text   │  │  Discourse  │ │Epistemology│ │Postcolonial/ │
    │         │  │   analysis  │ │ knowledge  │ │ postmodern   │
    │         │  │             │ │ structure  │ │  discourse   │
    └─────────┘  └─────────────┘ └────────────┘ └──────────────┘
```

Intertextuality & context	Cultural grammars	Ethnoscience/ ethno-archaeology	Writing culture
Ethnographic semantics	Discourse system	Etic/Emic approaches	Local/folk models
	Discourse semantics		Decision models
Symbolic/cognitive anthropology	Content analysis	Thick description/ Gane theory code/decode	Goal structures & motivational systems

Figure 4.1 The scheme of interpretive anthropology.

definition of ethnobotany: 'the study of plants employed by indigenous peoples'. Furthermore, in the 20th century this discipline is redefined as 'the interaction between researchers and plants'.

From Geertz's view, Local Knowledge has three important features:

First, local knowledge is the knowledge contrastive with the knowledge in western world. Although there is no direct claim that Western knowledge is universal, this classification is to separate the knowledge in the West from other ones. It would be dissatisfying since some kind of local knowledge is always in contrast with something beyond the local. Such relationships, as well as the non-universal, non-scientific position of local knowledge, could indicate some ethnographic exception of the sense of 'local'. And in many academic and folk discourse, though ongoing science passes through lots of local positions, it still managed to keep itself as 'non-local' which are exactly the same one, while appearing in all positions. In all the places it has surpassed. In particular, the locality of knowledge is mentioned compared to its relationship with the Western knowledge.

Second, local knowledge also refers to non-modern knowledge in contrast with modern knowledge. Because of the Eurocentrism, not only natural knowledge from non-western nations could hardly be regarded as science paralleled with modern science developing in western world, but also in the continental philosophy, especially in France, those who claim for non-western knowledge still think of them as 'Ethno-sciences', or 'non-formal sciences'. Non-modernity may be equally understood in two dimensions: one is the historical dimension, for which local knowledge is not always functioning as modern knowledge. The other is contemporary dimension, for which local knowledge has played only limited role in local region, instead of modern society.

Third, local knowledge must be intimately related to the agents of local knowledge. In other words, local knowledge is not available if it is separated from the specification on knowing who, where, and which contexts to employ it. Universal knowledge has no need to ask for whom it belongs to and in what specific situations it functions. The agent might be a nation, or an individual. In short, it is not the rational 'individual' of the modern conception, which may involve some broader or subtle meaning.

In fact, local knowledge is closely linked with local area. In Geertz's anthropology, local knowledge is presented primarily as knowledge concerning to geographical and ethnic cognitive modes. Although it has strongly criticized the view of 'logocentrism', local knowledge has involved still some post-colonialist implications.

Proposing the notion of local knowledge is greatly significant since it has shown how non-western knowledge to get recognized. But this notion of local knowledge still has many problems, of which many scholars have been aware in recent years.

1 According to the popular view in natural and social sciences, only some belong to local knowledge. As is generally recognized, knowledge from minority peoples is local one. Therefore, some anthropologists claim there is a failure to interpret how universal knowledge is possible from the notion of local knowledge. In the anthropological perspective, local knowledge is studied only in the sense of minorities, while never including Western knowledge.

2 The second question is that, if 'local knowledge' is merely to name knowledge in a particular species, excluding 'universal knowledge' as another kind of knowledge, the most importance of claiming local knowledge is to remind universal knowledge not to ignore its little brothers. In that sense different kinds of knowledge still rank among each other, and local knowledge is ranked as a subordinate status. If Western knowledge is universal, non-western knowledge could be only local. Therefore, it actually stabilizes the priority sense of Western knowledge, compared to non-universal but only local knowledge.

3 It is misunderstanding to claim that local knowledge is unable to separate from their agents, namely, the local people, and that no local knowledge could have clear expression in propositions. In fact, knowledge-based economy in contemporary world has taught people to emphasize the role of agents of knowledge. Today, to acquire knowledge also needs to ask: who, where, rather than merely what? In addition, the link between knowledge and its agents cannot be narrowly understood as unable separation in a physical sense, but it should be interpreted as intimately involved with skills and contexts of agents.

In a word, for one thing, the key to understand 'local knowledge' is to correctly understand the 'local'. For another, there is an understanding in the opposite direction, that is, to ask whether there is universal knowledge, and whether the nature of knowledge is universal, or local.

For the first question, cultural anthropology has no longer set up knowledge completely in a given domain as 'Local'. Local knowledge is no longer attached to a certain location, but to a dynamic sense of cultural identity. However, people still attempt to link locality with the notion of space. Recently, anthropologists tend to connect space with time, which claim that location is a particular intersect of time and space among social relationships. However, there are still many misunderstandings correlated to 'local' and 'now'. Especially the term 'local', in fact, has to be interpreted according to various contexts. Moreover, people often understand the 'local' from the notion of space, which not only narrows and distorts the notion in anthropology, but also narrowly understands the locality in PSP.

As for the second question, the understanding in the opposite direction is actually a scepticism on universal knowledge, which has not been discussed in anthropology. It is the PSP that puts forward a reasonable answer for this question.

It seems to me that the most fundamental problem is not to specify whether or not local knowledge is a species of knowledge, but to ask whether the nature of knowledge is local or universal.

4.3 Local knowledge and its significance in PSP

PSP claims that locality is the nature of knowledge. Joseph Rouse has put forward arguments as follows:

> The standard model of scientific knowledge takes it to be knowledge of universal laws, valid at all times and places. These laws can be applied to particular situations, however, only by using various bridge principles and by determining the relevant facts about the situation that need to be included in this instantiation of the law. Thus, the problem is always how to bring universally valid knowledge to bear on local situations.... The new empiricism suggests an analysis moving in the opposite direction. In scientific research, we obtain a practical agency of locally situated phenomena.
>
> (Rouse, 1987, pp. 21–2)

I will suggest an analysis of scientific practice that reveals the local, existential character of the understanding it produces. Scientific knowledge is first and foremost knowing one's way about in the laboratory (or clinic, field site, etc.). Such knowledge is of course transferable outside laboratory into a variety of other situations. But this transfer is not to be understood in terms of the instantiation of universally valid knowledge claims.... It must be understood in terms of the adaptation of one local knowledge to create another. We go from one local knowledge to another rather than from universal theories to their particular instantiations (Rouse, 1987, p. 72).

'Understanding' in Heidegger's sense is always local, existential knowledge. In calling understanding local and existential, I mean that it is bound to concrete situations, embodied in an actual tradition of interpretive practices carried on from generation to generation, and located in persons shaped by specific situations and traditions. Understanding is thus not a conceptualization of the world but a performative grasp of how to cope with it (Rouse, 1987, p. 63).

The first paragraph above illustrates how knowledge is characterized in traditional philosophy of science. The standard view is that universal knowledge applies to local situations by bridge principles. However, the second paragraph argues that universal knowledge is not taken by us into local context. We can only grasp local and contextual phenomena in practice, so that it is the locality rather than universality that has ever been grasped. The traditional philosophy of science goes on a wrong direction. And then J. Rouse points out his understanding of locality in the third paragraph.

The conception of local knowledge in PSP is in a normative sense, which does not particularly refer to non-western knowledge. The locality mainly refers to the specific local context or status in producing knowledge and its justification, such as particular culture, values, interests, and the resulting sight and positions. Local knowledge in this sense is an epistemological view which excludes universal knowledge. Universal knowledge is only the product from the standardization of local knowledge.

Therefore, Rouse's notion of local knowledge, which is constructed in PSP, is very different from Geertz's notion. Of course, Rouse might be inspired from Geertz's thought. Among the three books written by him, Rouse has only one direct reference to Geertz's work, though it is cited in hermeneutics sense, rather than directly referring to Geertz's 'local knowledge'. The indirect reference shows that Rouse is not satisfied by Geertz's notion of local knowledge. Though he may be affected by works from anthropology, Rouse does not agree with the view about local knowledge in anthropology. Rouse makes it clear that his conception of local knowledge is inspired by Thomas Kuhn, who argues that scientific knowledge is contained into capacities on specific instantiations when lacking agreement on interpretation, and also absorbs new empiricist insights that the technology in scientific practice could be applied without any need to extend the theory on which the technology is based; In particular, Rouse also makes reference to Heidegger, who argues that skilful practice in local, physical, and social sense is important for understanding and interpretation (Rouse, 1987, p. 72).

As mentioned before, there is a big problem for Geertz's notion of local knowledge in anthropology, since local knowledge in that sense cannot be generalized, or have been ranked equally as universal knowledge. In anthropology, local knowledge is opposite to universal knowledge. That opposition cannot be eliminated within anthropological research. Western anthropologists pay much attention upon non-Western knowledge, though certainly recognize local knowledge in some content, still make sense of local knowledge in contrast with science, or as a helpful complementary part for universal knowledge. Locality always means both negative and restricted position. Therefore, to demonstrate local knowledge from the perspective of non-Western knowledge, in order to show how it complements universal knowledge, would not in any case break the monopoly status of the universal knowledge, since it already assumes the rational paradigm from scientific knowledge. There seems to be no possibility to cross the gap between locality and universality in this way.

A complete resolution is a deconstruction of universality, which argues that universality does not exist in a primary sense, and the so-called 'universal knowledge' is a fiction, an ideal. A seemingly universal thing is in fact a product of the standardization of local knowledge.

Rouse accepts that solution, but still recognizes some functions of universal knowledge. For example, he points out, on one hand, that the application of science and technology is a kind of extension outside laboratory, which is the 'translation' of practice in order to adapt to local situation. On another, he believes that there is universality of scientific knowledge but only rooted in the achievement from local construction of practical skills in specialized laboratory (Rouse, 1987, p. 119). Recognizing the universality of knowledge, based on local constructions, may be a throwback to the anthropological perspective on local knowledge. Actually, the complete resolution, which argues for the deconstruction of universal knowledge, must adhere to the following basic ideas:

First, from the perspective of PSP, there is no universal knowledge, and scientific knowledge is also local. Locality is the nature of scientific knowledge. Scientific practices are local, contextual, and fit in a specific laboratory or situations. The knowledge gained from specific situations and contexts is local in its primary meaning. Universal knowledge is the result of translation by scientists, and therefore merely in a secondary meaning.

Second, the locality depends primarily upon science as a local practice. Scientists cannot leave specific laboratory to carry out research activities. Locality of scientific knowledge demonstrates mainly from the contextuality and indexicality of knowledge and its practice.

Third, though scientific knowledge can give out superficial universality, that is only resulting from standardization of knowledge. Seemingly universal knowledge is actually a representation for the standardization of local knowledge.

Fourth, the standardization of scientific knowledge often indicates itself as 'dislocalization'. However, it is merely to be a local extension or alteration to other positions, namely, a replacement by one locality with another.

Fifth, the dislocalization of scientific knowledge contains decontextualization and non-indexicality. But these manifestations are merely appearances, which results from the standardization of knowledge.

The first and second points are explanations of the local characteristics of scientific knowledge, which needs to describe scientific practice. The third, fourth, and fifth points are mechanistic explanations on how to produce local knowledge and why it looks like universal knowledge. Here we take up Rouse and others' works in order to argue for the first and second points above.

The standard view of scientific knowledge includes three basic ideas: 1) universal proposition is the standard form of scientific knowledge; 2) scientific explanation is to explain special phenomenon from general proposition. 3) Scientific knowledge is universal knowledge. On the contrary, the PSP, according to Rouse, maintains that scientific knowledge is local, contextual, and subjective indexical.

First, from the perspective of Existentialist hermeneutics, science should be recognized not as a network of universal propositions, but as a specific practice. There is no universal knowledge in primary sense; scientific knowledge is also local since it is also justified by particular scientists in specific situations in scientific practice. Without practice of science, knowledge is not only unable to be produced, but non-intelligible, non-transferable, and unjustified. For instance, the laymen without laboratory education are unable to understand experimental results and the relevant knowledge. N.R. Hanson argues that X-ray picture in professional hands can be explained, while ordinary people encounter X-ray picture without any understanding about the facts it represents. Hanson uses the example to support his claim on theory-ladenness of observation in science. It is also evidence of locality of scientific knowledge from the perspective of PSP. Since X-ray professionals have disciplinary training which ordinary people are lacking, they of course are in different background of local knowledge.

In fact, there are three significant aspects on such a point. First, it is irrefutable that scientific knowledge has experimental characteristics, which are not established by theory, but only figured out by instruments in laboratories. Second, instruments, and the microscopic world constructed by using the instruments, are the closest referents for those scientific claims. And finally, the knowledge upheld by scientists depends fundamentally upon their skills in employing instruments. Some knowledge cannot be obtained if there is no specific skill. Similarly, both Heidegger and Hacking argue that local practice and employment of instruments are the foundation of knowledge. According to Heidegger's existential ontology, activities of science are also grasps for Being-in-the-world in a '*zuhanden*' sense. Hall Effect, Josephson effect, and other experimental effects are not naturally found in the world, rather they are constructed by scientists with their grasping upon laboratory equipment. That's why Hacking (1983, pp. 227–30) argues for an important distinction between a novice and a veteran on whether to know how to act in such local environment as laboratory.

Some points from critics could be described as follows: though the origin of scientific knowledge is local, it does change to be universal once after its generation. The former local knowledge could have gone through decontextualization in order to be transformed into universal knowledge. For example, Newton's law of gravity has once been local knowledge when it generates from Newton's specific observation and calculation. However, since it has turned out as true under various conditions, the law of gravity has already come to be universal knowledge. How can PSP explain this law as local?

We have two main responses to those doubts. First, it is indeed a fundamental metaphysical position to hold knowledge is not universal, but it does have good empirical supports. People can really testify the relativity of knowledge in history of science. For example, according to Einstein's theory of relativity, Newton's law of universal gravitation has failed in some respects, which is recognized not as that universal as before. Knowledge is always generated in specific context, and thus it cannot be effective in all particular situations. Scientific knowledge has specific scope of application, which in itself is local and contextual effective. According to Cartwright (1999, p. 4), 'the impressive empirical successes of our best physics theories may argue for the truth of these theories but not for their universality.... So the laws of physics apply only where its models fit'. And thus, 'there is no universal cover of law (ibid., p. 6). Indeed, Cartwright suggests that all scientific theories are local knowledge, for instance,

> we are inclined to ask 'how can there be motions not governed by Newton's laws?' The answer: there are causes of motion not included in Newton's theory ... the wind is cold and gusty; the bill is green and white and crumpled. These properties are independent of the mass of the bill, the mass of the earth, the distance between them.
>
> (ibid., p. 32)

Second, the law of gravity, for example, when not involved in specific position in space, is just a mathematical form without any experimental sense, just like the mathematical form of the inverse square law of electrostatics. When it gets empirical meaning, the law should be applied to specific contexts, which is necessary to correlate with the environment and condition. For example, in employing the law of gravity, how would two objects move, such as coming to each other, moving in an opposite way, moving in an angle of 90° to each other, or just staying without movement? When those given conditions are different, it will have different calculations and situations of movement. Besides, the law of gravity functions never merely by itself, rather there may be other functions simultaneously. Therefore, on one hand, it needs to consider the specific combination of different functions; on the other, to let the law of gravity function by itself must screen off other effects, which in Cartwright's argument is to establish a nomological machine. If there are some difference on conditions and screening-off, gravity would give us different experience and situations of movement. Therefore, the empirical expression of gravity would be different under various circumstances. Thus, it is not universally valid, but correlated with specific contexts.

More importantly, the law of gravity is idealized when it is talking about mass point. Particles as mass points, as is known, are just idealized objects in classical physics. Similarly, in fact, there is the relationship of volume, temperature and pressure in thermodynamics, also the law of ideal gas. Ideal gas does not exist in reality. Some corrections are necessary on van der Waals law in order to apply that law to reality.

Let us focus on the evolution of high-energy physics from 1940s to 1990s, following A. Pickering's study, to put forward an example.

Pickering argues in his research that, from the 1940s to 1990s, physicists with different local theories were in competition in high-energy physics. Some of them failed, and others achieved a dominant position which has already been a so-called universal theory. Pickering has carefully studied two kinds of high-energy physics theoretical model in 1960s and 1970s (V-A [V-A theory by Feynman and Gell-Mann] and W-S [Weinberg-Salam Model]) which were in a competitive process, and he clearly demonstrates that the so-called universal theoretical physicists model is constructed in the competition and standardization. During the competition with each other, both theoretical models of V-A and W-S are not recognized as universal by theoretical physicists. Different physicists upheld different theoretical models, which thus should be thought of as local knowledge. From the view of PSP, the V-A physicists and W-S physicists hold their own local description of universe respectively. Though the W-S theory may obtain a superficial support from experiments finally, it seems that, by Pickering's inspection for over 29,000 photos of observation, the evidential supports for W-S post are actually far less than the supports for the V-A theory. However, physicists still accept W-S theory. Clearly, it is unable to explain that merely from the objective and independent view of epistemic universality. Moreover, if scientists merely characterize a scientific theory in a dominant position from a representationalist view, they will leave behind its previous local characteristics, and instead think of this theory as always universal. Pickering (1984, p. 409) argues, 'Each theory would appear tenable in its own phenomenal domain, but false or irrelevant outside it. There would be no realm of extra-cultural facts against which the empirical adequacy of different theories could be impartially measured'.

Third, it is most significant that the local view of knowledge is a critical and interpretive position in epistemology, which is opposed to the doctrines of universalism or absolutism. The locality of knowledge suggests that specific local contexts are always involved in the generation, formation, and transmission. Traditional philosophy of science in the past claims theoretical priority too much therefore it distorts the real image of science, ignoring the participating and opportunistic characteristics in scientific research, or even simply not considering the nature of contextuality in scientific theories. In fact, if we take into account the local context of scientific activities, it will be found that science is a kind of local knowledge. The characterization of the locality would suggest that science as local knowledge depends upon a practical grasp for equipment, technology, and the configuration of social roles, as well as the understanding for possibility of employment (Rouse, 1987, p. xiii), and that concerns on knowledge from scientific practice are to correct the traditional conception of science which has been excessively distorted.

Scientific knowledge is not primarily a network of universal propositions, but a kind of practice in its very beginning. Even on the level of representation of reality, the universal view of scientific knowledge is untenable since those universal statements are still conditional on specific contexts. Usually that is illustrated from the indexicality of scientific knowledge.

The indexicality of scientific knowledge is also a performance of locality. In this regard, we have also found a lot of evidences. According to Mertonian sociology of science, the naming of scientific discoveries, such as Newton's laws of mechanics and Coulomb's law, is a kind of recognition of creativeness in scientific reward system. From the point of view of PSP, that is a manifestation of the indexicality of scientific knowledge, which is not only the recognition of the privilege in generating scientific knowledge, but also illustrates the particularity in generation and justification of scientific knowledge. For example, we could not apply Newton's laws to the theory of quantum mechanics or relativity. That is the indexicality of scientific knowledge, which is intimately correlated to specific contexts.

The notion of indexicality was originally put forward for the expression by ordinary language, namely, indexical terms such as 'here', 'I', and 'today' with different referents in different contexts. C.S. Peirce, for instance, suggests that a symbol can have different meanings in different contexts, while the same meaning can be also expressed by different symbols. Peirce's notion of indexicality is taken from two aspects of sense. First, indexical terms must be related in use with specific contexts. Second, there is no specified referent for any indexical term if it merely takes on syntactic level. The indexicality of scientific knowledge would also blind scientists in linguistic sense, who may be confused to believe falsely that scientific knowledge does not need contexts

K. Knorr-Cetina argues that the notion of indexicality in anthropology indicates the specification of senses of words in particular time and space, and finally the specification of 'tacit rule'. Therefore, for no matter whether anthropologists or SSK scholars in laboratory studies, meaning is specified in a situational way, namely changing with the variety of specific contexts, and manifesting itself by the interactions among participants. That view of indexicality agrees to later Wittgensteinian theory of meaning, which is popular by its slogan as 'meaning is use'. Both of them emphasize the practical manifestation of meaning in particular contexts.

K. Knorr-Cetina (1981, p. 33) argues that,

> I will use the term "indexicality" to refer to the situational contingency and contextual location of scientific action. This contextual location reveals that the products of scientific research are fabricated and negotiated by particular agents at a particular time and place; that these products are carried by the particular interests of these agents, and by local rather than universally valid interpretations; and that the scientific actors play on the very limits of the situational location of their action. In short, the contingency and contextuality of scientific action demonstrates that the products of science are hybrids which bear the mark of the very indexical logic which characterizes their production.

Therefore, according to Cetina, the notion of indexicality roughly embodies in the following four aspects:

First, indexicality implies opportunism. In other words, the research work of scientists is like repairing something, which involves opportunistic characteristics. Similarly, scientists usually have important understanding upon their opportunities encountered in particular situations, and then take advantage of these opportunities to fulfil their programs. Opportunistic characters do not mean scientific work is unsystematic, irrational, or career-oriented. Instead, opportunism means indexicality in the way of production of knowledge by scientists.

Second, contingency is also contained in the research work of scientists, which is demonstrated in the existing restrictions on raw materials and equipment. For example, there are considerable limitations upon the existing laboratory instruments and director, especially on her academic background and training, which restricts the research approach for the entire laboratory. People tend to make use of existing equipment or respected leadership in choosing their own research interests. In those cases, the research topics do not rise from existing problems, but from the employment of existing resources. Even the famous Cavendish Laboratory has already changed its own research directions many times for those reasons. Meanwhile, scientists may change their original plan in order to use some existing instruments, merely because of some preference for some technical equipment. The contingency of research work does not mean irrationality or destruction. On the contrary, the reason why science has constructive capacity of new information is the uncertainty from contexts of scientific research, which has precisely led to the possibility for the averaging and translation of those contexts.

Third, local characteristics should not be ignored from specific laboratory operations. For example, there are differences among choices and usages of materials, components, or instruments in laboratory, as well as differences among measurement and experimental control. Those differences suggest that scientific method is a local form of practice without any non-localized universal paradigm. Scientific method is contextual rather than universal, and therefore it is a form of practice among other forms of social life, which have also been rooted in social actions.

Finally, the construction of knowledge is always accompanied by the choices made by scientists, which involve a series of communications and decisions. These choices are in the midst of some networks, one based on another, and then based on others furthermore. Every choice is orientated to some specific research objectives and costs, without any unified or general standard.

4.4 Why does science seem like universal rather than local knowledge?

Karl Popper believes that science is falsifiable, which has already implied that science is not universal but merely valid in certain conditions. Moreover, historicism claims scientific knowledge lies in hands of scientific communities, while no knowledge claim could be accepted by all those incommensurable communities. This indicates that, in a certain sense, knowledge has local characteristics. Therefore PSP is in a certain sense a natural consequence of historicist philosophy of science.

However, PSP has a very different view from critical rationalism or historicism. For instance, Joseph Rouse's criticism upon epistemic universalism takes two strategies: first, a general argument for the nature of knowledge would consider science in a non-representationalist way, but as some practical interventions into the world; second, it is necessary to reveal the mechanism how local knowledge transfers to a universal appearance.

In the first aspect, there are three main points worth noting.

First, science should be thought of not as a system of theoretical statements, like what traditional philosophy of science does, but as activities of cultural practice. Since science is a cultural practice, though very different from other cultural phenomena, there in no way means to claim science as a unique kind of cultural activities. Since there is no essential difference between science and other cultural events, the universality of scientific knowledge is also deconstructed, and therefore cultural relativism could be also applied upon issues about science.

Second, in all scientific practice, laboratory practice illustrates the most important nature of scientific knowledge, which therefore must be emphasized in science and technology studies. Rouse (1987, pp. 89–92) has argued that many lab discoveries have local characteristics with some reasonable explanations, such as Watson and Crick's discovery of DNA in Cavendish Laboratory. Latour and Woolgar (1986) give out an interpretation on activities of scientists in Science Laboratory. K. Knorr-Cetina (1981) puts forward some similar work on cultural studies of laboratory.

Third, scientific practice is local and isomorphic to the world. Thus scientific explanation makes sense only in local contexts. The so-called universality is precisely produced through standardization from locality of knowledge.

Many philosophers of science have stressed that scientific explanation needs contextuality which is not merely concerned with propositional knowledge, but also with the material environment including Lab and its instruments, which is referred to by Pickering as 'material force'.

Since activities of scientific research take place in specific contexts, science in its nature is necessarily local knowledge. The production, formation, and justification of scientific knowledge depend upon specific scientists and their local communities. Scientists should not be human beings without personality or inter-subjectivity, but actually local particular subjects.

Now we turn to the second aspects in order to elaborate the mechanism by which local knowledge transfers into a universal appearance.

As mentioned before, it is the standardization of knowledge that provides a universal appearance for sciences. Standardization is a dislocalized process, while it actually extends local knowledge from one context to another, which means that the local knowledge suppresses more and more other local characteristics. There are three processes in order to remove local characteristics from scientific knowledge: decontextualization, dislocalized, and nonindexical properties, which are intertwined with each other.

There is a very clear distinction between universalistic and localized notions of science. Universalism implies that scientific knowledge has experienced a real process of decontextualization. Localized notion insists that the decontextualization of scientific knowledge would not make itself really universal, but merely standardize itself in order to extend to other local positions. Rouse is fully aware of that distinction, and also realizes the importance of standardization for scientific knowledge. He argues,

> [s]cientific knowledge does often appear shorn of contextual reference, and the ability to extend scientific capabilities outside the laboratory has been a hallmark of modern science and has been especially prominent in shaping its cultural image. We cannot regard the foregoing account of science as fundamentally local knowledge to be well established unless we can adequately account for the apparently widespread evidence for its transcendence of the local.
>
> (Rouse, 1987, p. 112)

There are two features of scientific knowledge which seems to strongly support it as universal knowledge. One is that science can be applied outside laboratories by technology; the other is that science literature is presented and communicated in a decontextualized way. Both can be criticized in order to reveal how those local characteristics will not be eliminated thoroughly from scientific knowledge in its standardization.

An important kind of standardization is to put forward standards for 'scientific questions, tools, procedures and results'. Tools are usually designed to accomplish specific tasks for specific goals. However, once they are produced in particular contexts, tools could become improved and promoted for a more general purpose. In particular, they could be manufactured later in accordance with some standard sizes and styles, rather than based on everyday practices, such as manufacturing hammer or cutting clothing in accordance with standard size. Thus tools really began to lose references to particular persons, tasks, or situations. The standardization 'involves both transforming the things themselves to make them applicable outside their original setting and developing more exoteric interpretations that make them accessible to the nonspecialist' (Rouse, 1987, p. 113). Therefore, tools are standardized, and the product from research described through a standard laboratory could be understood by other researchers outside, which requires it to be repeated in similar conditions. The standardization does make people deny its local characteristics and referent relations. However, according to Rouse's view, standardization in fact merely disturbs its referential relations to local contexts, rather than eliminates them. Scientific knowledge can be applied in each specific situations, which enhances the intelligibility of knowledge. That process of standardization could be interpreted sufficiently in case studies of history of science.

Rouse's first case comes from Fleck's explanation of Wassermann reaction, illustrating that standardization is a learning process. In the initial stages of the experiment, the reaction only gets 15%–20% of the confirmed positive results for syphilis; while in the end, Wasserman reaction can specify 79%–90% of syphilis. How could those standardized results be possible? According to Frank's study, it is the adjustment of reagents and learning how to read results out that promotes the efficiency of Wasserman reaction, which ultimately becomes a standardized test for this field (Rouse, 1987, p. 115).

The second case involves studies on thyrotropin releasing hormone (TRH), or thyrotropin releasing factor (TRF) in the late 1960s, which is taken from *Laboratory Life* by Latour and Woolgar. According to Latour and Woolgar, only a handful of laboratories and specialized equipment could distinguish TRH (F) before 1969, while this laboratory work has been carried out more than a decade. Latour and Woolgar (1986, p. 148) argue,

> The advantage of having situated TRF in the relatively restrictive context of analytic chemistry became obvious as early as November 1969. To find out what TRF was before this date would have entailed a laborious search through a complex mesh of forty-one papers, full of contradictory statements, partial interpretations, and half-baked chemistry. After November 1969, however, eight syllables enabled the rapid spread of news, and thus raised the possibility of a radical change of network structure. A tiny group of specialists might have concerned themselves with the same problem for years, simply by citing a relatively small number of papers. Now, however, a considerably larger public could use the eight syllable formula as a fresh starting point for their research. The three amino acid formula also had the substantial advantage that it could be used to order as great a quantity of the substance from any chemical company as money was available to pay for it.

Before 1969, our knowledge of TRH is local knowledge, since there are only controversial statements before any facts are established. After 1969, it appears to be universal knowledge. At that time, various arguments concerning TRH pass through continuous mutual criticism and collision in order to make one or more local representation ultimately to be recognized as facts. However, the knowledge acknowledged to be factual is still local, since it cannot remove all references to laboratories, though it may get rid of references to a specific laboratory. The referential relations of local knowledge which can be removed are the ones to specific laboratories or scientists. However, there are still referential relationships to laboratory work or general scientific practice which cannot be deprived from. In other words, even such referential relationships are generalized, and contain more and more potential uses, they still require some special intelligible situations. In fact, according to Latour and Woolgar, knowledge of TRF (H) still has different significances, such as scientific knowledge or technique, in variety of contexts. As Latour and Woolgar (1986, p. 110) claim,

> TRH exists as a 'new recently discovered substance' within the confines of networks of endocrinologists. Its treatment as a nonproblematic substance is confined to a few hundred new investigators. Outside these networks TRH simply does not exist. In the hands of outsiders and once devoid of its label, TRH would be merely thought of as 'some kind of white powder'. It would only become TRH again through its replacement within the network of peptide chemistry where it first originated. Even a well-established fact loses its meaning when divorced from its context.

And Rouse (1987, p. 116) makes further comments that 'if laboratories and the activities associated with them were to be destroyed, synthetic pyro-glu-his-pro-amide will lose its significance'.

Scientific knowledge is local. It is the extension of technology that makes those knowledge claims able to be applied outside the laboratory. The extension of technology consists of two parts: the first is the standardized application of science into technology; the second is the further development of technology itself required by the development of science. Joseph Rouse suggests that in order to extend knowledge which scientists actually know (e.g. skill, trick, etc.) beyond their local context, one must not only refine and adapt the programs, tactics and tools themselves, but also reconstruct the situation in which the skill knowledge actually is applied (Rouse, 1987, p. 118). Thus, it is actually our reformation of world according to science, which makes the local knowledge of science extend beyond specific laboratories.

Usually, such standardized reformations are often undiscovered. We know, however, they are real, as long as see the rise of so many standardization sectors: measure, observation, survey, etc. The special or more general experiment centres created such situations. For example, by the division of time zones, the standardized time measurement was extended to everywhere. Such standardization of knowledge will bring out the uniformed technological standards firstly. An example is the transformation of the standard of time, from sandglass and mechanical clock to caesium atomic clock, with the gradual unification of the various local time measurement standards around the world. The measure of time is not only more precise but also more standardized. There is a good example to illustrate how this local knowledge about time was standardized, which is the story that Japanese docked their time measurement with that of West.[1]

The network of time measurement which is used nowadays is the system of International Atomic Time (TAI) which was built on 'about one hundred atomic clocks in Europe, U.S., Australia, etc.' (Wu Shouxian [吴守贤], Qi Guanrong [漆贯荣], Bian Yujing [边玉敬], 1983, p. 123). It, however, happened in only half of a century. Before that, the network of time measurement was always built on the rotation (shown as the apparent diurnal motion of sun) and the revolution (shown as the cycle of seasons) of the earth for thousands of years. Indeed, TAI itself was made according to Universal Time (UT) and Ephemeris Time (ET), the two systems of time measurement according to the two periods of the earth running.

However, the network which divides the time 'evenly' is not as unalterable as it is usually imagined. It's shown as a 'necessary' choice for human being to measure time by that network. There was another different network of time measurement in the world for a long time before the 'even' network which matches the occupation of Newtonian science and industrial civilization in all over the world. That network was also built on the two periods of the earth running, but divided the period of the apparent diurnal motion of sun according to another rules, by which every period of the apparent diurnal motion of sun was divided into daytime and nighttime at the moments of the appearing and the disappearing of sun on the horizon, and each half was divided into six kokus (刻) evenly. This network of time measurement was used in Ancient Egypt, and Japan until the sixth year of Meiji (1873).[2]

This time system is called the seasonal time system or the variable-hour system (不定时法) by the Japanese. It looks stupid for us who have been used to the fixed-hour system. Because in this time system, not only a daytime koku is not as long as a nighttime koku for most of a year, but only the length of every koku will change with the seasons. But maybe, in a more unprejudiced view, the seasonal time system isn't more unreasonable or more complicated than the fixed-hour system. Indeed, it follows the basic time reference, the apparent motion of sun, more closely than the fixed-hour system. Exactly, it divides a day into kokus directly according to the apparent motion of sun. Thus, the definitions of the time units below 'day' are unified with the definitions of the time units above 'day' in this system. On the contrary, in the fixed-hour system, only the time units above 'day' are divided according to the apparent motion of sun strictly. And to divide the time units below 'day', people must depend on another standard, a water clock or an equatorial sundial which is deliberately placed at an artificial angle to the ground.[3] In the view of religion, furthermore, the seasonal time system has got a closer relation with great Apollo obviously. And it conforms more to the ancient innocent habit of working from sun to sun as well.

100 *Local knowledge*

It is not the strangest that the seasonal time system exists yet. The strangest is that people can customize the measurement standard according to their measurement system in some cases. An example is the clock. Though the European clocks which were designed on the basis of Newtonian mechanics (which is built in the fixed-hour system) and which indicated the time in the fixed-hour system were introduced into Japan, Japanese did not give up their seasonal time system for these clocks, rather they refitted the system to indicate the seasonal time. It is a painstaking task to design a Japanese clock, or in the Japanese word wadokei (和时计), to make it work in the seasonal system. And it was much more cumbersome to use such a clock than a normal clock. To make the Japanese clock function normally, the user needed even a professional technician to move the weights on the oscillating foliot in the morning and evening every day at the beginning, since the lengths of hours were different at daytime and at nighttime.

Nevertheless, Japanese did not give up but tried to transform the mechanism of the clock to reduce the labour. They invented the double-foliot type (二挺天符式) which had two oscillating foliots which worked respectively at daytime and nighttime (see Figure 4.2).[4] A control gear called as count wheel (数取车) was connected with the principal shaft of the clock. It turned every 6 kokus to switch the working foliot. Thus, the indicator on the dial of the double-foliot Japanese clock could rotate at different speeds in the daytime and the nighttime automatically. In addition, each of the foliots was carefully divided into 24 teeth which corresponded to the 24 solar terms. To move the weights on the foliots a tooth every solar term, the clock would indicate the *correct* seasonal time at any moment in a year.

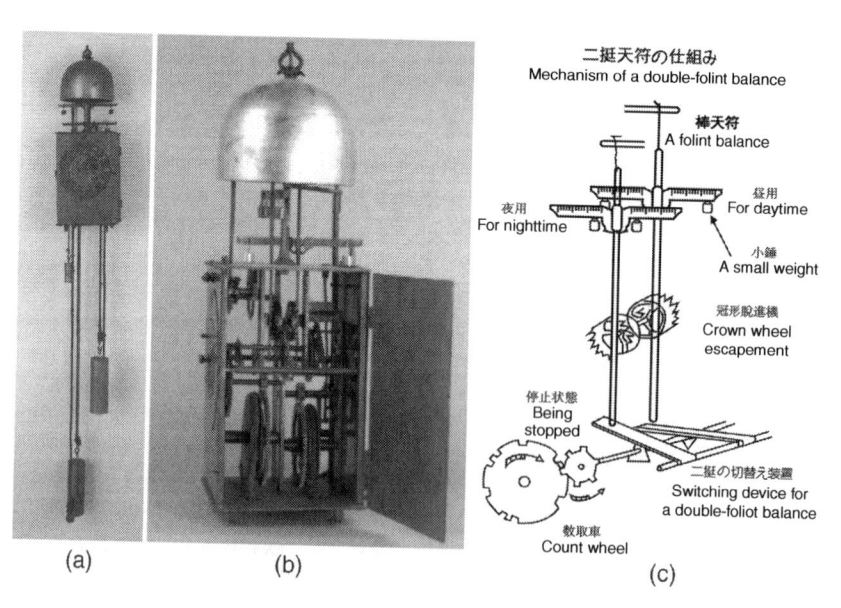

Figure 4.2 The double-foliot Japanese clock.

Moreover, Japanese did not transform the Japanese clock by this only way. They gave more solutions, such as the rotating dial type (割駒式) with a movable dial on which the interval of each hour could be adjusted; the circle-graph type (圆グラフ式) with a graphical plate and an indicator which expanded and contracted automatically in a year (see Figure 4.3, right); and the corrugated panel pillar clock (波板式尺时计) with a dial panel which was replaceable with the solar terms and a plumb bob that doubled as a time indicator.

More meaningfully, Japanese gave up the seasonal time system and switched to the fixed-hour system eventually not because of its non-scientific status or the cumbersomeness of the former, but because of, as some Japanese scholars suggested, 'the wholesale transplantation of Western social structure' in Meiji Japan.[5] And only by the edict of Meiji Mikado who was keen to bring Japan up alongside the European powers, the switching was finally achieved in all of Japan.[6]

New empiricism makes new contributions for the demonstration of 'mathematical and experimental science' as local knowledge in philosophy of science. For instance, Nancy Cartwright argues that modern science satisfies truth and valid conditions in local sense, namely, counterfactual conditionals by inferences *ceteris paribus*. She believes that modern science cannot be valid without nomological machine which is usually illustrated by models in scientific practice. By nomological machine and *ceteris paribus* conditions, modern science can be put in laboratories simply as a pair of factors, in order to manipulate and intervene the causation as anticipated by scientists. Thus, the

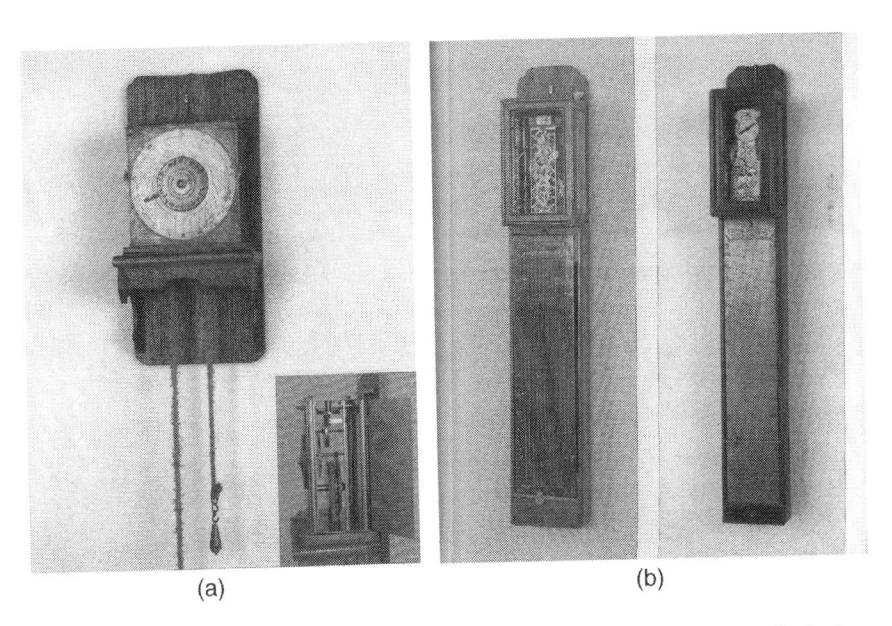

Figure 4.3 The dials of the corrugated-panel pillar clock and the circle-graph clock.

laws which satisfying nomological machine and *ceteris paribus* conditions are established where the nomological machine is effective. So that law could be transplanted among laboratories by setting up the same nomological machine in different local situations. Cartwright (1999, p. 46) claims, 'with a few notable exceptions, such as the planetary systems, our most beautiful and exact applications of the laws of physics are all within the entirely artificial and precisely constrained environment of the modern laboratory'.

Today our living world has already been transformed artificially by science and laboratories. We almost live in an artificial world. For instance, wild flies are not suitable for laboratory research, which need to be transformed through generations for the purposes of laboratory work. Mice have been transformed generation after generation, and have even become a kind of patented product. The mice for laboratory work are not natural but manmade. Therefore, technically what scientists study is not real nature, but an abstract, originally natural but already transformed as artificial. 'Natural' condition holds merely under non-counterfactual conditions. Therefore, what are thought of as natural are actually artifacts. In our opinion, Cartwright has specified local characteristics and conditions for modern science. The reason why it is local is that the condition is only suitable in laboratory, while it is invalid outside (with some exceptions on issues such as planetary systems). Modern science does seem valid in a general sense, merely because we have moved its satisfying nomological machine from one place to another. When laboratories extend to the extent of worldwide, laws of nature which are originally valid in laboratories now suit for all over the world, and therefore appear to be general. It is through this massive transformation of nature by laboratory that science is qualified as universal knowledge and adoptable for the whole world.

Above we have already discussed the local issues of scientific and technological knowledge from the view of PSP. Local knowledge is one of the main points of view in PSP, by which it is entirely separated from traditional philosophy of science. This outlook of knowledge provides important challenges to the widely accepted view which claims scientific knowledge involving essential universality. From the view of PSP, scientific knowledge is essentially local, which seems to be a really shocking point. The main value of such notion of knowledge would be to put all kinds of knowledge on an equivalent level of assessment. Science from the western tradition, for instance, would not be prior to other local knowledge from any non-western tradition. The only difference among them is on local conditions, rather than any distinctions about their grades or values. People are not totally free upon their choice for local knowledge, which is subject to the constraints from opportunities of resolving problems, effectiveness of resolutions, and other contextual factors.

Admittedly, scientific knowledge from the Western tradition does better jobs in its standardization than from other nations and traditions. Because of the advantage in standardization, it forms a relatively consistent scientific language and methods which are standardized and recognized as some terminologies followed commonly by scientists. However, we should not forget that all of these are created from works in laboratories and nomological machine.

In short, Rouse's 'local knowledge' is such a kind of knowledge: first, the generation and defence of knowledge has its specific scenarios, which includes historical conditions, cultural background, values and interests and other groups. Second, knowledge is something created and developed with things being generated, rather than anything fixed, universal, or effective at any time or occasions. Third, the 'locality' does not mean 'closedness'. Locality is open to all participation and practices. Local context is always changing in a dynamic process. Based on such a notion, it's not difficult to understand that 'scientific knowledge is local, and science is a kind of practice'. This is not to say that scientific knowledge is generated entirely subjective, while it is a result from practical interaction. 'Interactiveness' can be understood by 'local contextualization, world guiding, and dynamic practice' and other factors in a comprehensive structure. These factors not only affect each other, but also are affected by their own components, while this in turn affects the component of its internal and external factors (Ren Yufeng, 2007, p. 121).

Notes

1 This example is quoted from *Measurement, Relation Reality and A Network of Relations: A Question Brought Out by A Measuring Rod* (June 2006) by Su Zhan as his agent's degree thesis in the Department of Philosophy of Beijing Normal University, hereinafter referred to as Su's thesis.
2 See (1) SASAKI Katsuhiro (佐々木勝浩), 2003, 'History of Mechanical Clock in Japan' (日本の機械時計の歴史), *National Museum of Nature and Science News* (国立科学博物館ニュース), (414): 6–8; (2) HASHIMOTO Takehiko (橋本毅彦), 2003, 'On the Accuracy of Japanese Clock' (和時計の精度をめぐって), *National Museum of Nature and Science News* (国立科学博物館ニュース), (414): 9–10; (3) ODA Sachiko (小田幸子), 1994, *A Pictorial Record of Japanese Clocks Preserved at the Seiko Institute of Horology* (セイコー時計資料館蔵和時計図録), Tokyo: Seiko Institute of Horology (東京: セイコー時計資料館), pp. 3–7.
3 There are two types of sundial, horizontal and equatorial. The uniform scale on a horizontal sundial can't indicate the fixed-hour time precisely because the shadow moves faster in the morning and evening than noon in most locations except the equator. To solve this problem, people put the dial in the equatorial plane and put the gnomon in the direction of the Earth's axis. Thus an equatorial sundial is made, which can indicate the time precisely. But it must depend on a series of complex computations, astronomical observations, geological surveys and a whole set of theories to make out the angle of the equatorial plane. Therefore, the fixed-hour system may not be as natural as we think. It depends on a whole set of theories and artificial conventions.
4 The figures come from, ODA Sachiko (小田幸子), 1994, *A Pictorial Record of Japanese Clocks Preserved at the Seiko Institute of Horology* (セイコー時計資料館蔵和時計図録), Tokyo: Seiko Institute of Horology (東京: セイコー時計資料館), pp. 6, 8.

5 Hashimoto Takehiko, 'Introduction[J]', *Japan Review*, 14 (2002), pp. 5–9.
6 See (1) SASAKI Katsuhiro (佐々木勝浩), 2003, 'History of Mechanical Clock in Japan' (日本の機械時計の歴史), *National Museum of Nature and Science News* (国立科学博物館ニュース), (414): 6–8; (2) ODA Sachiko (小田幸子), 1994, *A Pictorial Record of Japanese Clocks Preserved at the Seiko Institute of Horology* (セイコー時計資料館蔵和時計図録), Tokyo: Seiko Institute of Horology (東京: セイコー時計資料館), pp. 3–7.

5 Knowledge and power

In the past few decades, with the development of the philosophy of science, the interaction between scientific knowledge and political power has become a hot topic in both philosophy and politics. Michel Foucault's analysis on the power in modern society and the theory of knowledge broadens the traditional boundary of the studies on politics and power, and it reveals how the micro power works. In philosophy of science, Rouse argues against that science is independent from any relations with political power. His understanding on how the modern science works depends on the introducing and developing Michel Foucault's theory on knowledge and power, with the dimension of power added in the perspective of PSP. According to his idea, political analysis of scientific knowledge is understood and deepened in philosophy of science. As knowledge and power have been highly combined in both ancient and modern China, analysis of the relationship between the two makes great sense to Chinese philosophers of science. By expounding the impact of power on scientific knowledge – power's dimension – this chapter elucidates its role in the generalization of local knowledge. In other word, it helps to explain Rouse's idea that power plays a role in the generalization of local knowledge and in the standardization of scientific practice. This analysis can be applied also to the science as practised in China.

5.1 Alliance of knowledge and power

Traditional philosophy of science rejects the dimension of power, or there was no dimension of power at all. Michael Mulkay, a sociologist of scientific knowledge, maintains that, traditionally, science is thought to be realistic and objective, and cannot be influenced by the observer's preferences and aims. The nature can be more or less reflected as the real world shows. Science is thought of as a series of academic activities which carry on an elaborate explanation of the things, processes and relations. So, traditional philosophy of science tells us that scientific knowledge originates from nature; it reflects real nature and objective experience. It is the true representation of being, which can be testified and defended according to the experimental evidence and logic rules. Anybody, no matter who he is, as long as he is thinking rationally by applying correct parameter, won't be influenced by any bias, and therefore can obtain objective knowledge and truth. As a result, scientific knowledge is not influenced by social factors with power included. Power has little to do with knowledge.

As Rouse said,

> Knowledge acquires its epistemological status independent of the operations of power. Power can influence what *de facto* is known, but its being known, and what it is for it to be known, cannot be subject to the influence of power. That is, power can influence what we believe, but considerations of power are entirely irrelevant to which of our beliefs are true, which of these are known to be true, and what justifies their status as knowledge. It is generally believed that knowledge is best achieved within an inquiry freed from political pressure, but that ultimately an epistemological assessment of that achievement must not refer to the intervention of power, either in support of or in opposition to knowledge. Similarly, power may or may not serve knowledge or draw upon it; it remains power all the same. In their constitution as power and as knowledge, power and knowledge are (in principle) free from one another's influence.
>
> (Rouse, 1987, pp. 12–13)

Therefore, only free from power can the original nature be found. To make scientific knowledge objective and free from the influences of human and social factors, Merton articulated four science norms: namely, universalism, communism, disinterestedness and organized scepticism. Indeed, the norms prevent human and social factors from interfering and intruding into the ideal norm.

An idea is hidden in the above views that politics and power are the factors ruining autonomy of science, bad for science research. However, in the 1960s, this idea of scientific knowledge and rationality came to meet challenges from every field.

Frankfurt School attacked the so-called scientific rationality at first. They thought that this kind of scientific rationality has a deceptive aspect isolating power. In this view, scientific rationality has actually ideological status in modern society, and therefore justifies human manipulation upon the nature, in favour of which is the slogan *Knowledge is power* by Frances Bacon. They also believe that, in order to liberate people, it is necessary to break the dominating power from science rationality.

With the rise of historicism in philosophy of science, man began to reflect on science and the relation between science and power. These considerations suggested that scientific knowledge is not pure and free from society. On the contrary, it is influenced by social dimension. The social dimension is deeply rooted in scientists' practice, which shakes the root of the traditional idea that knowledge has nothing to do with power. Thomas Kuhn and Paul Feyerabend held such point of view.

In his book *Structure of Scientific Revolution*, Kuhn paid attention to the history of science development and attached importance to the social function of science. These efforts revealed that the social and political factors should have importance attached to them when discussing epistemology.

Kuhn introduced the relativist concepts such as paradigm and incommensurability into science, and thus deconstructed the view that science is unified, integrated and general. Kuhn thought that natural science at any stage is based on a set of conceptual system, passed on from his ancestors to the modern researcher. The conceptual system is a historical product, rooted in culture, in order to get access to which the researchers should be trained. Only by understanding the explanatory methods historiographers and anthropologists applied can the laymen understand the conceptual system. Therefore, the paradigm becomes a norm to the academic community.

Different science communities have different paradigms (belief, standard of value, technology and practical skills). What is more, the rival paradigms are incommensurable. The change of paradigms give scientists different views on what they study. After scientific revolution, the scientists faces another world. Thus, during the scientific revolution, the tradition of conventional science have been changed, and the scientists' perception on the surrounding should be retrained. Under some special circumstance, he must be involved in a gestalt change. After doing this, his research world cannot be commensurable with the world he live in. Due to different paradigms, researchers see different questions and different researches, which let the world they study be also different (it is implied that the world between researchers is the world with local characteristics). So Kuhn introduced the relativism of epistemology into science.

Paul Feyerabend goes even further. He strongly criticized the superior statues (which is just the points of views that science is of ideology by Frankfurt School) that science endowed itself. He puts forward an anarchist epistemology. According to his theory, the achievement in science just like heliocentric theory, both are not victories of rationality. He declared that scientific rationality as reasonable criteria is only a fairy tale. Science is actually a heliocentric career, since knowledge does not represent the real world, nor get closer to the truth. Instead, it is an ocean of knowledge, increasingly extending, inconsistent, and even incomparable.

All theories about scientific knowledge are post-modernist. Postmodernists think that knowledge contains not only orthodox western science, but also non-western knowledge, techniques, dubious humanities, and beliefs and witchcraft with equivalent function to science.

In Foucault's opinion, the fictions like Telliamed have never gained the statues as science, but been labelled as mystery. Though most of them did not satisfy the established criteria of science, and went even further away from the criteria, they are still knowledge. To list all into the category of knowledge is a revolution to the traditional knowledge category. The Strong Program of Social epistemology regards science as a social institution. Social constructivism stresses micro-sociology research on scientific knowledge. The social factors such as scientific and artificial products, publicity technique, scientific research skills, the power conflict in laboratories, the power conflict between laboratories and fund administration department, will be overwhelming factors in scientific knowledge. Not only is knowledge rich in social factors, but is also itself a mixture of these factors.

Unlike Feyerabend, Foucault argued against modern knowledge by applying the methods of archaeology and pedigree. In his studies of the archaeology of knowledge, and pedigree about punishment history and insane, he declared that in traditional theory of knowledge, power is contradictory to truth, which is completely a fancy, an idea by typical intellectuals. In fact, the effort to separate power from truth is both an ideal and a cunning device for intellectuals. The idea actually hides the danger from seeing and presents a false knowledge landscape. However Foucault expounded the relationship between knowledge and power. That is, knowledge and power join in a plot.

Influenced by Foucault, in his book *Knowledge and Power*, Rouse investigated the partial challenge raised by pragmatism and new empiricism to the traditional knowledge/power theory, and then he criticized it by borrowing Foucault's opinion on knowledge/power. The acceptance of scientific theory completely depends on theoretical agreement to proved knowledge, as the traditional philosophy of science thought. However, Rouse did not think so. Knowledge, which was regarded as a presentation in traditional philosophy, is not only a representation such as text, idea, diagram, but also a live interactive mode, or the way to communication with the world.

This model compasses specific context. Only in the specific context can the representation be understandable, and they link with other practices with some meaning. Rouse also thought that the presentation hides the natural link between knowledge and power. He thinks the traditional idea about the relation between political power and scientific knowledge was misleading, which causes us to ignore the way that power works, and misunderstand scientific practice and its political influence. 'Perhaps a revision of our understanding of power will change the configuration of knowledge and power more fundamentally than have the recent developments in epistemology and the philosophy of science' (Rouse, 1987, p. 2). It indicates that in cognitive practice there is some power functioning, and thus also a new perception of the relation between knowledge and power. However, it is just the first step. The perception on the relation between knowledge and power by Foucault goes even further. On the one hand, power produces knowledge, since knowledge is produced, accepted, and communicated in an institution of power. On the other hand, knowledge works in the form of power, as a standard to judge other things and a norm of regulation. As a result, legitimated knowledge spreads the influence of power. As Foucault suggests, knowledge and power are combined. Knowledge is a form of power to record, disseminate, and replace something. It is knowledge that strengthens and regulates others. And it is the basic function for power to employ, occupy, distribute, and conserve knowledge. Therefore, in this sense, SSK also deals with the internal relation between knowledge and power. These regulations also indicate that the academic system has never lacked powerful relationship in academy. Academic knowledge contains the clue to academic power. Power is always an internal dimension of knowledge.

By investigating the characteristics of power in science, Rouse goes further than Foucault. After deep investigation and consideration, Rouse puts forward a claim that power shapes and constrains the field of possible actions of persons within specific social context. Here power has to cover practice, in which everything is related to power and shaped by power. If so, science has to be covered and shaped by power. But could it be arguably plausible?

5.2 Another dimension of scientific rationality: the power characteristics in practice

Strictly speaking, political philosophy is not a part of philosophy of science, or a core concept at least. However, the argument in this section is not in pure political sense. For one thing, this part embodies how Rouse's practice norm is produced and the logical thinking works (concerning to this part, we will discuss in the following chapter). For another, it sets scientific rationality free.

In the beginning, rationality was regarded as universality, necessity, and normativity. The understanding was formed with the rise and development of positivism in philosophy of science and damaged by the decline of positivism. Thomas Kuhn also accepted the rationality in science. He claims that rationality in science does not lie in the neutral observation or division of the subject and object, but in the collective value of the science community. Therefore, rationality represents accuracy and consistency.

Many philosophers' contribution to the question of rationality has been a failure, and consequently the rationality of science gets in risks, which decreases the significance of philosophy of science. In fact, deeply indulged in foundationalism and absolutism, man cannot find a way to characterize rationality.

Rouse is one of those philosophers. To avoid foundationalism and absolutism, and to suit himself in thinking about the nature of scientific knowledge, he puts forward an assumption on how to understand science based on practice. He breaks the routine that theory is superior in the traditional way. By combining behaviour and context, he points out that the previous explanation of Kuhn (Kuhn II) will lead to a theory-dominated thinking, while the understanding of real Kuhn (Kuhn I) should be based on a practical interpretation. However, if we understand science basing on Kuhn I, can we avoid the rationality crisis? Concerning the question, Rouse's answer to the practical interpretation is that we find the answer only in practice can. The rationality of science is not a discussion about theory, but practice. In another word, the rationality of science is not the rationality suited to formal logic, but to the nature. It embodies the local understanding that scientific practice brings us. Therefore, Rouse provides a new framework to solve the problem.

But how can we understand the rationality of science in terms of practice? Why is power a dimension to explain the rationality?

First of all, in Rouse's opinion, power and knowledge are internally related.

> Power does not merely impinge on science and scientific knowledge from without. Power relations permeate the most ordinary activities in scientific research. Scientific knowledge arises out of these power relations rather than in opposition to them. Knowledge is power, and power knowledge.... Power as it is produced in science is not the possession of particular agents and does not necessarily serve particular interests.
>
> (Rouse, 1987, p. 24)

Thus, the proposal does not neglect the traditional power, such as political or legal power. The influence that the traditional powers impose on science can be well understood only by understanding how the micro-power works.

Second, Rouse accepted all the characteristics of 'discipline' put forward by M. Foucault. His notion is that the acceptance of knowledge is a process to accomplish discipline. The process of finding knowledge also has the characteristics of discipline. You cannot find the problems in discipline unless you accept the discipline of finding knowledge. Knowledge itself should be understood as practice and the knowing-how behaviour. Practice itself has the power of sense to shape the external object as its will, and meanwhile the shaping has brought power in. As the previous paragraphs have mentioned, Rouse stresses context and shaping in the nature of practice, which are more important than its actors. Here comes a response that

> all interpretation (which includes all intentional behaviour, not just discourse) presupposes a configuration or field of practices, equipment, social roles, and purposes that sustains the intelligibility both of our interpretive possibilities and of the various other things show up within that field. I will be characterizing power as a characteristic of this field or configuration rather than as a thing or relation within in.
>
> (Rouse, 1987, pp. 210–1)

Therefore, 'to say a practice involves power relation, has effects on power, or deploys power is to say that in a significant way it shapes and constrains the field of possible actions of persons within some specific social context' (Rouse, 1987, p. 211). The reasons why practice is powerful is that power is one of the characteristics of practice. Power here does not subject to political or judicial sense, but to some internal discipline. Since practice has the characteristics of power, power in some circumstances defines and controls man's potential behaviour in a meaningful way. Meanwhile, power is not only the controlling discipline but a positive restriction. 'Such an account would emphasize the productive character of power rather than repression and distortion; it would focus upon its engagement with actions and practices rather than beliefs; and it would describe its local, decentralized, and non-subjective deployment' (Rouse, 1987, p. 209). Therefore, only the productive power can construct laboratories and extend its control beyond laboratories. The nature of power is the restriction in local context, so man can neither pursue it deliberately nor defy it totally.

Finally, science and power develop in the network of power woven by practice. Science develops, through practice, active links which interact with each other. In other word, they both have the characteristics of seeking being attended. So it is with finding and defying. The cognitive subject and the aims of discipline will evolve with practice. Power is the acting force in order to realize the relationship. Not only science but also power develop in the network of power woven by relations, which are required by the scientific practice relation.

5.3 The dimension of power in scientific knowledge and its characteristic presentation

Power has been a core notion in political philosophy. The classical theorists on political thought insist that state power should be regarded as owned and manipulated by the individual, social classes, political parties and groups, and churches. This power includes the drafting and implementation of laws, rules, as well as the formulation and implementation of policies. The fight for power takes place in the dramatic political events – social movements, general elections, court debates, law-drafting and implementations. Foucault identified the manipulation of law in politics as juridical power. In other word, it is called 'economical model of power', rooted in economic areas and as an extreme forms of powers.

Foucault analyzed the notion of power in a particular way. He thinks of the traditional notion of power as dealt with the complex power relations in a simple and macro way. In order to grasp how the social power works, it is necessary to understand power from a totally new perspective. He emphasizes that power is neither an entity nor any property, but a relation and interwoven network existing everywhere. What is more, it is not central but multiple, coming from everywhere.

The alliance of power and knowledge is deeply rooted in social relations. Institutions such as army, school, and factory are all subject to discipline in the relation of power immerged. Power is not only an oppressive external control, but also a kind of productive practice, producing knowledge, inspiring activities, introducing discourse and thoughts. Obviously, Foucault endowed the power with a broader interpretation. Rouse attempts to apply and develop Foucault's theory of knowledge/power to understand scientific practice. He maintains that scientific practice, especially in the laboratory and its reconstruction of the world, cannot separate from power. He initiates a new branch in philosophy of science, namely political philosophy of science. From Foucault's theory of knowledge/power and Rouse's philosophy of science by, we can understand the powerful dimension of scientific knowledge that to some extent power affects scientific knowledge.

5.3.1 Scientific knowledge is constructed by power

Foucault is the philosopher who first regarded knowledge and power as a whole. He realized that the production of scientific knowledge is a process which combines knowledge and power, when he expounded the process in which knowledge and power cannot separate from each other. Human science itself is produced together with a powerful institution. In the process of recording, investigating and analyzing, the power institution has built a vast archives hall, on which the human science is formed. Vice versa, human science provides tools and explanations for the better working of power. The relation between human science and power is not extrinsic but mutualistic symbiosis. In the traditional philosophy of science, power has nothing to do with scientific knowledge. However, for Foucault power has entered the inside of the scientific knowledge, and the employment of power strategies is the process of shaping scientific knowledge. Foucault is more interested in how science could be changed into power institutionally. If science is regarded as a series of processes, it is far from enough to point out the errors, and to disclose the mythological truth.

Rouse introduces Foucault's idea that scientific knowledge is constructed by power into the philosophical studies of natural science.

Traditional philosophy of science indicates that scientific achievements are distinguished from factors of power, while Rouse analyzes scientific knowledge from the perspective of power and knowledge by Foucault. He holds the idea that scientific concepts and theories are the products of practice, for example, of constructing, isolating, introducing, tracing and recording the micro world of laboratories. Scientific knowledge and activities are related in a local way. All scientific knowledge is produced by some laboratory, research plan, local community, and research skill.

> The so-called generalization of scientific knowledge is just a transition from one place to another. The transition is understood as approaching another place, while the so-called de-contextualization is actually standardization. That is to make a standard, so that all the local scientific researches should follow some local standards and make the scientific knowledge under such as standard scientific knowledge.
>
> (Wu Tong, 2006c, p. 138)

Power is involved in both constructions of some context, in which such local knowledge is produced, and the standard of the local knowledge. In other words, local knowledge itself is the outcome of power construct. The process of power working is that which is accomplished in discipline and along with production of power. It is power that produces knowledge.

5.3.2 *The process of scientific knowledge creation contains power*

Scientific research involves an activity of searching in the specific contexts in which such practices take place as skills, practice and tools including theory models. Laboratories and instruments are more meaningful than the theory-dominant philosophy of science. Rouse argues that a laboratory is a locus for the construction of phenomenal micro-worlds. Power permeates in the process of construct phenomena in micro-world. Rouse's argument begins by noting extensive parallels between the construction and manipulation of laboratory micro-worlds and the various 'power/knowledge' relations that have been at work in numerous modern 'disciplinary institutions': prisons, schools, hospitals, armies, factories, and so on. Tactics for the construction, manipulation, and control of phenomena within the laboratory must be recognized as parts of a network of power relations running throughout modern societies.

In laboratory practice, the function of laboratory space comes at first. Is the power hidden in the closure and divided of the laboratory space? The closure and division are the premises of experimental science. Do the premises embody some power relation? Rouse (1987, p. 222) thinks that 'these partitioned spaces are also structured to make possible the surveillance and tracking of what goes on within them' and hence it is the monitoring relation of power in Foucault's sense.

Moreover, 'creating an appropriate, familiar and readily usable experimental system is a highly valued achievement in science' (Rouse, 1987, p. 222). The achievement divides the scientists and researchers in the laboratory into staff of different classes according to their capacity. Therefore, the master of the theory-function service and practice operation indeed involves power. The main strategy of practice in laboratories is to carry out research. First of all, if the isolation is a strategy of power, how does it operate in effect? In the laboratory, the objective system is constructed in the known context, and separated from the other influential factors so as to manipulate and trace them. The construction of micro-world is the internal isolation of the factors in the laboratory.

Intervention is a second strategy of power. The aim that isolated micro-world is constructed practically is that we can manipulate it in a specific way. Scientists are not content to present things in the isolated context. What is more, 'scientists aim to create new domains for investigation by subjecting things to influences and interactions that would not occur (or would not be manifested as phenomena) without their intervention' (Rouse, 1987, p. 102).

The third strategy is tracing. Tracing experiment concerns the control of the whole experimental process in the beginning. Tracing is not only the result of experiment monitoring but also the normal operation of thing monitoring. Tracing what happens to the micro-world in the laboratory is to classifying, encoding, recording, and conduction.

The fourth is standardization. There is an interesting corresponding relationship among the standardization of materials, procedures and instruments, which is manipulated in the discipline found by Foucault. In scientific practice, standardization has the function of normalization. Standardization adjusts the researches carried out by various scientists and scientific groups, deals with the abnormal results, and helps to form a reliable judgment.

The fifth strategy is the production of signs. 'The most important power relation embedded in the laboratory is the production of signs' (Rouse, 1987, p. 224). For example, 'scientists have induced mute and hidden things to "speak" to us of themselves: radioactive labelling, bubble chambers, X-ray crystallography, the various forms of chromatography and telescopies, and so forth' (Rouse, 1987, p. 224). The signs show how things present, construct, move, and transfer in micro-world in the laboratory. In the laboratory, these data are produced in volume.

Experimental practice disciplines the subjects being practised. The laboratory is the strictly closed and isolated space, which is closely monitored, traced, interfered and operated. In such a space, 'these constraints upon the materials and processes that occur within its constructed micro-worlds cannot be sustained unless there are also constraints (largely self-imposed and self-monitored) upon the persons who work within them' (Rouse, 1987, p. 237). It can be seen that

> the laboratory, like the clinic, the asylum, the school, the factory, and the prison, serves as one of the 'blocks' within which, according to Foucault, a 'micro-physics of power' is developed and from which that power extends to invest the surrounding world.
>
> (Rouse, 1987, p. 107)

The decomposition and isolation of material which makes the research results possible have to be partly extended, if the research results get reliable support from other places. It is not disputed that the standardization of circumstance is often promoted greatly. Even in the present experimental study, standardization has also the function to stabilize results. However, the substance purified and measured must be prevented from contaminating; machines must be kept clean with raw materials provided, maintained and properly operated; the temperature, humidity and bacteria number in the surrounding must be controlled. Even afterwards, the isolation should be kept to some extent. For example, the newly produced toxic substance should be prevented from going into underground water, air, or food chains. Once pollution happens, something must be done to clear it. Besides, the maintenance and appliance of measurement standard, the purification, isolation and use of experimental materials, the construction, and employment of devices would ask for an extension of following activities similar to the experimental activities, such as tracking, recording, numbering, filing, and looking up related documents, which is necessary for the scientific practice and science outcome extended beyond laboratories.

Experimental practice and its extension agree to the discipline practice and institution that Foucault attempted to describe. There were no normalized discipline and restriction put forward by Foucault. It would not be imaginable how laboratories are operated, and how practices and materials extend. It is important for science, from educational training of scientists to all kinds of monitoring, normalizing, and all the restrictions to provide for the scientists' study and reliability. The construction of micro-world in laboratory and the strategies of monitoring and controlling must be regarded as parts of the power network in modern society. Meanwhile, when scientific practice extends beyond laboratories, it is necessary for power to supervise practice, since they impose many controls and restrictions from laboratories to the world.

> Those who are attracted to Foucault's account of our social practices and institutions as traversed by capillary relations of force or domination will thus have to come terms with the forms of power/knowledge that invest the natural sciences and our dealings with the physical/biological world. But even those who are dissatisfied with Foucault's analysis cannot really escape dealing with the sciences in terms of power. For the extension of the concept of power to apply in some form to the nontraditionally political institutions of prisons, hospitals, schools, and factories has been much more solidly established than has Foucault's specific treatment of power.... Once that extension is accepted, the relationship between these forms of power and what goes on inside scientific laboratories and discourses has become an important issue for political thought.
> (Rouse, 1987, 244–5)

Therefore, laboratories 'as a space of stringently enforced enclosures and separations, of strict surveillance and tracking, of carefully controlled interventions and manipulations, ... laboratory practice imposes a detailed discipline upon those who engage in it. This discipline is not normally noticed, because it becomes routine and ingrained in scientists and technicians, who have long since internalized it' (Rouse, 1987, p. 237).

5.3.3 Scientific discipline derives from power

The disciplining of scientific knowledge can be dated back to the 1700s. According to Foucault, the disciplining of technological knowledge appeared in the 18th century; but knowledge is not singular but plural, distributed in different places, by social status and educational attainment. In the beginning, knowledge was put into endless controversies, since mastering secrets of technique knowledge at that time was equal to mastering fortune, and in further the independence among knowledge meant individual independence. With the development of productive force, the fights among technologies become more and more dramatic. It is a fight focusing on knowledge; however, it is not a fight of brightness against darkness, nor one of knowledge against ignorance. In fact, it is a fight against those who attempt to monopolize that knowledge which is closely related to power.

The disciplining of knowledge, in the 18th century, brought about an expansion of knowledge in all fields. In other words, the internal organization of knowledge had its own criterion to exclude false and fake knowledge, to normalize, homogenize and hierarchize the content of knowledge. Finally it came to an internal organization around the concentration of some generally acknowledged truth. And therefore, knowledge was collated as discipline. The whole field among disciplines consists of science. The science evolved from knowledge discipline to monitoring discipline as knowledge, to normalize knowledge system, and to make it a field with solid base and strict organization.

Thus, the disciplining of scientific knowledge comes to a consequence that science becomes a part of culture, and separates itself from others due to its uniqueness, in which sense Foucault redefined science. Another consequence is that science completely separates from philosophy: knowledge included in philosophy is no longer a totality. Science takes the place of philosophy and carries out the assignments of sorting and grading knowledge.

The disciplining of knowledge generates the form of discipline censorship, conducted by some institutes whose aims are to supervise knowledge. Censorship does not involve any narrative content or the agreement with truth, while merely focuses upon narrative rules. It is crucial to know who speaks and whether he is qualified to speak. On one hand, it results in at least considerable freedom for narrative contents. On the other hand, at the level of narrative procedure, the regulation becomes stricter and stricter, with broader meanings and coverage.

The disciplining of knowledge does not stop until now. A typical example of disciplining is the specialty adjustment in higher education of China. For instance, the catalogue of specialty has been modified three times since 1954, when the national-uniformed catalogue was released, and in 1998 a new *Undergraduate Specialty Catalogue of Higher Educational Institutions* was formed. In the contemporary catalogue, all scientific knowledge is divided into 11 disciplines with 92 sub-branches and 274 specialties. The specialties excluded from the catalogue should be strictly examined and approved. In the latest specialty adjustment in 2005, Chinese Ministry of Education classified Marxist theory as one of 72 first-level disciplines. Besides, it reviews the disciplines, laboratories and engineering centres in order to classify them as national, ministerial, and provincial levels. All these demonstrate that power has intervened into knowledge.

As we have already seen, Foucault and Rouse's analysis of the power in scientific knowledge reveals the cognitive process that political and legal power has transformed into the socialized power concept. According to their analysis, power no longer merely means political or legal power, but as a broader concept, a relational network of power. To understand the modern natural science from the broader notion of power, we are about to find

> power does not merely impinge on science and scientific knowledge from without. Power relations permeate the most ordinary activities in scientific research. Scientific knowledge arises out of these power relations rather than in opposition to them. Knowledge is power, and power knowledge.
> (Rouse, 1987, p. 24)

Now I take an Arabian scholar's theory and case study to investigate the close relation between the micro-political power and academics in Arabian academic community, which metaphorically illustrates the micro-political power means to academic research. First, scholars are able to apply educational capital like education background, family background, specialty, and contribution to maintain and to promote their own status. The political power a scholar has is related to the specific specialty on the behalf of him and to the social cultural structure where he stays. In addition, it is related to the types of various capitals as well as cultural, symbolic, social and economic values that he can control. M'hammed Sabour has classified the degrees of esteem that a scholar obtains in the Arab academic community into four degrees:

> People on the first level are the employees who pass the national examination and are hired in some academic field.
> People on the second level are the staff who are entitled to social impact and prestige, namely the ordinary scholar.
> People on the third level are the scholars who are able to maintain their status and protect themselves by academic means, namely famous scholars.
> People on the fourth level are the scholars with sovereign status, namely authoritative scholars. (In the Arabian world, due to the unstable political situation, scholars can be intervened or suppressed because of different opinions. However, the scholars on the level can be exempted for his authority in science.)

What is more important is how an Arabian scholar finds his place in the levels and what method and strategy he applies. Firstly, make use of the science prestige raised by one's own academic outcome and uniqueness. Secondly, join some union by getting close to some class or political belief. Thirdly, participate into administrative task, which is to participate into expert politics. If an expert can put the function to use, for example, becoming the leader of scholar or the member of some committee, he can obtain very high status and get close to official institutions. Some status can be guaranteed by combining with the official institutions (Figure 5.1).

As is expected, scholars think diligence and competence are the qualities to strive for a place in academic community. However, they are fully aware of the role that social capital, political participation and union play in upgrade the status. In Syria, there is only a party *Arab Baath Socialist Party*. Most people think it is important to join the political union.

Therefore, to find a place in academic organization or the administrative institutions in university is one of most effective ways for scholars to guarantee his diversified personal power (Compare with the academic institutions or universities, there exist the same cases). Just as diagram 5-2 illustrates, though the academic endeavour is widely accepted in Arab academic community, social capital, political participation and union also play a very important role in upgrading social status.[1]

Fourthly, it is particular for Arab countries to promote scholars' social status by using social capital and social relation network. In the Arab world, these relations include personal relation, relation of independence and petticoat influence.

Finally, M'hammed Sabour drew his conclusion as follows (see Table 5.1).[2]

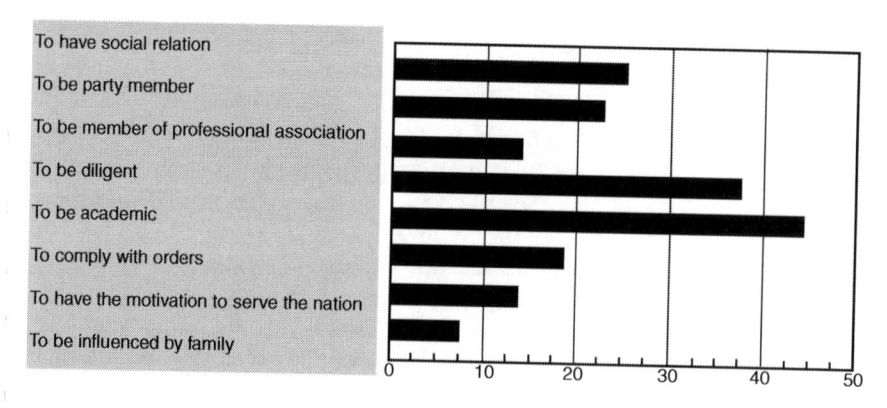

Figure 5.1 Arabian scholars' estimation of how to upgrade personal status.

Therefore, the relation between science and technology and political power highlights two dimensions. For one thing, science and technology are taken as macro-activity and social institution, which involves in an influential relation with politics. For another, Micro science itself (scientists in scientific research) and political power influence each other. In the former dimension we can often apply the relation between the political dimension of national governance and the development of science and technology. The latter dimension is reflected in scientific community and modern social organization and administration. The political study of science and technology in the micro-environment is carried out both inside and outside laboratories. Rouse puts forward that the power is not judicial power but the shaping of laboratories. 'To say that a practice involves power relations, has effects of power, or deploys power is to say that in a significant way it shapes can constrains the field of possible actions of persons within some specific social context' (Rouse, 1987, p. 211).

Laboratories train scientists in accordance with standard disciplines, and transmit them into a part of society, and therefore the administrative methods extend out. The manipulation in laboratories, including the measurement of objects and strategy, also becomes a part of power network throughout society, while making the contemporary society be along with science and technology.

5.3.4 *The impact of science on social power is stronger with the expansion of laboratories*

Rouse argues in detail for the influences of scientific knowledge on the power relation outside of laboratories. In the first place, the direct influence that science imposes on daily social practice and political practice is the transferring of new materials, methods and device from laboratories to the outer world. 'The stimulation, protection, regulation, utilization, and remedying of the effects of these

Table 5.1 How the Arab academic community gained its place and status

Method and strategy	Combination (position)	Authoritative influence	Dissuasion	Exemption
Academic achievement, creativity	To be appointed according to academic title. To get academic status according to culture and knowledge one obtained.	To get official status according to organization rules. To be a former member in academic community. To be accepted culturally. To play a communicative role.	To be academic authority To be possessor of scientific power Adopt retaliatory response in face of challenge	The respected and trusted status in science Highly productive creative Outstanding scientific achievement Symbolic capital Science prestige
Officials holding administrative position	To be an official, a member of officialdom	The outstanding role of organization. specialized knowledge and skills	Gatekeeper of administrative group Control and supervision	The privilege of decision and authority Right to act freely
Political union and political belief	To accept the existing political paradigm To be active or positive in combination To do a political tag	Get legitimacy and influence by participate organizations go-between's role	Pen support by officer's swords Both parties are able to dissuade in paradigm, but both are awesome	Inexcusable status, whose power varies from strong to weak
Social relation (social capital)	To be a member in a social association	Influence and prestige due to authoritative relation and social relation		Free from harm by virtue of social capital

120 *Knowledge and power*

translations of laboratory practices, materials, and procedures now occupy a large proportion of our overtly political activity' (Rouse, 1987, p. 227).

Second, the construction of micro-world and the regulations in laboratories reconstruct the outer world of our laboratories.

> The networks of equipment and practice that have their origins in the carefully constructed, controlled, and surveilled micro-worlds of the laboratory may well impose more expensive and stringent burdens upon us than do other kinds of tools or procedures. There may even be systematic constraints that the remaking of the world to resemble laboratory micro-worlds may impose upon us.
>
> (Rouse, 1987, p. 230)

Third, the extension of laboratories result in two effects: a) Technological system and social organization's interaction are becoming more and more complex and more closely related. b) The natural environment is simplified and controlled more and more seriously by man. 'Scientific knowledge is not something distinct from our practical ability to manipulate and control the phenomena it interprets' (Rouse, 1987, p. 234).

Finally, scientific understanding of the world and practice method also imposes a very important power impact on everyday world. For instance, the comprehension that separates the world and the methodology that comes from the experimental practice and scientific isolation have imposed a very interesting influence. Concerning to those issues in personality disorder and self-consciousness, we are sometimes regulated but also might sometimes be beneficial from the regulation in the calculating of academic resources.

5.4 On the relationship between social knowledge and power in Chinese society: investigations at both macro and micro levels

5.4.1 The relationship between knowledge and power in traditional Chinese society

To grasp the political characteristics of traditional Chinese society of China, the social characteristics should be understood at first. According to Asian production mode, the classical economists thought that traditional Chinese society is a typical oriental society dominated by water conservancy and agriculture economy. However, Max Weber thought that traditional Chinese society was one of the samples of bureaucratic states. The power which is dominant in society is the government's official growth system and its administrative means. Some scholar thought traditional Chinese society is not only a society of highly united of politicalized education but also that of knowledge and power. Some scholar thought that the situation that government officials are teachers has never been shaken.

There are four characteristics in the micro framework of knowledge and power. First, knowledge is focussed and monotonous, with limited Confucianism as the dominant influence. Second, knowledge highly depends on power and power is above knowledge, being intercessor of knowledge, while knowledge cannot supervise knowledge substantially. Third, subjective status is not stable and form independent social classes, in spite of imperial examination system. And finally, the integration of knowledge and political power strongly influences education in making Chinese traditional education too much political. Education is education and vice versa.

In addition, though in western society, politics also permeates the policies of education, science and technology, the western are claimed as a democratic society in which reversed opinions and viewpoints forms a tension, which is some conditional and balanced relationship, in order to keep the autonomy of cognitive dimension in science and technology from influence by political powers. However, the politicizing of traditional Chinese society did not allow science and technology and experiment to be developed in the autonomy of the institutions of science and technology. This makes the development of science and technology deprived of autonomous support. However, scholars have made empirical and theoretical thought that diversified cyberspace is prerequisite for society, economy, science and technology. The lack of diversity will drive the development to a dead end.

In 1840, western powers began to invade China, and China had to give up close-door policy of exclusion, resulting in big impacts on relationship between old knowledge and power in the traditional Chinese society. Under such circumstances, many patriots began to seek for the ways to save the nation. Science and technology was widely accepted under the double pressures, namely the rise and decline of a nation or culture, and they were localized in believing that science will save the nation. Someone thought that a Chinese journal *Science* launched in 1915 is the beginning of natural science research in China. Besides, natural science research was influenced by policies of instant gratification and science confused with technology, which struggled under some pressures. Obviously, the similarity of policies in different times transcended the differences of regime change in reflecting the strong influence from traditional culture.

A case is applied to illustrate the change of relationship between knowledge and politics in China. Recently some scholar have studied the change of the translated publication in Late Qing from the perspective of knowledge politics and attempted to reflect change happening to the relationship between knowledge and power at that time. The translated publication is the normal channel to communicate knowledge between different systems. However, to expect the prosperity in Wei-Jin Dynasty and between the Ming-Qing dynasties, the publication translation only prospered in Late Qing. For instance, Guo Jianming believed that it was not pushed by cultural law but by needs generated from politics, and He summarized the characteristics of the change by comparing the translation before and after the Sino-Japanese War of 1894–1895.

Between Opium War and the Sino-Japanese War of 1894–1895, the main characteristics of translated publication are three features. The aim of translated publication is to extend the traditional knowledge system; the translation was conducted by the coorperations between western missionaries and the government; the theme of translated publication was about western natural science, especially the technology.

After the Sino-Japanese War of 1894–1895, the main characteristics of translated publication were that knowledge had got rid of political control and attempted to promote the transition of politics. To be exact, the translation took place outside of the system. There are four features. The aim of translated publication is to extend the boundary of traditional knowledge and to lead it to transform to heterogeneous knowledge; with the transformation of the aim, the translators change differently and consist of folk translation institution and churches, led by new intellectuals; the theme of translation includes western philosophy, social science, and natural science; the translated books came from Japan instead of from Europe and America.

After 1840, the idea that political structure should be influenced by science had been prevalent for a while. Some elites who were learning western science realize the importance of science for the country. They put forward the slogan that *science save China* to the practice by themselves. Although they failed, it has been established that the pursuit of science could be applied to enhance society. The knowledge of the necessary combination of science and politics also influenced intellectuals.

However, on the one hand, due to the long-term turmoil and wars, the backward of Chinese society made the endeavour fail to influence politics. On the other hand, the practice of Chinese Communist Party (CCP), which represented the poor and came to power by revolutionary, broke down the idea to some extent that by politics could be established by scientific influence. Since the practice of state political power regulating science appears efficient, science and technology seem to be developed when they are subject to political power. Therefore, it imposed a very strong influence that the new political notion could change the old society. A dramatic change happened to the structure of science and politics after the founding of People's Republic of China. Since the appeal to autonomy was changed by the appeal to national aim and its politicization, the qualification as autonomous politics and social rights for the Chinese scientists and engineers would be attached to new politicalized subject and the political process.

5.4.2 The relationship between knowledge and power in the development of science and technology, from the founding of the PRC to the Cultural Revolution

After founding People's Republic of China, CCP asks the intellectuals to admit the Marxist ideology, which intensifies the traditional priority of political power to science. There are some other important reasons why Chinese society is deeply politicalized at that time.

First, traditional China is a politician-centered society. The highest pursuit of Chinese intellectuals is the personal morality, collective identity related to family, and serving motherland after their graduation. In fact, it is the basic characteristic of politicization and society. It is reasonable that the culture will continue no matter how China is revolutionized and modernized. That factor is one of social and cultural elements when we account for the political environment of science and technology development.

Secondly, after the founding of the PRC, the Chinese intellectuals of science and technology were basically judged as the ones who had the same viewpoint with bourgeoisie intellectuals, and were *attached* to the bourgeoisie, the landlord class and imperialism, so that they ought to be transformed into the proletarian intellectuals. Because of their class consciousness and identity, scientists were conscientious to accept the transformation so as to reduce the conflict between autonomy requirements and national scientific ideology. But this judgment mixed with the process that the state speeded up the completion of the modernization, forming a basic political requirement to revolutionize modernization, which is the basic starting point for observing and thinking about problems, and for the decision of many scientific and technological policies and intellectual policies after 1949. After the founding of PRC, through three major transformations, the original understanding of intellectuals should have been reversed, since the majority of intellectuals, especially scientists undertook the responsibility for the national fate. In the 20th century there was no ivory tower for Chinese intellectuals to live a calm life. Concerned to economic development and national defence construction, the majority of Chinese scientists made many contributions to make the nation rich and strong.

However, from a political point of view, the leading class of CCP felt worried about what Chinese intellectuals and scientists were doing. During this period, the intellectuals' painstaking research for a scientific theory and the effort to do academic research were regarded as the behaviour alienated from CCP and morality. There is political and academic context in which intellectuals were kept keen vigilance in thinking and politics, though their skills were made use. The CCP's intellectual policy also constituted an important part of the political context. According to research by some scholars, the CCP interprets the intellectual issues from the perspective of social politics. Whether intellectuals were qualified depends on their class attribute and political direction. Marx's theory of class analysis constituted the starting point of this interpretation. As a group not engaged in or not directly involved in any material production, intellectuals had to attach to some class, due to not having independent interests. In the communist countries, intellectuals were part of proletariat. However, the intellectuals who came from the old society still had to make class position and the transformation of the world view. Only by discarding bourgeois political position or world view and by accepting Marx's world view and proletarian political position, could they be proletarian intellectuals.

The thought was the core viewpoint that the Communist Party of the Soviet Union and the CCP held on intellectuals.

That understanding is the key point of the CCP for the policies of intellectuals. The viewpoint, position, and methodology are historically reasonable and legitimate, since CCP is a revolutionary party before 1949. After 1949, CCP needed to rebuild the country. However, from 1949, until the Culture Revolution ended (the thought tended to go further due to inertia), the viewpoints on intellectuals were fixed, which results in negative effects. At the present, CCP leaders are attempting to correct the mistake by regarding the intellectuals as a part of the working-class and by highlighting the role that the intellectuals have played in the construction of modernization, which greatly free those who work as successor of science and technology and change their political statuses for the development of science and technology.

Third, the traditional system of knowledge clashed fiercely with western one. Though the knowledge system needed to absorb the modern western science and technology largely and fleetly, the political dimension tended to suspect and to reject the western culture. It requires that the party in power should strengthen Marxism and Chinese identity. Thus, the process was closely related to be instituted combined with politics and power, which were also a process to enforce political powers.

Taking university as an example. Before 1949, it is the lofty mission that universities served state during the period of the Republic of China. However, the government lost its control over university, which created freedom and autonomy for universities. For example, Cai Yuanpei who was influenced by the western academic freedom and university autonomy served as leaders of educational administration or universities, leading to professor managing and university self-determination. At that time, university self-determination was best displayed. But after founding of People's Republic of China, power strengthened the intervention over universities. In the 1950s, CCP established the leadership over universities. The internal institutes of universities copy the government mode, and universities with multi-disciplines were transformed universities of art and science or institutes. Moreover, intellectuals were transformed to the cadres and professors from administers to ordinary labours. In 1958, the idea that *education should serve proletarian politics* was put forward, and the government frequently impacts universities. During the Cultural Revolution, the intervention from government came to its summit.

However, there were a conflict between the appeal for the politics of science and technology and the development of science and technology, including the following two aspects.

First, after the establishment of PRC, the state had to aimed at national industrialization with some slogans such as 'catching up UK an US in 15 years' and 'realizing communism in advance', which reflects the overestimation on the political institution and international status. Therefore, to improve the technology to meet the state requirement would greatly depend on the national level of scientific research. However, the national economy could not meet the requirement from scientific research without solid construction and accumulation of disciplines. Therefore, it is necessary to accurately interpret the policy of *science and technology serve the practical production,* which should not require that every aspect of scientific research has to point at problems in production. In order to serve the practical production, scientific research must solve the problem how to develop itself. Without concerning the problems of science itself, especially from the long-run period, science will lose the foundation to serve the practice. Thus, the conflicts at that level were caused by the simplification and practicability of science and technology. And the conflict is intensified by the rejection of scientific community to simplification and politicization.

In effect, the Party's ideologically calling for the combination of intellectual and workers had its sound consideration, and the practice contributes to development of science and technology, which combination of revolution and modernization. As we know that in 1950s, China was a very poor country with an extremely larger illiteracy and half-illiteracy population. It was the combination of intellectuals and workers that had rapidly changed the situation of illiteracy and half-illiteracy population. Therefore, in the planned ways, mass movements and concentrated resources were the great events accomplished, among which the ways and practices at that time were legitimate and reasonable.

At the second level, there are the conflicts among the notion that science is autonomous in its community, the pure notion that scientists do researches just for science, and the epistemology that science and technology policy is utilized respectively. The conflict had got radical because CCP combined those notions with political values. The attempt to conserve the authoritative status of community to scientific researches was regarded as the political confront with the Party's leadership. In the eye of CCP and new political power, the conflict between scientific community and them is obviously not an issue how to develop science and technology. From the utilized aims, the conflict concerned whether the state could apply scientific resources as much as possible to fulfil state target. Politically, the conflict lies in whether the state could reconstruct the society of scientists into a community with the ideology of socialism and communism in the realization of the aims of the state and society. The discussion about science policy in its political value means that the scientists are losing the resources to discuss on the scientific issues equally and thus to disintegrate the counter force when accomplishing the established aims and policy by the new state.

However, it is very wrong to strongly intervene and even eliminate the scientific autonomy by political powers, which cause serious damages to science and technology and hence to the scientific abilities on which the state aims depend. For one thing, there will be no theoretical and methodological roots for the application of science without any comparatively independent status of science and technology. For another, the emphasis of the service of science for production will fundamentally deny the independent value of scientific experiment and laboratory work, and hence fundamentally damage the base of scientific research.

Especially, the political damage on the development of science and technology becomes fundamental and potential, since it functions as an internal part of the self-consciousness of scientists and engineers, imposed by lots of political movements and ideological learnings, which later becomes very important since there is believed to be a conflict between the autonomous requirement of scientific community and the political and economic aims of the state. If some scientists claim for self-determination, the epistemological conflict could be identified as political conflict).

In conclusion, in a society with highly centralized power, such as China, it is essential to recognize the ways in which power permeates and then to avoid an over-powerful influence of politics.

Notes

1 In fact, the situation exists everywhere. In China, intellectuals join CCP and democratic parties in high proportion.
2 I believe that his research can also be transferred to the Chinese science and technology context, especially the political power context, because influence of these social relations in western modern society is more or less implicit, but in eastern society, especially in China, the traditional Chinese interpersonal relations still play an important role a society of acquaintance.

6 The contextual normativity of scientific practice

6.1 Naturalism and its problems in philosophy of science

The issue discussed about in this chapter will be the normativity of scientific practice, which is definitely correlated with naturalism in philosophy of science. It is helpful to make clear Rouse's idea about naturalism. For him there are two kinds of naturalism: the metaphysical one and the meta-philosophical one. Rouse attempts to clarify issues of normativity from the view of metaphysical naturalism, while promoting some kind of conciliation between naturalism and normativity from the meta- philosophical one. Arguably his philosophy of scientific practice (PSP) would be characterized as naturalism in the metaphysical sense.

Naturalism has already become an influential philosophy of science since 1960s, due to works from W.V.O. Quine and T. Kuhn. Quine put forward a naturalized epistemology, which took science as a kind of natural phenomena, and epistemology of science as an empirical science of science, just like studying other natural phenomena. In *The Structure of Scientific Revolutions*, Kuhn abandoned normative commitments of logical empiricism and critical rationalism, and turned to an empirical study of science, from historical, social, and cultural views, rather than a transcendental study of normativity.

However, philosophy of science is supposed to face with problem of normativity when naturalism has become a basic position. The problem of normativity could be characterized as incompatible relationships between the descriptive and the normative. There are three basic projects for naturalized philosophy of science. First, the radical naturalism, such as Quine's approach of naturalized epistemology, would deny any kinds of normativity. Second, some naturalists would attempt to make some conciliation between naturalism and normativity. L. Laudan and P. Kitcher have made lots of attempts here. Third, those philosophers who claim weak naturalism would reduce normative notions to definitions of empirical sciences. However, further inquiries and understanding would elaborate two intimated notions: first, how the notion of nature would be used in naturalism and anti-naturalism, which is very important to understand naturalism, since it is the notion of nature that determines what naturalism means. Second, since it is also intimately correlated to the understanding of normativity, the theoretical motivation of naturalism and anti-naturalism would also be essential for our understanding. And it seems that the second question should be answered first, which will then extend to the answer of the first question.

6.2 The history and development of normativity

In the early 20th century G. Frege and E. Husserl gave refutations on psychologism in logic respectively, which also implied defeats against naturalism in philosophy of science. For Frege, logic is a normative science; while Husserl thinks of phenomenology as the strictest science, which would make distinctions of essential structures of meaning from empirical psychology. Under their profound influence, early anti-naturalists, such as Russell and Carnap, argued that the logical forms and structures of natural science could be distinguished from its empirical contents. According to that view, the normative notions such as validity, objectivity, and rationality could not be clarified within merely empirical facts.

It seems indubitable that the premise of earlier anti-naturalism would be a kind of dualism between nature and normativity essentially,[1] or a particular conception of nature in opposition to normativity. Nature is conceptualized as the facts studied by empirical sciences, while normativity should be confined as those logical forms or structures lacking empirical contents. Anti-naturalists justify the distinction through an argument that non-empirical norms, as logical forms or structures, are necessary in a logical or transcendental sense, while empirical regularities are contingent. Therefore, according to the earlier view of anti-naturalism, normativity is founded on logical and transcendental necessity, which is considered as basic principles of explaining normativity in science.

However, that view has faced with serious challenges in the later 20th century. For example, Saul Kripke refutes the traditional equivalence between apriority and necessity, or the one between posteriority and contingency. Instead he puts forward notions of a priori contingent proposition and posteriori necessary proposition. Kripke's idea has implied that there is a new approach for explaining normativity, namely, the naturalized explanation. Following that approach, normativity could be derived from causal or natural necessity, which is a kind of necessity *posteriori*. Nevertheless, though it can explain why we should follow the norms in some circumstances, the necessity *posteriori* would not be always obvious in the historical or material world. It may not be available for our cognition, since it depends upon some non-historical, non-material contexts, which could not make effective ruling on the cognitive subjects situated in a particular, historical, and material context.

As Rouse (2002, p. 13) argues, '[A]ny attempt to ground normativity in necessity must be able to show how the alleged necessities are both authoritative and binding upon materially and historically situated agents'. However, the naturalistic approach could only explain the authority of norms without any mention of its power. Therefore 'the problem of manifest necessity poses a difficulty for any attempt to account for normativity on the basis of necessity' (Rouse, 2002, p. 71). That difficulty would confuse our actions rather than give them ruling power. Rouse (2002, p. 14) argues in that sense that the explanation of normativity from necessity would 'thereby account for normative authority at the expense of normative force', though indeed it can explain authorities of norms.

Both accounts of normativity have been criticized by Joseph Rouse since 1990s. For Rouse, naturalism could not be realized thoroughly by setting normativity either on necessity *a priori*, or more precisely, logical necessity, or on necessity *a posteriori*. Normativity cannot be merely understood as something necessary beyond our interactions with the world, no matter whether the necessity is interpreted as nomological or not. What Rouse would like to argue is a perspective from coherent naturalism, involving two fundamental principles. One is so-called Nietzsche commitment, which refutes any supernatural force or object imposed upon science. And the other is Quine commitment, denying any arbitrary philosophical confinement imposed upon science. Consequently, Rouse would not accept any systematic distinctions between nature and normativity or between spaces of reasons and causes, since any distinction of that kind would let each other be unintelligible or mysterious. Therefore, the dualism of nature and normativity, as the foundation for anti- naturalism, is not only necessary but also hopeless for naturalized philosophy of science. And Rouse's coherent naturalism would like to be an attempt of integrating nature and normativity in order to reconsider them, while abandoning the traditional opposition between nature and normativity.

From that new perspective, Rouse realizes that works from Heidegger and Neurath would be extremely helpful. Heidegger criticized Husserl's transcendental phenomenology, while Neurath refused the approach of rational reconstruction of scientific knowledge, claimed by Carnap. Both of them denied any structure of necessity assumed for the normativity of our interactions with the world. Following their arguments, Rouse argues that science should be interpreted as some practices progressing in causal interactions with the world. Nature reveals itself in those interactions. Therefore, nature is the object revealing itself in practical understanding, rather than the objective nature in traditional context of materialism, in opposition to normativity of practice. Scientific practices could not be merely regarded as non-natural activities, but also as recognizable patterns of the engagement of actors and their contexts in the world (Rouse, 1996, p. 134). Those patterns are 'identifiable by their normative accountability rather than by any performance or dispositional regularity' (Rouse, 2002, p. 12). Therefore, norms are involved in scientific practice itself, which is the normative accountability of scientific practice.

Insofar according to Heidegger's main idea, it is our participation within the world that interprets the world. Participation is interpretation for us, and meanwhile it regulates ourselves and our practice. The identity of practice, as some kind of repeatable pattern, is guaranteed by that normative accountability of practice. And the normativity lies in practice, rather than stand outside or opposite to the nature. Nature and normativity should be integrated within practice as two indispensable aspects. All those above have been claimed by Rouse's coherent naturalism. The thesis of naturalism is ontological rather than epistemological. Coherent naturalism would entail no pre-existing notion of science or nature, but 'locates us in the midst of the scientific and technological practices that continue to reshape what it is to be nature, and how we can understand ourselves and our possibilities as natural beings' (Rouse, 2002, p. 360). In terms of that argument, nature involves norms, while norms are derived from practices within which agents promote intervention upon the world, and where nature would be incorporate with norms.

Rouse argues that the problem of normativity would be dissolved in his coherent naturalism, since not only philosophical naturalism thoroughly dominates that approach, but also the social world of scientific practice encompasses nature. The dualism between nature and normativity, held by most of naturalized philosophies of science, is eliminated by Rouse's approach, while transforming the understanding of normativity into the normative accountability of scientific practice. And therefore, the questions remaining would be: how could the normativity of scientific practice be understood clearly? Especially, how to understand normative authority and force, as two basic aspects of normativity in practice? That depends not only upon the grasp of temporality, which is the essential feature of practice, but also upon agent's grasp of contemporary circumstances in practice.

6.3 Normativity, modality, and importance in scientific practice

Rouse's conception of practice integrating nature and normativity has already made a reconstruction of the notion of scientific practice, as well as a reconstruction of the relationship between nature and normativity.

According to that reconstruction, practice would not be in opposition to nature, but reveal itself as actions in some meaningful, social, historical, and material contexts. Those

> are not just patterns of action, but the meaningful configurations of the world within which actions can take place intelligibly, and thus practices incorporate the objects that they are enacted with and on and the settings in which they are enacted.
>
> (Rouse, 1996, p. 135)

First and foremost, practice is a kind of context in that sense. And also, guaranteed by the essential character of temporality, the practical context should be in an unfinished and reconfiguring process, facing with the future. Contemporary practices have been lying in a particular space-time, as a local activity, while also involving extension to the future, so as to make agent participate into cognitive practice from her understanding concerned with the past, the current, and the future. Every practice in the current is stable for the time being, not determined by its history. Practice is a dynamic process, while reconfiguring the past, constructing the current, and projecting the future. The most important thing for understanding normativity in practice should be the recognition of temporal essence of practice, which opens scientific understanding already upon various real possibilities for the future. The clarification of real possibility can be obtained from causal interactions between bodies and surroundings. Therefore, the principle of practical normativity could be thought of as that 'the normativity of scientific and everyday practice thus comes from being in the open-ended contingencies of a historical-material situation rather than in a relation to something "outside" or "beyond" it' (Rouse, 2002, p. 76).

It is argued that there is no appeal to any metaphysical or supernatural assumption in order to understand the normativity of practice. That is a naturalistic understanding. However, practical circumstances are still philosophical black boxes for understanding normativity, unless the following questions could be made explicit: How could the normativity from practical circumstances determine scientific practices or understanding? And how can we explain the authority and origin of force of the normativity? Probably the answers should be seated on some fundamental characters of practical circumstances, such as intrinsic causality and temporality, and on explaining in further the stakes and normative forces of them.

Philosophical naturalism insists on explaining normativity from natural causality. Causality and laws of nature are foundations for understanding the notion of normativity in a scientific way. Nevertheless, the relations between causality and normativity would be interpreted in various ways according to different conceptions of causality.

When he interprets the notion of scientific practice, Rouse argues that causality should be elucidated as some causal capacities shown from the interplays between agents and the world. He also claims that the nature consists of causality in an intrinsic way. Thus, any understanding of natural world must be laid in the intrinsic activities of causality. That would be essentially different from traditional notion of causality. According to Rouse, traditional naturalism has already been confused by a 'God's eye' view, upon which the notions of causal interactions and laws of nature depend implicitly. Therefore, it is a dualistic understanding of causation, which is interpreted as external to practical circumstances. And that is unacceptable for philosophical naturalists, who would prefer to avoid those mysterious structures of relationships brought by the dualistic understanding for scientific practice. In order to avoid a supernatural notion of causality, it is necessary to ascribe causal significance to scientific practice, as the whole of practical circumstances, rather than to objects independent from practices. Practical circumstances are the whole actuality of intrinsic causality. The commitment of putting the philosophy of practical causality as the primary is the radical core of philosophical naturalism held by Rouse.

The subjects and objects of inquiries are ontological derivations from phenomena of intra-action in practical circumstances. Consequently, there is no clearly ontological distinction between subjects and objects. Agents are figured out as cognitive subjects in the practical intra-actions. Objects are also configured in the same process. The boundaries of causal intra-actions in scientific practice are not properties of systems in separated states, rather they are determined by holistic configurations in phenomena of intra-actions. In a word, the totality of practice depends on its particular causal intra-actions.

Rouse's argument derives from Heidegger's notion of practice. According to Heidegger, elements must be excluded from practice unless they could inter-connect as a causal totality. For example, an agent can only be involved in a practice if she connects with circumstances by some causal intra-actions. And an agent, as a constitutive part of practical circumstances, must be able to involve some causal intra-actions with other constitutive elements. Therefore, Heidegger's view has already implied that causality holds somewhat normative authority for agents and other constitutive elements of practical circumstances.

Rouse argues that the repeatable pattern of practical circumstances depends upon standards of constitutive elements and modes. Causal capacities, constructed from causation among constitutive elements, belong to objects in phenomena rather than in abstraction. For instance, there are two kinds of chemical which function as constitutive elements of laboratory practice. They reveal some causal dispositions respectively in some common circumstances, even though they may not involve in an actual chemical reaction. Therefore, finally the dispositions are identified as constitutive and internal causal relationships. Practical circumstances are different in terms of particular causal structures and dispositions, according to different constitutive elements. Furthermore, those structures and dispositions should be interpreted as parts of holistic phenomena, rather than any independent objects. In that sense, the phenomena of practice are not distinguished by various regularities, but by repeatable patterns which is defined only in some particular causal structures and dispositions. Therefore, since they make clear those norms on correctness of repeatability, causal intra-actions indeed provide normativity in practice.

It is the responsibility of agents in practical circumstances to account for their interactions with the world, including the holistic phenomena and modes of actions in which she is involved. In a world, particular causality of practical circumstances determines the intelligibility of agent's actions. The correctness, appropriateness, and insightfulness of the actions of agent could not be explained unless based on particular causal intra-actions in practical circumstances. Therefore, it is also the causal intra-action that contains normative authority for agents in practice.

Now turn to the issues of normative force and temporality of practice. According to the traditional notion of practice, it is logical or transcendental modality that determinates possibilities of practical understanding, which has already been challenged by Heidegger's insightful philosophy. Heidegger put forward the notion of temporality, which fundamentally acknowledged *Being-in-the-world* as an understanding of practice. And in further he denied any possibility of independent understanding of theoretical inferences beyond the *Being-in-the-world*.

The temporality of practice has already promoted us beyond our current understanding, in order to keep it open for space and time, and also to retain variety of possible understanding in the future. For Heidegger, those possibilities are described from the causal intra-actions between *Dasein* and *Being-in-the-world*. However, for Rouse, the possibility of understanding in the future would be characterized from agent's embodiment, namely, interactions between bodies and environment, in the reconfiguration of environment. Therefore, things that matter here are not theoretical or transcendental possibilities, but real possibilities. The notion of real possibility, defined stricter than alethic or nomological possibility, is indispensable for understanding scientific practice, especially its temporal aspect for the future. The modal characters of understanding in scientific practice are essentially distinguished from the traditional notion of practice. Real possibility, as illustrated from norms of causation in particular practice, is the only conception of modality acceptable for coherent naturalism. For example, 'philosophical naturalists should take the normativity of causal intra-action as the basis for understanding alethic and nomological modalities rather than the reverse' (Rouse, 2002, p. 24). Furthermore, Rouse (2002, p. 334) regards 'real possibility as the most important modality for understanding scientific practices more generally'.

But how can we clarify the notion of real possibility? Hans-Jörg Rheinberger figures it out as 'epistemic thing', which, though not totally ascertained, is responsible for systematic inquiries in science.

Epistemic thing

> are material entities or processes – physical structures, chemical reactions, biological functions – that constitute the objects of inquiry ... they present themselves in a characteristic, irreducible vagueness. This vagueness is inevitable because, paradoxically, epistemic things embody what one does not yet know. Scientific objects have the precarious status of being absent in their experimental presence; they are not simply hidden things to be brought to light through sophisticated manipulations.
> (Rheinberger, 1997, p. 28)

Rheinberger's conception of epistemic things puts forward the fundamental sense of the real possibility. What makes difference on real possibility from impossibility is the thing which real possibility matters, referring to practical arrangement in particular circumstances, namely, material arrangement and those agents with different competence of experiment. The practical illustration of real possibilities has always involved things 'what is at stake', which is reconstructed by the evolution of practical circumstances, while 'are not what nature is, but what it is to be nature' (Rouse, 2002, p. 340). Therefore, what is at stake does not depend upon discoveries of objects and properties, but upon clarifications of those epistemic things which constraint us. For the stakes of scientific research, 'what was disclosed was not some definite object, however, but a field of not-fully-determinate possibilities' (Rouse, 2002, p. 343). Insofar as it is concerned, John Haugeland emphasizes that it is necessary for understanding science to participate into stakes of scientific practices, which should be accounted for, rather than identified, by us.

> We are accountable to what is at stake in our belonging (causally *and* normatively) to the material-discursive world: our fate is bound up with what is at issue and at stake in our practices, although those stakes are not yet definitely settled – indeed, that is part of what it is for them to be "at stake".
>
> (Rouse, 2002, p. 25)

It is not surprising that we are configured as who are only in the causal interactions with the world. The most significant things in practice matter so much for us as participants, agents, and cognitive subjects of practice. And 'our constitutive normativity involves responsibility (through both holding ourselves responsible and being held responsible by others) to coordinated, intra-active performances, issues, and stakes' (Rouse, 2002, p. 359). That makes essential difference with the traditional conception of normative force which claims that the force derives from agent's subjective commitments. Now we encounter constraining forces from what is at stake in practice. Agents could get normative force if they are localized upon some effective causal structures or relationships, since the normative authority for agents also depends upon causal intra-actions in practice. And for the same reason, 'what is at stake in practices only acquires normative force through causally efficacious intra-action' (Rouse, 2002, p. 359). That is the main conclusion on normative force for Rouse's coherent naturalism.

6.4 Concluding remarks

Rouse's notion of the normativity of scientific practice involves two basic presuppositions: philosophical naturalism and a practical conception of science. According to those assumptions, natural world studied by sciences has to be intimately connected with the social world of meaningful practice. Facts and norms, isolated in metaphysics, could be integrated together ontologically in practice. Therefore, normativity arises intrinsically from practice, and the normativity of scientific practice especially derives from the practice itself.

Based on his notion of normativity, Rouse also has clarified how scientific practice could have norms. And his conceptualization should appeal to the causal intra-actions in practice, e.g. causal structures and forces, in order to satisfy requirements from coherent naturalism, which, claimed by Rouse, argues that causality in practice should always be taken as fundamental principle in accounting for normativity. Finally, Rouse's aim would be to figure out the construction of the world from the normativity by causal intra-actions. 'Causality must be understood as always already normative, and normativity as always already causally efficacious' (Rouse, 2002, p. 183).

Rouse's argument has salient advantages, which substantially eliminate lots of philosophical disputes on normativity, in terms of his two insights. For one thing, Rouse recognizes philosophical naturalism as an ontological issue, rather than a mere epistemological issue, claimed by traditional naturalism, which would provide solid foundation for integrating nature and normativity in practice. For another, Rouse reconstructs the notion of practice as a whole of causal interactions between agents and the world, which could be modelled as repeatable patterns, and also as normatively accountable for the phenomena of practice as a whole. Consequently, the accountability of practical normativity could be founded upon the causality internal to the practice itself.

However, there are also significant faults on Rouse's view. First, what he has rejected is an epistemological issue merely in a narrow sense, namely, the inquiry for certainty abstracted from metaphysical notions of cognitive subjects and objects, while still presuming that '"knowledge" demarcates a coherent, surveyable domain of inquiry' (Rouse, 2002, p. 140). It is plausible to argue that Rouse does not reject epistemology thoroughly, but pursues for epistemological issues in a broad sense, which would incorporate subjective attempts for truths and thoughts of cognitive responsibilities at all. Therefore, since he would not plan to abandon pursuits for epistemology, it seems unavoidable for Rouse to be endangered by trapping in a dualism between the ontological and epistemological issues, which could both be thought of as indispensable parts of his approach.

Second, though he sets normative authority and force upon causal structures internal to practical circumstances, Rouse's accounts for causal structure as foundation of normativity are mainly theoretical rather than practical. As a result, Rouse's argument cannot make a sufficient clarification for understanding normativity in particular circumstances. The problem seems to be that it is not enough for providing a theoretical account for the normativity of scientific practice. Rather causal intra-actions should be put into discussion about particular, historical, and material practices of sciences, in a more intensive and explicit way, in order to open the 'black box' of causal intra-actions, and in further to promote understanding of normativity of scientific practice.

In conclusion, the normativity of scientific practice could neither be *a priori* nor been brought about on somewhat transcendental foundation, rather it derives actually from the practice itself. Practice needs norms in order to promote itself, while norms could also influence practice in an unexpected way. Those existing norms could be challenged from various aspects in the further practice of science. Therefore, norms are forced into interacting with practical circumstances, which would let them always changeable, since particular circumstances also vary at all time. And scientific practice would not be successful or fruitful unless norms are always revised and originated according to new problems arising from the changeable circumstances. In a word, norms should evolve through practice, rather than follow irrevocable regulations.

Note

1 The problem of naturalization of normativity could be thought of actually as the naturalization of rationality. For instance, H. Putnam argues against naturalism because of a supposition that rationality cannot be naturalized. Cf. R. Almeder, *Harmless Naturalism: The Limit of Science and the Nature of Philosophy*, Carus Publishing, 1998. Rouse attempts to revise contemporary naturalism in order to reconstruct the notion of nature, which could make conciliation with normativity in scientific practice. Therefore, it is necessary for him to eliminate the dualism between nature and normativity, and in further the disputes upon traditional problem of normativity, in order to realize the ideal of coherent naturalism.

7 Philosophy of scientific practice and naturalism (I)

Naturalism is an unavoidable issue for Philosophy of Scientific Practice (PSP). In Joseph Rouse's interpretation, PSP should also promote a kind of naturalism, which is not even a traditional one. It is worthwhile to notice that naturalism has never been a unique position, but only consists of various naturalistic approaches. This chapter seeks to clarify the evolutionary history of naturalism first, and then attempt to elaborate the relationship between naturalism and PSP.

7.1 Naturalism: issues in history

Naturalism, as a philosophical trend, has its origin in the natural philosophy in ancient Greece. In a specific way, it refers to those claims which insist on explaining nature from intrinsic causations, since modern science was born in 17th century. And in that sense natural scientists could also been referred as 'Naturalists'. However, naturalism has never become an influential school of philosophy until 1930s. Naturalistic philosophers, such as F.J.E. Woodbrige, Morris R. Cohen, John Dewey, and Sidney Hook, have already been pioneers of Naturalism in United States. They are also taken as pragmatists or critical realists, who emphasize continuity of naturalism and materialism and significance of scientific method, and who refute idealism and transcendentalism in epistemology.

Contemporary naturalism is founded on criticisms of the analytical tradition by Frege and Russell. The mainstream of analytic philosophy claims that traditional issues in epistemology could be reduced to issues on logic, conceptual and grammatical analysis. Besides factive and belief conditions, knowledge should satisfy a third requirement, namely, justification. The analytic philosophers believe that the assignment of philosophy is to clarify how knowledge could be justified. Therefore, psychology and other empirical sciences could not be involved in the enterprise of epistemology, since the appropriate terms of epistemology are logical and linguistic, rather than psychological ones. Analytic philosophy recognizes transcendental reflection as a necessary approach, in order to provide reliable foundation for knowledge.

Those doctrines have been defeated by contemporary naturalists. Naturalism has two important starting points in contemporary epistemology and philosophy of science. One is 'naturalized epistemology' firstly claimed by W.V.O. Quine.[1] Quine put forward a serious criticism against foundationalism in traditional epistemology, and also argued that it should be eliminated or replaced by a new kind of epistemology, which is involved as a chapter of psychology, and even of natural science. That radical notion of naturalism encourages and promotes a great development of cognitive science. The other is Kuhnian historicism. Kuhn defeats the philosophical project of 'rational reconstruction' by facts in history of science, which makes him believe that the history of science should take the first and foremost role in philosophical reflections for sciences. Kuhnian commitment of naturalism[2] influences others in two aspects:

1 Refuse any philosophical analysis prior or external to history of science;
2 Involve social and cultural elements into theory of knowledge.

Most of the contemporary theories of knowledge which involve historical, social and cultural dimensions could be traced back to Kuhn's historicist interpretation of science. The stakes of PSP should also be interpreted on a Kuhnian background. Following the approaches led by Quine and Kuhn, contemporary philosophy of science has already indicated a naturalistic turn. Today lots of various theories are referred to naturalism, such as Paul Thagard's computational philosophy of science, Ronald Giere's evolutionist naturalism, Steve Fuller's social epistemology, Stephen Downes's socialized naturalized epistemology, and Larry Laudan's normative naturalism.

Naturalism proposes a very different approach to traditional philosophy. For example, naturalism argues against foundationalism and autonomy of traditional epistemological rules, while emphasizing continuity between epistemology and empirical sciences.[3] It implies that methods of empirical sciences are helpful in order to promote our understanding of scientific knowledge. The continuity between epistemology and empirical sciences should be defined as a common issue for contemporary naturalism.

First, naturalism argues for a continuity of statuses of disciplines. According to naturalists, knowledge is just a kind of natural phenomenon, which should be studied in a context of empirical sciences. The theory of knowledge must be, rather than prior or external to, standing with empirical sciences in order to promote our holistic understanding of the world. '[T]here is only one way of knowing, the empirical way that is the basis of science', Therefore, 'from the naturalistic perspective, we should deny that there is any a priori knowledge' (Devitt, 1996, pp. 2, 49). On that sense, theory of knowledge has an equivalent status with empirical sciences. The only difference would be that, since knowledge as an object is too broad for studying, epistemological claims would be more abstract and general than scientific ones. For those reasons, naturalists deny traditional epistemology as a 'first philosophy' and abandon the purpose of justifying beliefs on an absolute foundation. Now it is free to employ achievements from psychology, cognitive science, evolutionary biology and sociology for construction of theory of knowledge.

Second, naturalism insists on a continuity of essence of disciplines. Traditional epistemology is a normative enterprise, which concerns essentially on justification of beliefs and rules. But sciences should be descriptive. Empirical sciences could only focus on the truth of the world, or 'saving the phenomena', rather than any evaluation such as rationality and goodness. Since they emphasize the continuity between epistemology and empirical sciences, naturalists would to some content discard or reconstruct the normativity of epistemology, which transfers theory of knowledge immediately as somewhat descriptive empirical science. For instance, Giere (1985, pp. 339–40) argues that

> [T]he general problem faced by a naturalistic philosophy of science, then, is to explain how creatures with our natural endowments manage to learn so much about the detailed structure of the world-about atoms, stars and nebulae, entropy, and genes. This problem calls for a scientific explanation.

And that would be a descriptive explanation of the human ability of cognition, which could also be agreed on by sociologists of scientific knowledge, considering knowledge as phenomena in our social nature.

However, it is very controversial to eliminate or discard the normativity of epistemology, since judgment of normativity is indispensable in both theoretical and practical reasoning. Therefore, the problem of normativity would be a real difficulty for the naturalistic position.

Finally, naturalism claims a continuity of research methodology. Traditional epistemology would like to take conceptual analysis, transcendental reflection, and intuitive method as its main methodology. According to naturalists, theory of knowledge should involve scientific methodology, such as observation, experimentation, induction, and examination. Epistemology, as science, should be an a posteriori enterprise: the problems of knowledge should be resolved in some empirical way. Contemporary naturalism would be an opponent to metaphysics by emphasizing methodology from empirical sciences. Naturalists think of the difference between the empirical and metaphysical methods as one between self-criticism and dogmatism. Scientific methodology would be taken as a refutation of dogmatism, while philosophical propositions involve incompatible and inconclusive position of epistemology.

7.2 Naturalism and philosophy of scientific practice

Due to Quine's great influence, naturalistic philosophy of science is usually dominated by interdisciplinary cognitive science. However, the essential problem of that approach is to insist on cognitive individualism. As a kind of natural phenomena, cognition or knowledge could not only be characterized from point of view of individual cognitive subjects, but also be altered by social and cultural institution. Therefore, it is not surprising that sciences studies have already evolved from social and cultural studies of science to studies of scientific practice.

Naturalistic epistemology pays attention on social dimension of scientific knowledge. For example, the strong programme of Sociology of Scientific Knowledge (SSK) would like to explain history of science from social factors as the first and foremost determinates. The normative and rational reconstruction of science, which is claimed by traditional philosophy of science, has been replaced by a more empirical and naturalistic sociology of knowledge. Barnes (1982, p. 5) argues

> Sociology is a subject with a naturalistic, rather than a prescriptive or normative orientation; it simply tries to understand the convictions and the concepts of different cultures as empirical phenomena. External evaluation of the convictions and concepts is irrelevant to this naturalistic concern; all that matters are why they were actually sustained, at a particular time in a particular context.

David Bloor (1991, p. 157) also emphasizes the empirical status of sociology, 'In the main science is causal, theoretical, value-neutral, often reductionist, to an extent empiricist.... This means that it is opposed to teleology, anthropomorphism and what is transcendent'.

On issues of epistemology, social constructivism instantiates social dimension in naturalism, denies priority of philosophical approaches, and attributes it to sociology. From the view of social constructivism, theory of knowledge should turn from macro to micro level, and laboratory becomes focused location for these studies. Scientific practice has now become indispensable for theory of knowledge. For instance, Schatzki, Cetina, and Savigny (2001) collect lots of studies for scientific practices from philosophers, sociologists, and scientists, in order to clarify the importance of practice for human life. Schatzki et al. believe that the sociality could only be organized, replicated, and communicated by human practices. However, contemporary thinking in the last two decades has mainly focused either on individual mind and activity, or on social structure, system and discourse, which are both potentially challenged by contemporary turn to theory of practice.

The practical turn is also manifested in contemporary philosophy and sociology of science. Pickering (1992) describes clearly how that turn occurs. He argues that there are some potential differences between understanding science as knowledge and as practice. Pickering characterizes the focus on practice as reconciliation among philosophy, sociology, and history of science, which refutes any foundationalist presupposition assumed by social constructivism and philosophical rationalism, and which in further goes beyond the debates between constructivism and rationalism.

Joseph Rouse's PSP has instantiated how to realize naturalistic position in studies of practices. In his *Knowledge and Power*, Rouse (1987, p. xi) remarks that science should be understood as a kind of practice, in terms of which is unavoidable to associate epistemic dimension with political dimension of science. In *Engaging Science*, Rouse suggests some deeper understanding of the political dimension of scientific practice, which is constructed in a broad notion of practice, in the sense of that no foundationalist theory of knowledge could determine. Rouse also puts forward a model of cultural studies of science, which would specify our understanding of science by connecting with specific cultural contexts and by transferring to new contexts. According to Rouse's view, there are six issues in the cultural study of science: (1) anti-essentialism about science; (2) a non-explanatory engagement with scientific practices; (3) an emphasis on the locality and materiality of scientific practices; (4) even more so on their cultural openness; (5) subversion of rather than opposition to scientific realism or conceptions of science as 'value-neutral'; (6) a commitment to epistemic and political criticism from within the culture of science (Rouse, 1996, p. 242). That could be taken as a serious criticism of foundationalism in epistemology from a naturalistic position. However, in *How Scientific Practices Matter*, Rouse explains how normativity could be compatible with naturalistic characters of scientific practice. Therefore, it is very clear in his three books that Rouse has made conciliation between naturalism and normativity for PSP.

In conclusion, contemporary epistemology has studied scientific knowledge from an individual or social view to a practical view. All of those views are naturalistic, which understand activities of knowing in a bottom-up way, on contrary to the top-down way in the traditional epistemology.

Would practice give us deeper insights for understanding science? That question could be fruitfully answered from a comparison of Rouse's PSP and social epistemology claimed by Steve Fuller and Stephen Downes. To certain content, socialized naturalized epistemology has already focused on practice compared to Quine's naturalized epistemology. For example, Downes argues that scientific practice consists of experimentation, theoretical extension, proposing and testing hypotheses, repeating and commenting others experiments and papers, and learning new results from others papers. Though the sociality of scientific practice is usually neglected by naturalized epistemology in philosophy of science, it is still necessary to realize that the activity of knowing in science should be those activities causing actual knowledge.

However, there are some essential differences between social epistemology claimed by Fuller and Downes and Rouse's PSP. According to Rouse's view, the thing matters here is that the notion of practice in PSP consists of micro-instrumental interactions in inquiry activities, as well as social interactions, while social epistemology merely takes activities of knowing in science as a common kind of social interactions. Besides, Rouse thinks of social practice as a background, on which scientific practice interacts with other parts of culture, and also changes with alternation of time and location. Therefore, compared with Fuller and Downes, Rouse prefers to understand scientific practice within the whole cultural practice, with emphasizing the intervening and participant characters of practice.

Another important point for Rouse's PSP is the comparison of SSK with feminism in *How Scientific Practices Matter*. The aspects of familiarity include: (1) recognizing scientific knowledge as results from social interactions; (2) criticizing traditional epistemology and philosophy of science as a distortion of scientific image; and (3) acceptance of political participation in science which makes it temporal and changeable. Rouse argues that feminism puts forward a practical direction which involves a dialogue between science and its interpretation. Feminists suggest that those dialogues consist of not only the objects of study but also the scientists themselves, since they do intend to connect scientific study with actual scientists rather than to separate them. Science incorporates lots of interconnected parts, rather than being explained as a whole. The goal of scientific inquiry could not be characterized as representation or reflective reconstruction, but as a changeable enterprise faced to future. That is an extension of naturalizing society itself.

7.3 PSP and problem of normativity for naturalism

According to A. Kertesz's comments, epistemology in a traditional sense should have essential characters, as follows (Kertesz, 2002, p. 270):

1. Belonging to the discipline of philosophy;
2. Assuming several transcendental suppositions, concerning with standards of rationality, for reflections of scientific studies;
3. Evaluating scientific knowledge as correct or wrong, rational or irrational, and justified or not, based on standards of rationality;
4. Recognized as the unique approach for dealing with foundations of science, which is also the 'first philosophy' approach.

Naturalism could be also characterized as follows in a corresponding way:

1. Epistemology does not belong to the philosophical discipline, which should deploy scientific method;
2. Thus, epistemology could be studied in a posteriori way;
3. Epistemology is not a normative enterprise. It is supposed to describe or explain results from scientific studies, rather than to evaluate them.

Kertesz's understanding of traditional epistemology could be elaborated as Transcendence (*TD*), Normativity (*NT*), and the Character of First Philosophy (*FP*). *NT* distinguishes traditional epistemology from empirical sciences; *TD* puts forward forms of origin and existence of *NT*, by emphasizing that epistemic norms are based on abstract and transcendental pure reason rather than any empirical facts; While *FP* explains relationships between epistemology and empirical sciences, namely, the fundamental status of epistemology to empirical sciences, or the guidance of epistemic norms for practices.

It is worthwhile to notice that *NT* has maintained a critical role in traditional epistemology. Also, the most serious criticism against naturalism would be its attitudes for problems of normativity. Undoubtedly, Kertesz's notion of naturalism, influenced by Quine, has already been a radical eliminativism of normativity, which is refused not only by proponents of traditional epistemology, but also by most of contemporary naturalists.

Traditional epistemologists could give an argument like this:

1 Epistemology is essentially normative;
2 On the contrary, Science is a descriptive enterprise;
3 Naturalism claims the continuity between epistemology and natural sciences, and the applicability of scientific method for epistemology;
4 Therefore, naturalism has actually deprived an essential part of epistemology;
5 Conclusion: epistemology could not be naturalized.

(Janvid, 2004, p. 36)

Most of contemporary naturalists, except those radical eliminativists, would also maintain the premise (1) and (2) in the argument. However, it is on the interpretations of those two premises that naturalists have made great differences from traditional epistemologists. It is supposed to admit that, no matter what positions naturalists would take, the argument put forward by traditional epistemologists should be taken seriously.

Therefore, it seems interesting to discuss about what PSP would say on the problem of normativity, since PSP has also claimed itself as a naturalistic approach. According to Rouse's interpretation, the naturalistic position of PSP would emphasize the intrinsic causal necessity for the purpose of resolving problem of normativity, which is distinct both from Quinean dependence upon natural necessity and from those claims of social-historical necessity from Kuhn and social constructivists. As a philosopher of scientific practice, Rouse puts all contexts relevant to science into the notion of practice, which therefore also considers normativity of practice as constructed in a piecemeal way during the configuration of practical situation. Practices in that sense would not be opposed to nature, but activities filled with social and historical meaning. And they

> are not just patterns of action, but the meaningful configurations of the world within which actions can take place intelligibly, and thus practices incorporate the objects that they are enacted with and on and the settings in which they are enacted.
>
> (Rouse, 1996, p. 135)

Practice is also a dynamic process of evolution, or an opening progress of history. From Rouse's view, it is the intrinsic causality involved in practical activities that provide normativity of practice. As the matter of fact, causal structure of practice is merely the circumstances encountered by actions, which are so objective that the participants could only accept them in order to make success, and which are so necessary for spectators to make intelligible meaning of actions of the participants.

Furthermore, there is no abstract or formalized norm from the view of PSP. Only in particular circumstances that laboratory, scientist activities, and methods and results of scientific research can be made intelligible. Any attempts for justifying norms of scientific practice would be problematic, since abstract or formalized focus on norms could not be meaningful at all. Justification of norms could only be functional in localized social network, which involves common practices, instrument, and problems, as well as common beliefs or values. Those strategies of persuasion for scientific claims should be founded only on the common things of the localized network, and also be revised by their feedbacks. The standard of justification is changeable. New results or interpretations of existing results determines which of scientific claims are acceptable, and which of others refutable. That is a very significant point argued by Rouse (1987, pp. 120, 124):

> Scientific claims are ... established within a rhetorical space rather than a logical space; scientific arguments settle for rational persuasion of peers instead of context-independent truth.... [T]here are no generally applicable standards of rational acceptability in science. There is only a roughly shared understanding of what can be assumed, what can (or must) be argued for, and what is unacceptable for any given purpose and context.... It is therefore a shared reference to 'what one does' in this shared field of practices that provides the basis for this variety of assessments and standards.

The notion of normativity in PSP is supposed to satisfy several requirements. First, it is constituted by which is internal to practices rather than pre-existing external to them. Second, it is a dynamic notion which means that there is some evolutionary process for constructing normativity. Finally, it depends upon the interventions made by subjects into the objects of practices. In terms of that justification of normativity, there is no limitation from universal rationality upon innovations of science, which recognizes the variety of local contexts and the evolution of dynamic process, and then guarantees a complete development of innovative potentiality. It would be helpful, arguable from PSP, to re-conceptualize normativity based upon causal necessity internal to practice.

However, there would still be some problems which make that proposal as an untenable position. Since PSP emphasizes contextuality in scientific practice, and its solution of normativity depending upon the internal causal structures of contexts, those problems would also focus on that particular emphasis and solution.

First, it matters so much on how to choose relevant context in order to explain the rationality of scientists' behaviours. Practice is so complicated that most of practical elements are co-existing. It seems impossible to consider all of elements to explain rationality, while some of the elements are too hard to mention or explain, such as some tacit knowledge. Therefore, it really matters much to consider which elements as relevant or irrelevant. in most cases we actually also recognize those elements relatively obvious. However, it is very hard to grasp the value of relevance or the content to be obvious in practice. On that problem, Rouse could only put forward an explanatory strategy rather than any elimination.

Second, it seems that Rouse would argue that some of those relevant contexts would be critical or indispensable for practice. He tends to believe in some 'stakes' in actual practices.

> We are accountable to what is at stake in our belonging (causally *and* normatively) to the material-discursive world: our fate is bound up with what is at issue and at stake in our practices, although those stakes are not yet definitely settled – indeed, that is part of what it is for them to be "at stake".
>
> (Rouse, 2002, p. 25)

Since that would be in opposition to the hermeneutic principle and logic of emphasizing locality and particularity, he does not or also unable to point out what the stake is.

Another potential problem is: could those stakes in various scientific practices be the same? Since the stakes of the same experiment could be different even just because of difference of locations, the interpretations of practices would be no explanatory power if they are not equivalent. Therefore, those explanations of practice could not tell us how it should be thinking and doing. On the contrary, if those stakes are equivalent, then the question would be: what is it? Is it the norms of theory and action constituted by logical analysis from positivists, or some theory of knowledge? Then PSP cannot be recognized as a whole groundbreaking position. It is at most an attempt to integrate naturalism, historicism, and social constructivism, which gives another philosophical explanation of scientific practice, in terms of its special notion of practice.

7.4 Naturalism and normativity: an open question

PSP stands on a position of naturalistic epistemology, in opposition to the traditional one which is mainly characterized as foundationalism. However, PSP has also been distinguished from naturalism in an individual cognitive aspect, such as Quine's naturalized epistemology, and from naturalism in a social cognitive aspect, such as SSK and social epistemology. It is worthwhile to notice that PSP could put forward lots of inspiration on understanding science today. Meanwhile, it is also faced with some urgent problems which may be common for most of naturalistic theory, for example, the incompatibility between the descriptive and the normative.

As far as it is concerned, PSP would base normativity upon the causal structure of necessity within practices, which emphasizes the special significance of practical contexts. However, compared with other relevant theory of knowledge, especially those paying heavy attention on transcendental normativity, the role taken by contemporary PSP would be more deconstructive than its constructive meaning. Therefore, it is still an open question on whether and how normativity could be involved in the naturalistic theory based on scientific practice.

In conclusion, it seems worthwhile to pay more attention on the problem of representation of science, since representation, as well as practice, is necessary for science. And the discursive practice could also be a kind of representation described from a point of activity. It is supposed that the incompatibility problem between the descriptive and the normative could not be eliminated without an effective clarification on relationships between practice and representation.

Notes

1 Most of contemporary naturalists have been inspired from Quine's idea of naturalized epistemology. That is the main reason why naturalism is directly equivalent to 'naturalized epistemology' in most cases, which makes the notion of naturalism deeply confused. Actually, naturalized epistemology could only be considered as a specific part of contemporary naturalism, rather than the whole.
2 Kuhn does not claim his own position as a naturalist. Actually, he even doesn't emphasize the notion of naturalism in his books. However, many naturalists such as Giere's evolutionary naturalism, Barnes's strong programme, and Rouse's PSP, are indeed inspired by Kuhn's work. By a detailed analysis of the whole development of naturalism, it is clear, at least on the whole, that Quine and Kuhn have led two naturalistic approaches respectively. The former is an approach as empirical sciences such as employing psychology and neuroscience, and the latter is as empirical social sciences, such as referring to history, sociology, and cultural studies.
3 James Maffie (1990, pp. 281–93) puts forward six kinds of continuity between epistemology and empirical sciences. There are continuities in the senses of the epistemological, the contextual, the methodological, the analytic, the metaphysical, and the axiological one.

8 Philosophy of scientific practice and naturalism (II)

The hermeneutic character of practice, the normativity of scientific practices, and naturalism are three dimensions of the nature of philosophy of scientific practices (hereafter PSP). Naturalism refutes the foundationalism of transcendental normativity in Kantian epistemology. However, this also makes it discard transcendental constructivism, which is very brilliant part in Kantianism. This chapter will argue, accordingly, that Joseph Rouse's philosophy of scientific practice is supposed to have two aspects. On one hand, PSP in Rouse's opinion should be naturalistic, which characterizes the normativity in practices as developing from instrumental interaction with a physical world. On the other hand, PSP agrees with those reconstructed transcendental epistemology, which also thinks of normativity in practices as a 'quasi- transcendental reference', and also as an object for critical reflection in order to deeply reform patterns of practices in sciences. Therefore, PSP could also be identified by somewhat emancipatory interests of cognition.

Naturalism and transcendentalism are different approaches in philosophy of science. Psychologism, criticized both by E. Husserl and by G. Frege, is one of various naturalistic approaches. Logical positivists take 'methodological reconstruction' as the whole work in philosophy of science, which actually retain some aspects of transcendentalism in terms of methodological normativity for sciences. However, thanks to works from Kuhn and other historicists, naturalism which emphasizes descriptive approaches has been back to the mainstream of philosophy of science. It seems really necessary to put Rouse's theory into that historical background for a comprehensive understanding.

8.1 The Kantian notion of constructivism and normativity

It is Immanuel Kant who firstly put forward an ideal for transcendental epistemology. For him, transcendental reflection means a philosophical approach that could make a critically inspection for human ability of knowledge.

'Not every *a priori* cognition must be called transcendental, but only that by means of which we cognize that and how certain representations (intuitions or concepts) are applied entirely *a priori*, or are possible' (Kant 1997, A56 B80).

Transcendental constructivism (hereafter *TC*) emphasizes that the object of scientific knowledge should depend upon the synthesis and unification from understanding, rather than anything-in-itself (*Ding-an-sich*). The aim of transcendental reflection is to ascertain how the objective validity of knowledge could be constructed from some subjective conditions of thoughts. As for transcendentalism, the objective of knowledge itself should be a synthesized achievement by the subjective of knowledge, a construction from understanding. The manifolds from sensation could receive the unity as an object, only thanks to the spontaneity of understanding and the unity of '*I-think*' consciousness in Cartesian sense. Therefore, Kant argues that it is the transcendental apperception that makes empirical objects possible *a priori*, which implies that a necessary premise for the objective validity of knowledge would be the 'pure, original, unchanging' unity of consciousness (Kant, A107).

Transcendental normativity (hereafter *TN*) is also an important view from Kantian epistemology. To say something is normative means that it could be typically characterized as an action following rules. For Kant, understanding is a normative enterprise, which could be explicated as 'faculty of rules' (Kant, A126). Robert Brandom has already clarified Kant's big idea from his talk about 'commitment and entitlement'.

Human judgments and actions are

> what we are in a distinctive way responsible for. They express commitments of ours: commitments that we are answerable for in the sense that our entitlement to them is always potentially at issue; commitments that are rational in the sense that vindicating the corresponding entitlements is a matter of offering reasons for them.
>
> (Brandom, 2000, p. 80)

However, it is worthwhile to notice that Kantian *TN* has assumed a kind of foundationalism, which claims something remote from human practices of inquiry as most basic for the whole knowledge. We shall never forget that the pure unity of apperception, which in Kant's argument is such as a transcendental foundation for empirical objects and validity of knowledge, is also actually 'unchanging'. It is a stable foundation since human cognitive practices cannot make any alternation, while actual practices should be decided necessarily. It is the transcendental foundation that justifies rule-following actions of synthesis and unification from understanding, which is really a constitutive part in Kantian *TN*.

8.2 Naturalistic criticism of transcendental normativity

Naturalistic criticism on *TN* mainly comes from its unsatisfactory presumption of foundationalism. Naturalism implies refutation of any absolutely transcendental foundation for scientific practices. And the unity of consciousness or apperception, which is already elaborated as the foundation of *TN*, would be mere facts causally produced in practices, rather than any valid paradigm *a priori*.

The reason why Naturalism always connects with Physicalism is that it needs to argue consciousness or mind are all the same as physical reality, rather than any super-natural existence, e.g. in a Cartesian sense. That argument is necessary for naturalistic criticism since transcendental epistemology usually thinks of the mental as unchanging foundation for the validity of knowledge. Another argument which is also necessary for Naturalism is to claim continuity from philosophy to empirical sciences. Scientific practices would not accept any arbitrary regulation from transcendental methodology. However, the evolutionary mechanism of scientific practices could be characterized scientifically rather than philosophically, such as descriptions from a biological, psychological or sociological view of point. Rouse (2002, p. 3) defines those two necessary arguments respectively as 'metaphysical' and 'meta-philosophical' naturalism.

W.V.O. Quine has argued for Naturalism in both senses. For one thing, he insists a behaviourist conception of meaning, which claims that any intentional or intensional notion cannot be acceptable unless they are reducible to descriptions of observable behaviours. Quine (1960, p. 221) emphasizes that only 'the physical constitution and behaviour of organisms' can be an appropriate propositional attitude for a language of science. For another, since any statement cannot get a specific empirical consequence by itself, the principle of indeterminacy of translation has made the epistemology of 'rational reconstruction' impossible. However, in Quine's view, the future of epistemology would be found when it is naturalized as a part of psychology, which identifies the mechanisms between empirical inputs and descriptive outputs, in order to understand the relationship between evidences and theories (Quine, 1969, p. 83).

Donald Davidson also argues for Naturalism in both metaphysical and meta-philosophical sense. For metaphysical naturalism, Davidson put forwards the argument of anomalous monism, which concludes that, on one hand, all mental events are actually physical ones; and on the other hand, there is no deterministic law for predicting or explaining mental events, which means that mental event is not reducible to physical event, according to the anomalism of the mental. That view of non-reductive physicalism could let us properly talk about mental events such as belief, desire, or intention, without any worries on how they could be reduced to observable behaviours, while also avoiding any promise to dualism between mind and world, since the mental object of intentional expression should not be non-physical.

For meta-philosophical naturalism, Davidson argues that the relation between sensation and belief could only be a causal one. Sensation cannot be a truth-maker of, or justify, belief. 'No doubt meaning and knowledge depend on experience and experience ultimately on sensation. But this is the "depend" of causality, not of evidence or justification' (Davidson, 1986, p. 314). All activities of justification could only occur within linguistic practice, which also demands that our reason for a belief must be another belief, rather than any unchanging foundation out of the whole system of our beliefs. Therefore, the focus of philosophy here would be made a turn, and the search for a transcendental foundation for validity of knowledge, the classical goal of transcendental epistemology from Kant to logical empiricists, has already turned out as nonsense. It seems that Davidson's naturalistic argument would be more sufficient and persuasive than Quine's. Both of them have argued for a replacement of transcendental epistemology by Naturalism in a metaphysical and philosophical way.

However, it is worthwhile to notice that, Naturalism puts refutation to all over transcendental epistemology. Even if its target may be only *TN* with foundationalism, its hostile attitude would also swallow up the other aspect of transcendental epistemology, namely, *TC*. And that would make disastrous consequences that we will hereby see. We have already known that Kant argues for TC in his conclusion that validity of knowledge comes from the construction of cognitive objects.[1] Nature would be considered, without *TC*, as objective in a sense of thing-in-itself, which is not true, trapped in 'the objectivist illusion' (Habermas, 1971, p. 305).

TC in Kantian sense is mostly put forward as constructivism from transcendental consciousness. But that's hardly enough. The possibility *a priori* of knowledge has to be construction also from instrumental intervention, as well as from consciousness or mind. Therefore, K-O Apel (1980, pp. 46–8) would like to characterize it as '*Leibapriori*', in order to make distinction from the constructivism in Kantian sense. And transcendental epistemology for him should also be reconstructed as cognitive anthropology.

Two questions would be found immediately for the reconstruction of transcendental epistemology. First, what role would cognitive subjects take in the newly reconstructed transcendental philosophy? It should not only retain something from Kantian epistemology, but also make some significant changes. Second, why should the instrumental intervention be considered as a new kind of 'ground' or 'reason', rather than cause, of knowledge, just like things in the naturalistic way? We must justify the transcendental approach by Apelian new reconstruction.

C.S. Peirce has already put forwards such kinds of reconstruction. The synthesis of understanding and its unity of pure consciousness, by Kant, are main roles that cognitive subjects would take. While for Peirce, the so-called 'unity of pure consciousness' would only refer to the agreement on how to use and interpret signs, since human thoughts are usually constructed by making use of signs. Therefore, the community which could arrive at agreement about interpretations of signs could replace the role of apperception in Kantian transcendental epistemology, and also has been thought of as the possibility *a priori* of the validity of knowledge.

'Thus, the very origin of the conception of reality shows that this conception essentially involves the notion of a COMMUNITY, without definite limits, and capable of a definite increase of knowledge' (Peirce, 1998, 5.311).

Apel (1980, p. 81) argues that what Peirce has already done can be defined as a 'semiotical transformation' of Kantian transcendental logic, which identifies Kantian transcendental subject of knowledge as a community of experimentation and interpretation. In other word, the community which is taken as the possibility a priori of knowledge in a reconstructed transcendental epistemology would be actual subjects engaging in scientific practices. Scientific knowledge, as a result from those practices of community, has to presuppose the possibility of agreement on the interpretations of signs. The most interesting thing for reconstructing transcendental philosophy is that, the agreement could only be possible by instrumental intervention of the world. That is why *Leibapriori* is so important for a post-Kantian transcendental epistemology. There are three main arguments for that big idea.

First, according to Peirce (1998, 2.712), every valid inferential relation among signs, for instance, deduction, induction, and abduction, which is very famous by Peirce's work, are all commitments to some framework of instrumental rules. Scientific practices could not be held without making inferences like those patterns. However, it seems unavoidable to presuppose a referential framework of instrumental action when applying those patterns of inferences. The referential framework has already been characterized as a new kind of possibility *a priori* for the objective validity of knowledge, deriving from scientific practices.

Second, think about those phenomena described by scientific knowledge. Most of them would not even exist without instrumental, especially experimental, practices construction. Ian Hacking (1983, p. 230) believes that 'to experiment is to create, produce, refine and stabilize phenomena'. Since they are actually constructions from instrumental systems, those phenomena as objects of scientific practices could not be thing-in-itself. Here we cannot make a clear distinction between the artificial and the natural. It can be implied, then, every phenomenon in experimental constructions could not be taken as merely localized, but as involving something of which we could make sense. That would require experimental results to be repeatable at least in principle, which means that the phenomenon such as 'if x, then y' is observable not as a particular event, but as something general in a certain range. And it also makes constructions of agreement on interpretation of signs.

Finally, it is necessary for the community of experiments and interpretations to manipulate, rather than just think about, objects in the world, in order to make agreement on objective reality. 'Experimenting on an entity does not commit you to believing that it exists. Only manipulating an entity, in order to experiment on something else, needs do that'. Hacking (1983, p. 263) argues for experimental realism which claims reality of object from its manipulability, which provides evidential force for abandoning doubts on reality among experimental community.

In a word, for the newly reconstructed transcendental philosophy, instrumental practices would be identified as a necessary presupposition of making agreement on using signs, and also as a new kind of possibility *a priori* for validity of knowledge. Our objects of knowledge would not merely come from synthesis of understanding, more importantly, but from the configuration of world by instrumental practices. Undoubtedly, that is a reformed TC, even not in an original Kantian sense, but actually still refutes the illusion of thing-in-itself, and the totally wrong objectivism.

8.3 PSP as a naturalistic approach

Philosophy of scientific practices (PSP) has been brought about based on the disputes of naturalism and reconstruction of transcendental epistemology. Actually, we have to accept that PSP is really a hybrid position of both sides. For one thing, philosophers as proponents for PSP, like Joseph Rouse, would deny any so-called 'foundation' out of or even beyond actual practices for sciences. That would make a definite agreement with naturalists, who also repute *TN* with foundationalist supposition imposed on scientific practices. However, for another, Rouse also emphasizes that there is also no Nature without normativity, since scientific practices as interactions between mind and world should also involve 'normative accountability' (Rouse, 2002, pp. 11–12), which could be elaborated as serious *TC* criticism to objectivism.

It is worthwhile to say something more about Rouse's notion of normative accountability, which has been certainly influenced by contemporary anti-naturalists, especially philosophers in Pittsburgh. John McDowell (1994, p. 72) argues that there is a distinction between a pure disenchanted nature and a 'second nature' involving Sellarsian 'space of reasons'. It is the disenchanted one that would be object of modern sciences, such as brute causation without normativity. And it is the second nature that brings meaningfulness and intelligibility back to the pure nature.

Robert Brandom (1994, p. 27) claims that normative accountability of causal interactions is produced by discursive practices of 'giving and asking for reasons', since discursive practices involve implicit norms, which instantiate themselves as propriety, demanding to be followed properly by actions. And the products from perceptions and actions, which is considered as brought about from causal interactions of experimental community, could only be meaningful as the inputs and outputs of discursive practices. Brandom (1994, p. 234) gives us two reasons for this idea. First, the factual reports from observation and action has already contained conceptual contents, explicated in discursive practices of community, and therefore could be thought of as authorized reasons. Second, though factual reports cannot be meaningful independently, discursive practices could be meaningful even if there are no inputs or outputs – such as, say, factual reports.

Rouse's notion of normative accountability criticizes both McDowell's and Brandom's ideas. For Rouse, nature and normativity must be interpenetrated each other, rather than any dualism of causality and normativity such as the position that McDowell's argument would take. And Brandom is also wrong, since discursive practices, though Rouse agrees to its specific significance, could not be intelligible independent from any non-linguistic instrumental practices. Here Rouse (2002, pp. 260–1) argues for two points:

1. The causal interactions of instrumental practices, including perceptions and actions, are that which make inferential practices possible and meaningful;
2. The normative accountability of scientific practices, dependent upon discursive inferences, is indispensable for the configuration of world in instrumental practices.

Actually, 1) has already been got proofs from the reconstruction of transcendental philosophy. Discursive practices in scientific inquiry could be identified as a process of arriving at agreements on interpretations of signs among the community of researchers. Perceptions and actions should be presumed in order to bring meaningfulness to discursive practices, since we have argued, in the last section, that the community could not make those agreements without supposing interventions to the world in instrumental practices. However, an argument for 2) could be defined as Rouse's main contribution. Contrary to Brandom, Rouse argues not only for discursive practices as presupposition of the meaningfulness of perception and actions, but in further for the constitutive relation between discursive practice and instrumental practice, in which the world is configured by scientific practices as a whole.[2]

It has also argued that repeatability of experiments should be assumed as presupposition of agreements on sign interpretations. However, the problem is how to produce such repeatability from those localized experimental situations. Standardization seems to be a necessary strategy. Besides, the material arrangement of experiment would also make lots of importance. The first and foremost arrangement would be to identify boundaries between measuring system and measured objects. An experiment could be distinguished from others merely by arrangement on 'measuring x with y', even they make use of the same materials. Rouse (2002, pp. 275–6) illustrates, for example, 'the scattered light may be directed towards a photographic plate ... used to record the position, or the light may be directed towards a piece of equipment with movable parts used to record the momentum of the scattered light'. Therefore, it is necessary for the repeatability of experiment to fix material arrangements explicitly, since specific arrangements could guarantee meaningful phenomena produced from experiments, in order to become rules followed by different experimenters.

In a word, the replicability of experiment requires material arrangements to be transformed into some explicit norms of practice, which is rightly the significance of discursive practice for the whole enterprise of inquiry. A specific configuration of the world has already been determined by the notions involved in experimental reports describing material arrangements. '"Position" is what a detecting apparatus with relatively fixed components can measure, whereas "momentum" is defined by an apparatus whose components can move appropriately in relation to one another' (Rouse, 2002, p. 286). Those notions involve implicitly the normativity of scientific practice – not only the discursive normativity, emphasized by Brandom, but also those 'oughts' for guiding operations of experiments, which could be understood and followed by experimenters, who have already been professionally trained and have got certain experiences. Therefore, Rouse (2002, p. 287) would like to say, 'Concepts ... are the normativity of physical arrangements as reproducible'. It is the normativity of scientific practices that guarantees the repeatability of experiments, at least in principle; though experimental design has already been considered from resources available for actual experimenters, which would produce unavoidable characteristics of contextualization and opportunism.

In other words, normative accountability of scientific practice is derived from the specification and stabilization of material arrangement by discursive concepts. Compared to Kantian *TN*, the normativity of scientific practice commits to no place for foundationalism; while compared to Brandom's notion of discursive practice, it emphasizes that conceptual normativity participates those specific perceptions and actions in experiments, which is a necessary supposition for repeatability. However, Rouse would agree to Kant and Brandom, on conceptualizing commitment and entitlement as the essence of normativity. For example, making use of the notion of 'position' seems to commit an existence of corresponding system of measurement. To determine locations of particles would also be an action responding appropriately to experimenter's commitments, while changing their deontic situation, which entitles them to employ the notion and to act led by the notion (Rouse, 2002, p. 287).

If Rouse's argument is correct so far, normativity of scientific practice could not be independent from McDowell's notion of 'disenchanted nature', but internally constitutes causality in nature. It may seem to be some agreement with the criticism of objectivism, though Rouse would not like to lead hereafter to any transcendentalist epistemology, since here the normativity does not presuppose any foundation of knowledge which is unchangeable according to actual practices of inquiry. On the contrary, conceptual norm could not be intelligible unless related to scientific practice in which people are actually engaging (Rouse, 2002, p. 318). Therefore, Rouse's PSP should be taken as a naturalist position, which claims that discursive practice is just one part of the causal interaction with the world.

Rouse believes that PSP is essentially a naturalistic approach. Though it emphasizes the constructive role of normativity, PSP does not commit to any non-naturalistic attitude, no matter whether metaphysically or metaphilosophically. Nevertheless, it is the notion of scientific practice that makes normativity possible to be understood in a naturalistic way (Rouse, 2002, p. 16). Rouse insists a 'relational' notion of causality and normativity, which argues for non-reductive relation from normativity to some brute causality, as reciprocal constructive relationship. This would be an 'expressivist',[3] rather than reductivist approach. Causality and normativity are mutually intelligible conditions for each other, while without any specific explanatory connection (Rouse, 2002, p. 270). Therefore, Rouse should be a naturalist in Davidsonian rather than Quinean sense. While Davidson turns naturalists to be aware of the Anomalism of the Mental, Rouse would try to let them know why it is unavoidable to discard the familiar dualism of causality and normativity (Rouse, 2002, p. 23).

8.4 PSP as quasi-transcendental philosophy

Philosophy of scientific practice (PSP) seems certainly to involve characters of metaphysical naturalism, since it has already argued that normativity of scientific practice is derived from experimenter's causal interaction with the world. However, it is still doubtful whether PSP could satisfy the requirement from naturalism in the meta-philosophical sense, as Rouse supposes. There are three points to construct a reasonable doubt here.

First, Rouse argues that causality consists of normativity in scientific practice, which makes it unavoidable to commit to truthful conditions while employing propositions about causal knowledge. Then, how could PSP admit some truth-making relation for one hand, and insist on meta-philosophical naturalism for another, since Davidson has already turned it out that it is necessary to deny any truth-making relation for argumentation of naturalism in a meta-philosophical sense?

Second, it seems that Rouse has pointed out PSP as following Brandom's expressivist approach, which, while in Brandom's view, actually promises a serious criticism against naturalism.

Finally, the self-reflectivity would provide attacks on the naturalistic definition of PSP: does Rouse's relational notions of causality and normativity have to be a necessary condition *a priori* for scientific practice? Would the notion of nature without normativity also be unthinkable for Rouse, as well as the validity of knowledge without *Leibapriori* for Apel?

Rouse (2002, p. 174) could reply indeed that the temporal structure of normativity would make some difference, in order to distinguish PSP from any transcendental epistemology. It is absolutely different from the foundationalist supposition of *TN*, which claims the foundation of knowledge as unchangeable and external to scientific practice, by the temporal structure involved in PSP. In sum, norms could only rise from actual practices, while also regulating them. And it is indeed a position of meta-philosophical naturalism that characterizes scientific practice as regulating itself during evolution in time.

However, that reply can only justify that PSP would not be a Kantian transcendental philosophy, while it is not yet enough for distinguishing itself from those reconstructed transcendental epistemologies since Peirce. It is worthwhile to make a comparison with Habermas's notion of interest, which would serve as a proper counterexample. For Habermas (1971, p. 196), interest is 'the basic orientations rooted in specific fundamental conditions of the possible reproduction and self-constitution of the human species, namely work and interaction'. Interests rise from productive and communicative practices, involving an internal temporal structure, while could not be thought of as empirical relationship from psychology, as naturalism supposed to be, since interests of various kinds provide referential systems for the validity of knowledge, in a 'quasi-transcendental' (Habermas, 1971, p. 194) sense. Therefore, I would say that the traditional distinction between the empirical and the transcendental seems to be inappropriate for the notion of interest. Like Apel's notion of *Leibapriori*, the notion of interest attempts to deny the foundationalism without temporal structure, for one hand, and to reconstruct somewhat quasi-transcendental philosophy.

Therefore, it seems still reasonable to claim that norms of scientific practices, involving temporal structures and serving as referential system for the validity of knowledge just like what the notion of interest does, could consist of the same quasi- transcendental aspect as the interest and *Leibapriori*, which is also an absolutely denial of its role as a naturalistic approach.

Supposing he insists naturalism thoroughly, Rouse would argue that it is only an identification of some familiarity between PSP and the reconstructed transcendental philosophy, such as those works from Peirce, Apel, and Habermas. That would not mean any choice made by PSP in order to direct to transcendental epistemology. In Rouse's view, the relational notions of causality and normativity could be accomplished in the whole naturalistic approach.

But Rouse is wrong. It seems impossible for naturalism to support those relational notions, since a necessary condition for the argument would be to refute the objectivist illusion, which, as we have already argued, cannot be satisfied by naturalism. Naturalism, unfortunately, has already discarded transcendental constructivism (*TC*) when it puts forward persuasive criticism against transcendental normativity (*TN*). If it is a totally naturalistic approach, philosophy of scientific practice (PSP) would focus on the consideration of objects of knowledge as products from some causal mechanisms. That has accepted the mechanism as 'thing-in-itself', which is definitely an objectivist illusion refuted by *TC*. It is *TC* that emphasizes, only in terms of some specific presupposition of normativity, can we construct causal mechanism as 'thing-in-itself'. The emphasis has a significant implication: it is up to us, for example, the experimenters in actual practices of inquiry, to make an essential change on existing mechanisms, in order to improve scientific practices, since those are not 'thing-in-itself', but just construction from us. Therefore, scientific practice could get profound critiques from the view of *TC*, which could not be provided by naturalism.

In other word, it is only within a transcendental approach, rather than a naturalistic one, could PSP get those critical significances supposed by Rouse. For example, ' "Nature" is a thoroughly political concept and marks a difference in how we deal with things and people, and what sort of standing they have' (Rouse, 1987, p. 188). And he argues that the criticism from PSP is just like those from Feminist philosophy of science, which has already elaborated gender bias as a normative presupposition deeply rooted in scientific practices. Feminist philosopher believes that gender-relevant normativity constitutes the way of scientific inquiry for nature, for instance, those masculine discourses about the notions of objectivity and force, which lead the whole enterprise of scientific explanation and theoretical construction. Contrasting with the feminist criticism, social constructivism, as a naturalistic approach, pays more attention on social mechanism by which methodological normativity of science is produced. That is what a naturalistic approach can describe, however, as a reduction of practical norms to products of some facts in the sense of 'thing-in-itself'.

That which feminist philosophy of science aims at is, not to describe an existing mechanism, but to clarify the normative role of gender view in scientific practice. That kind of clarification would refute the objectivist illusion covered on the productive mechanism of knowledge, which makes it possible to 'revise norms of rationality or objectivity' (Rouse, 2002, p. 137). Therefore feminists, in the very sense, take not only positions of bystander for objective description or explanation, but actually as participants of scientific practices, though at most of time as critics of the existing mechanisms. Rouse recognizes 'engaging science' as an essential character for the cultural studies of sciences, including PSP. It is definitely a kind of emancipation for scientific practice to be released from repression and control of powerful relationships. And the first and foremost condition for emancipation lies in eliminating the objectivist illusion.

Since, as we have argued, it is in further necessary for refuting that illusion only by *TC*, which brings about reflective awareness on normativity, rather than any description of causality, PSP should also be a transcendental – at least a quasi-transcendental – epistemology synthesizing reflections on normativity and constructivism. And that is the main conclusion of this chapter. It is indispensable to obtain a synthesis of normative and constructive aspects for PSP in order to deny objectivist illusion and in further to emancipate practices from powerful relationships. The reasonability of criticism from naturalism is that *TN* cannot accept any foundation of knowledge external to the development of scientific practices. It seems that PSP and other reconstructed transcendental philosophy attempt to prove that, in order to promote revolutions and innovations of practices, normative aspects could still be thought of as 'quasi-transcendental systems of reference', as objects of critical reflection, even though they are considered as derived from scientific practice itself.

In sum, the cognitive interest of emancipation seems to be constitutive for Philosophy of Scientific Practices, which pays lots of attention in critical reflection on existing mechanisms of practices. Both transcendental normativity and constructivism are necessary for that critical reflection. The reasonability of naturalistic criticism lies on refuting appeals to foundationalism in Kantian *TN*. However, what Rouse and others' reconstructed transcendental epistemology are attempting to prove is that, even if they are rising just from those changeable practices of inquiry, the normative elements could also be thought of as 'quasi-transcendental systems of reference', which is the objects of critical reflections, in order to promoting innovation and revolution of scientific practices. That claim does not intend to deny those naturalistic implications of PSP. It is just an emphasis that, when he argues that 'the primary concern of feminist science studies is less with the present state of knowledge than its future possibilities' (Rouse, 2002, p. 147), Rouse would undoubtedly be benefited from an at least quasi-transcendental approach of epistemology.

Notes

1 Logical positivism is different both from Kantian epistemology and from Naturalism. For one thing, positivists define rational reconstruction as the main aim for philosophy of science, which accepts most parts of *TN*, and thus is against naturalism. For another, positivist approach for methodological inquiry has already discard Kantian framework, the dualism between the subjective and the objective, and also has accepted a kind of neutral facts, which is absolutely very close to Naturalism, while opposite to transcendental philosophy. See Habermas, 1971, pp. 67–89; Apel, 1980, pp. 77–8.

2 It is worthwhile to make a clarification on difference between Brandom and Rouse. For Brandom, discursive practice is something intra-linguistic, which is essentially distinguished from non-linguistic instrumental activity. Consequently, perceptions and actions could connect with discursive inferences only as their products, as the inputs and outputs. But Rouse (2002, p. 260) denies anything that could independently intra-linguistic: 'Speaking can thus never be a fully innocent representation of an unchanged state of affairs, but instead acts upon the field of possible sayings and doings of others. Words and the sentences they compose are an integral part of the world we live in'. It seems that Rouse (2002, p. 214) believes that representationalism is still remained in Brandom's notion of mind, since he still accepts a linguistic and non-interventional intermediate between perception and action, even Brandom (2000, p. 7) has already argue himself that no representationalist notion would be assumed in his inferentialist semantics.

3 Davidson (2001, p. 156) believes that there is an interdependent rather than a reducible relation between language and thought, which is conceptualized as a 'relational' position for linguistic meaning by Brandom (2000, p. 5), namely, that speech and belief could be meaningful only in terms of their interdependence. The meaning of notions in discursive practice would not be clarified until they are related to those implicit beliefs, especially beliefs about normativity. Brandom (1994, p. 649) argues for expressivism as his research program, which aims at elaborating implicit normativity hidden in daily discursive inferences, and also 'making explicit the implicit structure characteristic of discursive practice'. Rouse agrees with Davidson and Brandom on that point, and also extends the expressivist approach onto PSP, especially on the relationship between discursive normativity and natural causality.

9 Philosophy of scientific practice and relativism

The basic idea of relativism is that there is no universal and completely consistent knowledge standard, thus there is no universal and identical knowledge, neither for knowledge with universality. In philosophy of science, relativism generally reflects the position of no universal standards in philosophical problems. Traditional philosophy of science adheres to the position of theory-dominance, and then triggers some problems such as the 'theory-ladenness of observations/experiments' and 'incommensurability'. And this hinders the emergence and establishment of universal standards and entraps traditional philosophy of science into the predicament of relativism, which exactly results from the insistence for universalism and foundationalism). Whereas Philosophy of Scientific Practice (PSP) takes position of the practice-dominance, under which the theory-ladenness of observations/experiments has been criticized, and a more fundamental position of experiments has been redefined. PSP transform the 'paradigm' concept and make efforts to eliminate the 'incommensurability'. PSP abandon the representationalism and establish the view of 'local knowledge', which has avoided the relativism to some extent, and has made some stands of relativism be accepted.

9.1 The problems of relativism and its performance in the philosophy of science

Relativism is not a school of philosophy, but only an ideological element included in the thoughts of many schools of philosophy. The definition of relativism has various versions, which all advocate that there are no universal standards. For instance, cognitive relativism believes that knowledge has no universal standards; moral relativism thinks that morality has no universal standards; aesthetic relativism considers that aesthetic evaluation has no universal standards; and cultural relativism deems that no culture is superior to other cultures, etc. Moreover, these theories can be further subdivided. For example, some scholars divide cognitive relativism into ontological relativism and epistemological relativism. And then they subdivide epistemological relativism into rationality relativism, conceptual relativism and truth relativism. Ontological relativism denies that knowledge can understand the real world;

while epistemological relativism repudiates that there is only universal truth. Relativism of rationality denies that there is procedural rationality during the cognitive process; conceptual relativism claims that the conceptual systems of two theories are incommensurable; and finally, truth relativism repudiates that there is only universal truth. These theories of relativism are not separated, but they connect with each other. From the opposite perspective, relativism can be against absolutism, objectivism, rationalism, foundationalism, realism, universalism, and other concepts. These definitions and classifications can help us to recognize the conception of relativism.

Relativism has a long history. For instance, Protagoras in the early period of the west, and Chuang-tzu in the early period of the east are early advocates of relativism. However, in ancient, medieval, and modern philosophy, relativism has not been a major trend. After the huge Hegel system of rationalism broke down, many important modern and post-modern thoughts, such as nihilism, existentialism, hermeneutics, pragmatism, historicism, and deconstructivism, etc., contained the ideological element of relativism, either piecemeal or intensively, either obliquely or obviously. It is stated that the whole modern and post-modern cultures are enshrouded in the relativism. However, regardless of whether relativism is the shadow or sunshine, it is really a main characteristic of modern and post-modern culture.

Many philosophers who searched for the universality made every effort to overcome the relativism, though failed finally. If relativism turns out to be a 'problem', the problems of relativism have always existed, especially in modern and post-modern cultures. MacIntyre (1987, p. 385) said that

> relativism is one of the theories which have been overthrown for many times as the scepticism. If a theory contains some truths, those can't be ignored, you can never find a more significant sign as it is overthrown again and again in the history of philosophy. And the theory which can really be overthrown only needs to overturn just once.

This quotation contains complex feelings with both respect and frustration to relativism.

Some people may retort that we are in the age of science when people believe in scientific rationality, so relativism can only have impact on humanities. However, there is no 'scientific rationality' that can be applied to all fields in the history of science. And it is not enough to explain the development or summarize the characteristics of science only with rationality. Many scientists do not believe or follow scientific rationality, which even can't be defined by them. Thus, the prosperity of science and technology in our age can't lead to the denial of relativism in theoretical aspect. And it is not surprising that the problems of relativism also exist in the philosophy of science.

In the philosophy of science, relativism demonstrates that there are no universal standards in some philosophical problems. For instance, in terms of meaning standards, relativism believes that there are no standards of meaning;

in terms of scientific explanation, it agrees that any model of scientific explanation is unilateral; in terms of methodology of science, it may claim 'anything goes', which means that there is not only one kind of rational methodology for science; in terms of the development model of science, it considers that every development model of science does not conform to the true history of science; in terms of science demarcation, it thinks that there is no clear boundary between science and witchcraft; and in terms of scientific realism, relativism is absolutely the anti-realism.... In fact, the cognitive relativism and cultural relativism mentioned in the previous parts both have been embodied in the philosophy of science. Ontological relativism and epistemological relativism are included in cognitive relativism; rationality relativism, conceptual relativism and truth relativism are included in epistemological relativism –all of which are reflected in the philosophy of science. Furthermore, they are concerned with many issues, such as the close connection between 'incommensurability' and conceptual relativism, and the strong relation between scientific realism and ontological relativism, etc. Therefore, the philosophy of science is a very active field of relativism thought.

But we can't help to question closely, are the answers based on relativism proper or not to these problems? It refers to rational standards but relativism doubts rationality itself, so this question has no results at all. Should we accept relativism? How we are going to explain the fact of prosperity of science and technology? Could we see it only as a miracle? It surely makes philosophers awkward and causes the stagnation of philosophy of science. And it throws us into introspection: should we overcome relativism within theory aspect? We should discuss relativism in another way instead of from the perspective of theory, which is followed by the PSP rising in 1990s. The advocates of PSP abandon the 'theory-dominated' stand of traditional philosophy of science and take the 'practice-dominated' stand to have provided a new science view, which has resulted in the emergence of new research perspective in many philosophy of science problems. In the sight of PSP, relativism has been perfectly avoided at least.

In the following part of this chapter, we will review how traditional philosophy of science produces a lot of relativism problems in its own development, and why they can't be solved; then, this paper will introduce how PSP establishes its own science view to avoid the relativism or make relativism no longer be a problem to traditional philosophy of science.

9.2 The traditional philosophy of science and relativism

Logical positivists intend to transfer philosophy into strict science and endow science with rigid logic specification, which means only after their combination can philosophy of science become a category of philosophy. Logical positivism shows the first common standard in philosophy of science – namely, verifiable principle. This significant principle is: with cognitive meaning, only one statement can be verified by a series of observational propositions. When the principle is applied to scientific development, the development of science will be observed and verified as the process of scientific theory accumulation in series. While it is applied to science demarcation, the statement that can be verified by observing propositions becomes scientific knowledge, and those cannot be verified appearing nonscientific.

Verifiability principle is close to the simple understanding of people towards science. Besides, having undergone a rational and frenzied period, the development of all subjects can form its core. Thus it can be recognized that philosophy of science starts from this mode. However, this principle is logically inoperable. First of all, with limited observation, the general theoretical propositions cannot be verified completely. Secondly, the observation and the experiment have loaded the theoretical presupposition. In order to solve the first problem, Carnap developed the inductive logic that proposed the suggestion that choices of theory acceptance or rejection should be decided by the level of probability. Whereas later, Popper pointed out sharply that under a certain number of evidence, all theories' probability appeared zero, whether those theories were of science or of pseudoscience. In order to solve the second problem, Ayer adapted the verifiable principle to be: when and only when the statement with cognitive meaning as well as other auxiliary hypotheses are conjunctions, the observing proposition can be deduced, which means that the observing proposition cannot be deduced solely by auxiliary hypotheses.

On this subject, Church, a logician, put 'Church formula', whose form is ($\sim O1 \wedge O2) \vee (O3 \wedge \sim S$). Any statement that applies this formula as its auxiliary hypothesis can satisfy the requirements raised by Ayer. Thus, the adaption of verifiable standards from Ayer has failed. Other than logical problems, there are so many accepted scientific theories without verification or principal verifiability, such as the theory of the black hole. With its great mass, a black hole can absorb all materials even including light rays so that it cannot be observed, which results in the unverifiability of the theory of black hole. Therefore, verifiable criterions of logical positivism are abandoned out of the philosophy of science field. Once criterions are lost, philosophy of science will have to face the challenges from relativism.

Popper put forward the falsifiability standard through brand new thoughts. Whether the theory can be regarded as scientific lies in whether it can be falsified but not be verified. The more contents one theory contains, the more falsifiable it appears to be, thus the theory becoming more scientific. Scientific research always starts from one problem on which scientists propose hypotheses constantly and then refute them boldly in return. Hence scientific knowledge develops in the mode of 'Hypothesis1 – Falsified – Hypothesis 2 – Falsified ....'

However, there are some points in Popper's falsifiability standard that are not logical or do not conform to the history of science. 'Duhem-Quine thesis' indicated that all statements are the conjunctive propositions integrated themselves with auxiliary hypotheses. Thus, we cannot verify a statement directly, nor can we falsify it. If there is any counter example, the auxiliary hypotheses can be regarded to be wrong whereas the statement itself is not, which makes the falsification impossible. Besides, there are lots of examples that violate the falsifiability standard, one of which can be the discovery of Neptune and Pluto, being the outcome of scientists' refusing to falsify. But the development of science is not carried out by the mode that Popper proposes. Therefore, Popper's falsifiability standard has fallen into dilemma, making science of philosophy get into the panic of relativism again.

The two failed experiences in logicism trigger people's doubts: is it reasonable to artificially bring history of science into the so called regulated logical frame? Or regardless of the image of philosophers, should scientists seek history of science for scientific features as strict as historians do, no matter whether these features can be found? As it becomes more people's preference, historicism school, represented by Kuhn, develops.

Kuhn revealed that, according to the history of science, the mode of science development is: 'pre-science period – regulated science period 1 – scientific revolution period 1 – regulated science period 2 – scientific revolution period 2 ....' While in regulated science period, main scientists constitute a scientific community, whose members can share one paradigm, and scientific revolution is the process of one paradigm substituted by another.

'Paradigm' serving as the core concept of Kuhn's theory, there appears different definitions on it according to the academy. For example, is the main content of paradigm theoretical or practical? Is paradigm defined in accordance with scientific community or the other way around? While many problems remain controversial, there exists one point for sure, that is, the incommensurability of paradigm, which leads to relativism.

They are not commensurable and there is no standard in comparison both between modern science and ancient thought (as between modern astronomy and the *Tianwen* of Qu Yuan), or between scientific theory and superstition (as between modern astronomy and astrology). While one scientific theory A is different from nonscientific theory B in paradigm, one nonscientific theory B differs from another one, thus there exists no difference between science and non-science. Whether a scientific theory correctly represents the truth of the objective world can only be decided in paradigm, which is not universal.

Despite it advances or not, the development of science lies in the substitution of paradigm, whose process represents the transformation of social mentality. Although Kuhn does not accept the conclusion and continues to seek for the standard in paradigm's mutual comparison from predictive ability, universality, simplicity and others, these respects cannot serve as the common standard. Later, Lakatos, a historicist, mainly compares the predictive ability in distinguishing advanced program and backward program, which has not made big progress. However, Feyerabend had completely accepted relativism, considering scientific approach is 'anything goes'. The mode of scientific development is a process of natural selection. As with religion, myth, and witchcraft, science is one of the various kinds of ideology. As historicism develops to this step, the altitude towards relativism gradually turns to acceptance from rejection.

After the sociological approach is introduced into philosophy of science study by strong program, relativism tendency is more obvious. They put forward 'impartiality', and believed that truth and falsehood, rationality and non-rationality, and success and failure should be treated without prejudice. And both parties of the dichotomy should get explanation. Also, they proposed symmetry principle, and considered that this explanation is symmetrical, and the reason of the same type can explain the correct belief, as well as the false belief (Cai Zhong, 2002, p. 21). Based on requirements of the two principles, a completely different or even contrary theory can be interpreted as true and correct. Scientific and technical knowledge is not of a rational and logical extension of existing knowledge, but the product formed by different social, cultural and historical processes. Scientific and technical knowledge is an accidental consequence, without inevitability or universality, which is an obvious view of relativism.

Feminists believe that, since women have a unique cognitive style, women's participation in science can change the situation of science. If so far it is females not males that are leading science, science would not be the way as it is, but people still believe in and rely on it. This claim also contains the thought of relativism.

In short, in the traditional philosophy of science before the 1990s, the thought of relativism had become a mainstream. Although there are many philosopher of sciences, like Putnam, Shapere, and Laudan, who offered many sorts of new standards to save the rationality of science, but like many antecessors, they could not completely overcome the problem of relativism. Traditional philosophy of science has been in trouble, presenting a stagnant state.

Through the process that traditional philosophy of science is integrated to relativism, it can be found that the standpoint of theory-dominated is the fundamental reason. Main schools in traditional philosophy of science deem science as a representational scientific theory system and a network of concepts. It includes two aspects. Firstly, science is a theory system and a network of concepts. In traditional philosophy, research topics such as meaning criterion, scientific explanation, scientific methodology, science development model, demarcation of science, scientific realism, set scientific theories and logical relations of scientific arguments as the main study objects. Although philosophers of science have recognized that the scientific study is a kind of practical activity, they just stay in the recognition in common sense, and never improve the practice property of science to a height for deep study. The non-theoretical factors in scientific research activities, such as individual scientists, research institutes and instruments, experiment operation means, and social situation on research, are always mentioned just as appendages to scientific theories, not the focal object of philosophy of science. Although the 'paradigm' concept of historicists already includes the classical experimental case and other factors, with the tendency of focusing on experiment, it still puts more emphasis on basic scientific concept, presupposed background knowledge, core scientific laws, fixed deducing process and other theoretical factors. Second, a scientific theory is a representational system of the objective world. It is related to the problem whether the representational system and representational object are consistent. If consistent with the objective world, the scientific theory is correct, otherwise it is wrong.

The standpoint of theory-dominated results in relativism from many aspects. Firstly, it raises the position of theory, but belittles the position of practices. It is inappropriately promoting the generation of 'theory-ladenness of observations or experimentsLadenness' and destroys the foundation of logical positivism.

Secondly, it is believed that all scientific theories are universal statement, and the correctness of universal propositions is universal. The right scientific theories are applicable everywhere, while the wrong ones can get nowhere. Therefore, the traditional philosophy of science surely will discuss questions about universality of scientific theories. For example, can scientific theory be universal? And how? In fact, it is also the same. The meaning criterion, scientific explanation, scientific methodology, science development model, demarcation of science, scientific realism and so on in traditional philosophy of science are all associated with the universality. To some extent, the opposite of universality is relativism. Once the answer to these questions turns out no, the results of relativism will immediately come out. Therefore, after Kuhn puts forward the concept of 'Paradigm', people will compare different paradigms and decide which one is true. At last it is found that they cannot be compared, and then the problem of 'incommensurability' in paradigms appears.

Thirdly, if the scientific theory is regarded as representational system of the objective world, many old and difficult problems of philosophy are involved, such as the problem whether the objective world can be recognized, the homogeneity problem of the objective world, the problem of scientific realism, and the problem of induction. These disputed and unresolved problems overwhelm the traditional philosophy of science. If the criteria, deciding whether the representational system and representational objects match, cannot be figured out, it will be integrated into relativism. Therefore, the standpoint of theory priority is the fundamental reason of relativism.

9.3 PSP and realism

Although the mainstream of traditional philosophy of science is theory-dominated, it implies the tendency should focus on practice. For example, Kuhn's concept of 'paradigm' already includes some classical experimental cases; In Kuhn's explanation for the reason of the scientific revolution, he began appealing to some reasons on social psychology. The field investigation method in anthropology is introduced to study scientific activities by SSK, and this study contributes to the view that science is regarded as a process; negotiations among scientists and even the political dimension are also introduced into the analytical framework of SSK. The naturalistic thought existing in the philosophy of science, believes that everything (including scientific activities) is a part of the natural world and can be explained in terms of the method of natural science. When scientific activities are researched by methods of natural science, the practical part cannot be neglected. The philosophy of technology has been concerned about practical components in scientific activities such as tools, instruments and equipment. All these reflect that scientific practice will be concerned much more. These turns were collected in the 1990s and absorbed ideas of hermeneutics of practice from philosophy in the Continent, forming the PSP school. At present, as mentioned above, the study approaches in PSP mainly are the neo-experimentalism and hermeneutics. The new experimentalism approach emphasizes on re-discussion on relations between theory and practice (especially experiment); hermeneutics approach introduces Heidegger's hermeneutics of practice into PSP, studies the basic property of philosophy of science from a theoretical point of view, and explores these important issues such as the type of scientific practice, the property of scientific knowledge, the relations between scientific practice and laboratory and power.

The school of PSP abandons the position of theory priority and adopts a practice-dominated standpoint. They no longer regard science as a network of theory, but the unique form of human culture and social practice. Based on this research position, all scientific researches are activities of human work and practice. Scientific researches serve as a way for us to contact with the world, which opposes to the view held by traditional philosophers of science that they are a way of understanding the world. Science is not only the surface feature of the objective world, but also the involvement and construction. We are living in a world constructed by scientific practice instead of a network of scientific concept. The formation and evolution process of scientific theory, like other aspects of scientific activities, are scientific practices. As a practical activity, scientific theory is also involved in the reconstruction of the world. In this way, the boundary between theory and practice is broken through putting theory into practice, after which we will no longer pursue the universal correctness of scientific theory, rather we begin to focus on problems as to the components of scientific practice, the way scientific practice constructs the world and the way to spread the results of scientific practice. Therefore, it is conceivable that there exists great difference in the research purport between the PSP and the traditional philosophy of science, providing many new research opportunities and avoiding many problems that traditional philosophy of science fails to solve in a certain sense. As to the problem of relativism, the PSP strives to avoid relativism through the criticism of 'theory-ladenness of observations/experiments', transformation of the concept of 'paradigm', and abandonment of representationalism.

Firstly, the criticism of the New-experimentalism of the PSP on the proposition of 'theory-ladenness of observations or experiments ladenness' is also a work that avoids relativism, which has been specifically analyzed in the ninth chapter. The New-experimentalists point out that the purport of the proposition has been magnified and strengthened in the application. The magnification brings about the result that the previous perception all observations depend on is 'theory' and the use of common language becomes dependent evidence of load theory. The strengthening incurs the result that observations and experiments will certainly rely on or be loaded with the strict theory, which means that the proposition has both a weak and a strong version. The weak version just mentions that there must exist some concepts and instruments related to nature emerging before observations and experiments, while the strong version claims that some systematic or systemized theories emerge before, influence or load the observations or experiments. The weak version, to some extent, unconsciously expands the concept of theory to anything like spirit, thought and conviction, on which sense all human activities are naturally theory-laden. Therefore, although the proposition can be established, it is not worth refuting and makes no sense, while the strong version is confronted with many counter examples and fails to be established. There are a lot of experiments and observations with no theory-ladenness in the sense of strong version in the history of science, such as the irregular motion of the pollen reported by Brown. Chemist Davy inadvertently observed that algae could produce a combustion-supporting gas in the sunlight. At that time, he did not know the nature of this gas, let alone knowing the theory. A lot of experiments not only do not load theory, but also can cause or inspire the new theories of high level or large scope. One case in point is Galileo's observations of the sky through a telescope, which stimulated the future development of astrophysics theory. As a result, the new experimentalists propose that experiments have a life of their own. In this way, the purport of 'theory-ladenness of observations or experiments' of the strong version is refuted, making the experiments and observations once again become the necessary and appropriate basis of theory. Of course, the New-experimentalism does not return to logical positivism, rather it makes this criticism from the perspective of PSP (Wu Tong, 2006a). Once the theory is constrained by observations and experiments, it will not be universally applied, and relativism will no longer be so bold and straightforward.

Secondly, the PSP reinterprets Kuhn's 'paradigm' concept because Kuhn put forward a scientific development pattern that quite conformed to the history of science and won the recognition from many peers. However, from the perspective of traditional philosophy of science, the incommensurability of 'paradigm' will lead to the problem of relativism. Therefore, the PSP makes a transformation on the concept of 'paradigm'. Through the analysis of the works of Kuhn, Rouse, representative of the hermeneutics approach, pointed out that the paradigm theory was not a recognized standpoint, but a conceptualized exemplary way intervened by special experience context.

Besides, it is the practical master of specific scientific achievements that can be used to lead the new study field (Rouse, 2004, p. 31). It is better to say that acceptance of the paradigm view is the acquirement and application of a practical skill rather than an understanding and belief of a statement, such as the application of concepts and mathematics skills, the use of instruments and equipment as well as the grasp of opportunities. This reinterpretation of the paradigm concept is not subjectively imposed by Rouse, and Kuhn himself particularly pointed out that they could agree on a paradigm, but would not agree, or even try to make complete explanation or perform rational work for it. The guidance on research from the paradigm theory would not be hindered even in the absence of the standard interpretation or the recognition of rules (Kuhn, 1970, p. 44). Therefore, it is conceivable that the paradigm theory cannot be reverted to the theoretical rules. But it can guide the operation of the scientific community if it is not theorized. In this way, paradigm has the characteristics of practice. A lot of practice components can be shared between different paradigms of practicality which is not incommensurable. Some instruments, operational methods and technical applications in the era of Newtonian mechanics can continue to be used in relativistic age. Although relativity now comes into vogue, many theories and technologies adopted by people in production and life still belong to the Newtonian mechanics system. Therefore, the transformation of the concept of paradigm has, to some extent, solved the problem of incommensurability.

Thirdly, the PSP no longer regards science as the only characterization system of the objective world. The world is not just the object we get to know but a stuff in which we live and that responds to our activities. The characterization of science on the objective world is not the mirrored type, but the intervention and construction that start from the construction of micro world by the laboratory. In the lab, the measures taken by scientists on the research objects like isolation, highlighting, manipulation, intervention and tracking are not passive reflection and characterization of research objects, but the active construction of them. What they will become after the construction subjects to various factors including scientists, research institutions and equipment, experimental operation means and social context.

However, these factors are unique to the laboratory, so the scientific achievements, the results of the construction, are not universal scientific theories but 'local knowledges' which are not required to comply with the original appearance of the objective world but are only responsible for the laboratory that has brought them into existence. If they can be applied in elsewhere, then it is not because they correctly characterize the objective world, but because scientists artificially copy the criterion of the original laboratory to the new place, repeat the same process and get the same result. This is not a universally practical process of theories on science but a standard transferring one. The spreading process of scientific achievements turns out to be a process continuously constructing the objective world. The world will not become what it is now without the continuous intervention and construction of the science. With the premise of no resorting to the representationalism, this scientific view successfully explains the fact of science, prospect and advanced technology of present world. Because of the rejection of representationalism, it avoids a series of philosophical problems faced by representationalism. It no longer needs to give the judgment as to whether representation system and object conform to the criterion, and therefore, in a certain sense and to some extent, avoid relativism.

Consequently, a conclusion can be drawn as follows:

The traditional philosophy of science adheres to the position of theory-dominance and takes science as theoretic system and concept network featured by representation, which immediately brings about some purports, such as, 'theory-ladenness of observations or experiments' and 'incommensurability'. Also, the philosophical problems caused by the representation hinder the attempt of the traditional philosophy of science to build a universal criterion and make the philosophy of science fall into dilemma where the traditional philosophy of science cannot form the true relativism unless it adheres to universality. PSP does not give priority to the theory but to the practice. In PSP, science is not regarded as the theoretic network but the characteristic form of human culture and social practice. Under the standpoint of practice priority, PSP criticizes the purport of 'theory-ladenness of observations or experiments', redefines the foundation of the experiment in scientific research and transforms the concept of 'paradigm'. And it eliminates the problems of 'incommensurability', abandons the representationalism and builds the view of science as local knowledge, thus avoiding or eliminating the problems on relativism.

9.4 Relativism and absolutism

Many scholars engaged in the philosophy research make their utmost effort to avoid and detest relativism and they all make clear to conquer relativism in all kinds of arguments. In fact, relativism is closely related with an important philosophical standpoint all the time, namely, foundationalism or essentialism. The former is opposite to the later relativism is bound to bring about anti-foundationalism and anti-essentialism. When we hold the standpoint and idea of foundationalism or essentialism, we always try our best to admit that facing the phenomenon refers to one reflection, one explanation or one instruction. It is universally acknowledged that if we have two ideas concerning one event, one of them must be wrong, which I think is correct. However, in fact when putting this cognition to the extreme, we are on the way to absolutism.

The term of 'absolutism' refers to different things in different fields. The absolutism in metaphysics opposes subjectivism and relativism, and admits the principles of ultimacy, eternity and objectivity as the origin and criterion of truth and value. Ethic absolutism claims that all rational animals obey essential and universal moral principle wherever they are. Moral absolutism is contrast to moral relativism denying the universal effectiveness of any moral principles. In the political theories, absolutism claims that the power and right of the government are absolute. When having conflict with the rights, benefits, demands, hobbies and wishes of the public or social groups, the government always takes up the advantaged position. Toulmin, a philosopher of science, said that the disputation of absolutism in ethics leads to the admission of various moral authorities from the beginning, each of whom admits its partial validity.

What is the harm of absolutism? In epistemology, while blocking different cognition roads, it turns out to be a dogmatism, which brings about rigidification and durance. Liu Huajie, professor of Peking University, once said that relativism took advantage over absolutism. He thought that absolutism was exactly supposed to be conquered (Liu Huajie, 2004, pp. 1–4). In ontology, absolutism relying on the realism prides itself on the only representative of the realism. It appears to be the representative of the truth by mixing up the objectivity with absolutism, the truth with universality, and the truth with absolutism. In fact, at present we all know that there are many patterns on realism, for example, Cartwright advocated metaphysics of multi-local realism, which is the realism subordinating to relativism. Therefore, relativism is not always opposite to realism and absolutism is not always accompanied by realism. With the standpoint of local diverse realism, relativism opposes dogmatism and realism base on the foundationalism, turning to be the monist realism of absolutism. In axiology, absolutism is addicted to the monist value, which is a monist dogmatism and closeness. It does harm to the political practice, denies the cultural diversity and value orientation of different groups, and serves as the theoretical foundation of centralism and totalism.

Of course, there is trouble putting relativism to the extreme. Relativism may be a defense of backwardness, an excuse for the weak and a paradox in logic (for example, the judgment that everything is relative is absolute. And this problem exists in all universal negative judgments.) However, in practice, relativism is a way with possibility and without contradiction. The most important is that we are not supposed to take it as an excuse; on the contrary, we should respect other cultures and interact with them instead of brushing against and ignoring them. As to the understanding of scientific practice and knowledge, the problems brought about by relativism are mainly about local nature and the way to spread the local knowledge, which is like the problem that whether to admit the universal rationality after the permission of the partial rationality. We have discussed this in other chapters and will not repeat it here.

After the discussion of the relation between PSP and relativism, we will introduce the parallel subjects of PSP (narrow sense), such as, philosophy of scientific experiment and philosophy of science of new empiricism. These subjects, to different extents, giving the enlightenment, nutrition and support to PSP (narrow sense), prove to be the supportive powers of its emergence and the companions of its development. Some views of these philosophies surpass the PSP; hence, it is necessary to discuss them. Without them, the practical reflection of the group is no longer the change and reflection of 'scientific practice'.

10 Partnering the philosophy of scientific practice

The philosophy of scientific experimentation

A book *What is This Thing Called Science* written by A.F. Chalmers, an Australasian philosopher of science, puts significant influence on teaching and studying philosophy of science in China. In the preface of its third edition, Chalmers points out that new experimentalism is one of the most influential schools of contemporary philosophy of science, which is definitely a high evaluation of new experimentalism in philosophy of science.

Generally speaking, the philosophy of scientific experimentation (PSE) belongs to the PSP. For that reason, there are two basic approaches in the research of PSP. The first is based on hermeneutics, which investigates and discerns the meaning of scientific practice, illustrates the meaning of the most important issues based on the notion of scientific practice, such as realism, observation and opportunities, laboratory research and local knowledge, in traditional philosophy of science. Meanwhile, it also explores content and meaning of the philosophy of science from the perspective of hermeneutics. The second approach is to specify stereotypes of scientific practice. For instance, philosophers could reconsider the relationships among experiments, observation, and theories by reinterpreting experiments and other activities in laboratories. Besides, some potential issues, such as the functions of scientific instrument, or the stereotypes of objective-thing knowledge, may also provide much empirical content for PSP. That could give evidence in arguing for multiple relationships between scientific practice and theories, without any necessity of theory-laden observation.

In a word, the first one is to provide normative justification, while the second one is an approach of empirical explanation.

If we try to understand the PSP in a narrow sense, the philosophy of scientific experimentation is not PSP. Despite having many similarities, they differ from each other in many ways, even in some basic points. Therefore, we may as well think of PSE as a fellow of PSP.

10.1 The significance of the philosophy of scientific experimentation

To review the claim that 'observations/experiments are theory-loaded' is of great significance to study philosophy of science. This claim once became the bane of positivism and an important theory viewpoint for the historicist philosophy of science. Hanson, a historicist philosopher of science, puts forward the theory-ladenness of observation/experiment. His idea was widely accepted by philosophy of science because it is better than the positivist idea, which is thus delivering a fatal blow to the positivism in philosophy of science.

Once the three of them are mixed together, questions such as the progress, comparison and truth of the scientific theory become puzzling because of the lack of necessary and reasonable foundation. Therefore, the fundamentalism based on the proposition of empirical observations loses its solid foundation. Besides, the flourishing of relativism is also related to that claim. Can we reconstruct the observation/experiment as a necessary and appropriate experimental basis for theory, without simply returning to logical positivism? This is an important question of philosophy of science. The new experimentalism starts from reexamining the history of science, provides a large number of cases from the history of science to demonstrate the existence of observation with no theory-laden. This provides an important theoretical basis to correct the extreme theory-dominant viewpoint in philosophy of science, and resets the viewpoint of the dialectical relationships both between experiment and theory and between accumulation and revolution in history of science. Therefore, the new experimentalism becomes a significant approach.

Second, traditional philosophy of science often confuses observation and experiment. In the positivist philosophy of science, experiment is not more important than observation since the role of the observation and experiment have few essential differences. The traditional philosophy of science only attaches importance to their representation named as observation statement. However, both of them provide evidence for theories, and they are the bridge between theory and reality. Moreover, experiment and observation are also used to testify theory. They are guided by clear theoretical assumptions about theory-loaded observation/experiment.

From PSE, the experimentation usually has a life of its own (Hacking, 1983, pp. 150, 165). Instead of being guided by theories, experiments are guided by hints about what is worthwhile to study and by how to carry out such hints. PSP attaches much importance to the experiment, since from its perspective, science is all about the way to act in the world, rather than the way to observe and describe the world, for 'the experiment always undertakes an important task of epistemology' (Knorr-Cetina 1995, p. 110). In this way, the experiment/observation and theory have important differences. As a result, the PSP obviously attaches more importance to the experiment. The target, design, and means of the experiment, instruments used, and the phenomenon obtained from the experiment, all of them are received a higher degree of attention.

Third, PSE also intensely criticizes the claim of the theory-ladenness of observation/experiment, which has deeply influenced philosophy of science. For example, Hacking, the pioneer of the new experimentalism, argues that,

1. Observation and its statement are not the same thing. In the past, logical positivism paid too much attention on the statements, but the activity of observation had not been treated in a deserved way.[1]
2. The relationship between theory and experiment is varying according to different stages of development, rather than keeping as totally unchanged.
3. The fundamental study of reality is prior to any relevant theories.
4. Classifying the matter according to the theoretical and experimental terms is a kind of misleading, since it disposes the theory as a class involving an essence, and treats the experiment as another class (Hacking, 1983, chapter 9–10). The new experimentalism suggests the existence of theory-free experiments and therefore rejects the claim of the theory-ladenness of observation/experiment by a large number of cases, in which observation is antecedent to and independent from relevant theories, such as the cases of observation of methane by Humphry Davy, of the discovery of unknown celestial bodies by Herschel, and of the non-theoretical significance of the Brownian motion.

The relationships among experiment, observation, and theory are very complicated. To generalize them into one conception might be an illusion, since if experiment can have its own life, so does theory. Furthermore, experiment and theory can also have a life that intertwined together like double spiral coil. The most important thing is that we should go deeply into the scientific experiments in laboratories to explore multiple relations rather than to give a simple explanation.

Therefore, laboratory is getting new status and significance in epistemology, which is a common conclusion from the new experimentalist approach in PSP, the hermeneutic approach of scientific practice, and the laboratory studies in SSK. For instance, Knorr-Cetina claims that the notion of scientific laboratory has replaced the significance of experiment in the history of science and scientific methodology. Laboratory context is made up of the practice of instrument and symbols, which roots in the skilful activities of science. Laboratory has now become an important theoretical concept in the philosophy of science, which is an important agent of scientific development. Knorr-Cetina even thinks of the proxy process as follows: laboratory is considered as a natural process in which things are brought 'back home', which makes the natural objects and conditions get 'domesticated' and 'Social audit'. The significance of laboratory is its promotion of the 'social order' and 'cognitive order' (Knorr-Cetina, Jiasanuofu, et al., 2004, pp. 112–13).

In traditional philosophy of science, laboratory is the place merely to produce scientific knowledge, with which the development of scientific knowledge and theory have nothing to do. This makes the laboratory much like the notion of space in the Newtonian framework, in which the development of experiment has nothing to do with the particular situations in the laboratory. But the new PSP argues for an important status of laboratory, which makes 'laboratory to experiment and theory' a bit like 'the time and space to the material under Einstein's relativity theory'. Any scientific experiment cannot be separated from specific laboratory situation, which is closely associated with scientific practice.

1. The laboratory which plays a decisive role for production and defence of scientific knowledge is the place and context to produce knowledge. Knowledge is local in a practical dimension which must contain some laboratory characteristics. After tremendous transformations, objects are no longer purely natural ones;
2. The role of laboratory is to isolate and control the objects, which make the researched objects much clearer;
3. The laboratory intervenes with instruments, equipment, and skills, the world in research, and the laboratory itself is part of the scientific practice;
4. The laboratory also provides an entire understanding of experiment. If there is no laboratory, scientific practice is not anything meaningful for the very beginning;
5. Laboratory also provides practical and cultural foundations for understanding the new scientific resources. What is the new understanding about the laboratory in PSE? That is exactly the new understanding provided by the PSP approach.

10.2 The research themes of PSE

At present, PSE focus on six core themes: (1) the material realization of experiments; (2) experimentation and causality; (3) the science–technology relationship; (4) the role of theory in experimentation; (5) modelling and (computer) experiments; (6) the scientific and philosophical significance of instrumentation. Here we briefly introduce the six themes and focus on instrumentation, tools, and others.

10.2.1 The material realization of experiments

In experiments we actively interfere with the material world. In one way or another, an experimentation involves material realization of experimental process (the research objects, the apparatus, and the interaction between them). The question is: what are the significance of the productive character of scientific experimentation on ontological, epistemological, and methodological issues about science?

10.2.1.1 Ontology significance

The philosophy of new experimentalism argues that the ontological significance of scientific experiment lies in an appropriate ontological description of experimental science, which needs to correspond to certain configuration. Since the traditional philosophy of science is theory-dominated, it does not give enough attention and recognition to experimentation, including the practice of experimental design, the investigation on the role of reproductive ability of experiments as a machine, the use of graphic symbols in experiments, the program of 'virtual observation', the roles of specialist in using instruments, and the role of the human spirit in the experiment. Again, those possibility, ability, and tendency should be involved into the ontological study of experimental science.

10.2.1.2 Epistemology significance

The interventionist character of experimentation engenders epistemological questions. David Baird offers a neo-Popperian account of 'objective thing knowledge'. He thinks that the knowledge is encapsulated in material things. Thus, non-contextual knowledge exists in material things. Some illustrations of such knowledge include Watson and Crick's material double helix model, Davenport's rotary electromagnetic motor, and the Watt's indicator of steam engine. Baird suggests that the notions of truth, justification, and delocalization are transferred to the thing knowledge in those cases. In the discussion of instruments, Baird offers a concept of thing knowledge. He thinks there are three kinds of thing knowledge: model knowledge, such as the model of DNA; working knowledge, an instrument or machine that performs regularly and reliably; and measurement knowledge, which refers to the cognition of things' degree, scope of the size, and measurement when we consider them. These distinctions also involve the division of Popper's world 1, 2, and 3 with the problem of interaction in metaphysics. We will discuss the characteristics of these different kinds of knowledge later.

10.2.2 Experimentation and causality

Theoretical and empirical studies of experimentation are helpful to investigate causality. On the causality of experimentation, at least three different approaches can be found. First, the role of causality in experimental practice may be analyzed. The second approach involves analyzing the role of experimentation in interpreting and testing causal claims. Causal inferences can only be justified through (possibly hypothetical) experimental interventions but not through 'passive' observations. The third approach tries to explain the notion of causality on the basis of the notions of action and manipulation.

10.2.3 The science–technology relationship

If philosophers keep neglecting the technological dimension of science, experimentation will continue to be seen as a mere data provider for the evaluation of theories. If they start exploring the science–technology relationship seriously, one salient way is to study the instruments and equipment in experiments. In his book *The Philosophy of Scientific Experimentation*, Davis Baird argues for the importance of thing knowledge on a par with theoretical

10.2.4 The role of theory in experimentation

Another core theme in PSE is the relationship between theory and experiment. That theme can be studied from two approaches. First, the role of existing theories within experimental practices concerns the claimed (relative) autonomy of experimental science. As previously mentioned, the view of theory-free of experimentation is put forward against the claim of theory-ladenness of experiment. In addition, Michael Heidelberger argues that causal issues in experimentation can and should be distinguished from theoretical issues. The same distinction is also in classification of scientific instruments. Heidelberger suggests that while experiments with 'representative' instruments are theory-laden, the use of 'productive', 'constructive', or 'imitative' instruments is causally based and claimed to be theory-free (Radder, 2003, pp. 138–51).

The second major approach to the experiment-theory relationship addresses how theory may arise from material experimental practices, or how to conceptualize the transition from the material process to propositional and theoretical knowledge. Even if experimental research is not merely a means to theoretical knowledge, experiment does play an epistemic role with respect to the construction of scientific theories. A balanced philosophical study of this issue may profit both from 'relativist' approaches of science studies (Gooding, 1990, pp. 180, 211–5) and from 'rationalist' epistemological approaches (Franklin, 1990, pp. 2–3, 160; Mayo, 1996, pp. 405–8). Especially, through critical research of the subjective Bayesianism, Mayo resorts to and describes the appropriate relationships among the experiments, observations and theories.

10.2.5 Related research on experiment, modelling, and (computer) simulation

It has gained important propulsion that related researches on Experiment, Modelling, and (Computer) Simulation are promoted in PSE. Over several decades, the scientific significance of computer modelling and simulation has increased greatly. Many scientists nowadays are involved in what they call 'computer experiments'. Apart from its intrinsic interest, this development invites a philosophical discussion of what is meant by these computer experiments and of how they relate to ordinary experiments. Evelyn Fox Keller and Mary Morgan deal with this topic in detail. Both of them offer a classification of computer modelling and simulation. The first is computer simulation which uses computer models to simulate the existing phenomena. The second is the real meaning of 'computer experiments', which is an experimental study on the object producing by computer programming. The third class is model phenomena that lack a theoretical underpinning in any sense, such as research on 'artificial life' (Radder, 2003, pp. 198–215, 216–51). Morgan even gives out the difference on the three kinds of classification of experiments (see Table 10.1).

Table 10.1 Types of experiment: all three with representing relations (Radder, 2003, p. 231)

		Ideal Lab experiment	Hybrid experiments virtually	Hybrid experiments virtual	Mathematical model experiment
Controls on	Inputs	experimental	experimental on Inputs; assumed	assumed	assumed
	Intervention	experimental	On intervention and environment	assumed	assumed
	Environment	experimental		assumed	assumed
Demonstration method		Experimental in laboratory	simulation: experimental/mathematical using model object		deductive in model
Degree of materiality	Inputs	material	semimaterial	nonmaterial	mathematical
	Intervention	material	nonmaterial	nonmaterial	mathematical
	Outputs	material	nonmaterial	non-or pseudo-material	mathematical
Representing and Inference Relations		representative of... ...to same in world Representative for... ...to similar in world		representation of... ...back to other kinds of things in world	

10.2.6 The scientific and philosophical significance of instruments

In fact, the study of scientific instruments is an interesting insight for PSE. There are some issues ignored in traditional philosophy of science, but now figured out from new directions and approaches, including a variety of features of design, operation, and wider uses of instruments; the importance of schematic and pictorial symbols in designing instruments; the perceptual and functional information that is stored in images or visual perception; the relationship between thought and vision; the role of reproducibility as a norm for experimental research; the modes of representation which instrumentally mediates experimental outcomes, and so on.

10.2.6.1 The faults in traditional philosophy and history of science: scientific instruments

Looking at a large number of works on the philosophy and history of science, in addition to theory-dominated tradition, there is an ignored subject: scientific instruments.[2] Actually, this is very natural in a theory-dominated philosophy of science. In the theory-dominated view, scientific practice is usually divided into two contexts: discovery and justification. For instance, Karl Popper argues that the discovery belonged not to methodology of science but merely as psychology, and the instruments do not work in justification. In other words, scientific instruments are though helpful to the scientific understanding, it is only up to play an auxiliary role to the discovery; As for the justification of scientific knowledge, the instrument does not make any sense. Therefore, it is enough for philosophy of science to merely study the construction of theory.

When scientific discoveries are studied in philosophy of science, the status of instruments, will be increased. And the important role of scientific instruments for discoveries start to get a better understanding. However, discussion on the role of scientific instruments is not enough, for it still lacks adequate rationality.

Without Rutherford's α particle scattering experiment, there is no accurate understanding about the nucleus discovered. If there was no particle accelerator, high-energy physics could not be emerged. However, for the epistemology of science, that is just a recognition of the important role of understanding scientific instruments, rather than the understanding belonging to the justification of knowledge. It only illustrates that the development of scientific recognition cannot exist without scientific instruments, while not illustrating how scientific instruments take roles in forming scientific knowledge, or how the nature of scientific knowledge relates to scientific instruments, which is more important.

In fact, the traditional view of science has ignored scientific instruments in these aspects as follows:

Scientific instruments are meaningless to debates upon realism;
Scientific instruments are meaningless to the justification of knowledge;

180 *The philosophy of scientific experimentation*

Scientific instruments are meaningless to the relationship between experimentation and theory....

But that would not be true.

10.2.6.2 Scientific instruments: their philosophical significance

Traditional philosophers of science argue that the use of scientific instruments is only 'seeing' in a position of sideholder. However, when citing the new experimentalism, Chalmers keenly points out that the observation is an intervention of practice. Hacking, who is hailed as a pioneer of the new experimentalism by Rouse, also points out that, the observation 'is learning by doing, not just passive looking' (Hacking, 1983, p. 189). Here, the important role of the practical intervening of the scientific instruments has been roughly sketched out in the process of observation and employment of scientific instruments. In addition, in his book *Representing and Intervening*, Hacking sets up one chapter to illustrate an example of using microscope in discuss the epistemology of scientific instruments, especially the realism and anti-realism debates.

Below, basing on Hacking's discussion, we use the microscope as a case to look at the role of scientific instruments to the realism argument.

The debate about scientific realism is not about whether there is a general entity, but about whether the theoretical entity predicted by science is real. The development of the microscope is often related to such debates.

For example, according to Hacking, Gustav Bergman denies that the object under the microscope is real. He argues that

> the object under the microscope is not true sense of the physical things, but they are given by language and illustrations fiction.... When I look through a microscope, what I see is a patch of colour which slowly swept through my horizons, like a shadow over the wall.[3]

However, Grover Maxwell promotes a kind of scientific realism. He denies the fundamental difference between observable and theoretical entities. And he makes use of the opportunity to propose a visual continuous spectrum, 'look through the panes of glass, look through the glasses, look through binoculars, look through the low-power microscope, look through a high power microscope, etc'.[4] Hacking refers to these arguments to claim that there are some things not visible sometimes, but later becomes observable because of some new technologies. Therefore there is no ontological significance between observable and theoretical entities (Hacking, 1983, pp. 187–8).

Most philosophers just inherited a sideholder's position in epistemology in the debate of scientific realism. According to Hacking, one should learn to see through a microscope in practice, not just by seeing as a sideholder (Hacking, 1983, p. 189). Hacking mentions two different views on the microscope in different times:

1. Abbe's research becomes explicable. It is demonstrated that microscopic vision is *sui generis*. There is and there can be no comparison between microscopic and macroscopic vision. The images of minute objects are not delineated microscopically by means of the ordinary laws of refraction; they are not dioptical results, but depend entirely on the laws of diffraction (Hacking, 1983, p. 187).
2. The microscopist can display a familiar object in a low power microscope and it let us see a slightly enlarged image which is 'the same as' the object. Increase of magnification may reveal details in the object which are invisible to the naked eye; it is natural to assume that they, also, are 'the same as' the object. (At this stage it is necessary to establish that detail is not a consequence of damage to the specimen during preparation for microscopy.) But what is actually implied by the statement that 'the image is the same as the object?'

> Obviously, the image is a purely optical effect.... The 'sameness' of object and image in fact implies that the physical interactions with the light beam that render the object visible to the eye (or which would render it visible, if large enough) are identical with those that lead to the formation of an image in the microscope....
>
> Suppose however, that the radiation used to form the image is a beam of ultraviolet light, x-rays, or electrons, or that the microscope employs some device which converts difference in phase to changes in intensity. The image then cannot possibly be 'the same' as the object, even in the limited sense just defined! The eye is unable to perceive ultraviolet, x-ray, or electron radiation, or to detect shifts of phase between light beams....
>
> This line of thinking reveals that the image must be a map of interactions between the specimen and the imaging radiation.
>
> (Hacking, 1983, pp. 261–3)

Sometimes philosophers should abandon the spectator's epistemology. Philosophers can intervene in microscope practice indirectly, since philosophers could 'enter' it through studying the works of scientists who engage in technical practice. Sometimes we should listen to the suggestions from scientists, especially in case of what is 'seeing' and why the similar structure revealed by various microscope. For example, the microscopist in 2) thinks that we see through a microscope only when the physical interactions of specimen and light beam are 'identical' for image formation in the microscope and in the eye. We often think of 'seeing' as observing something through ordinary optical microscope. However, the microscopist thinks that ordinary optical microscope works by diffraction; it is completely different from ordinary vision.

From the two quotations, the meaning of 'seeing' is different, since what is the 'seeing' also depends upon different understandings. In 2) there are two factors for the image: imaging and specimens. The image is the result of using optical microscopes or electromagnetic properties technology, but specimen is the object itself, which plays an important role. Therefore, when we are able to use a variety of microscope to 'see' an object, the observed entity have been very likely real.

In addition, different kinds of microscope depend on different optical principle. Thus, it may be theory-loaded: loaded with the theory of optics or other radiation that any statement about what is seen with a microscope. Hacking does not agree on the idea. He points out that, in order to manufacture microscopic you need theory. However, you don't need theory to use the microscope' (Hacking, 1983, p. 189). Especially for biologists, the real perception under the microscope has nothing to do with the physics. Observation and manipulation are loaded with no physicist theory, therefore it is completely independent from the cells or crystals to be studied.

Hacking enumerates the ordinary optical microscopy, electron diffraction microscope, the polarization microscope, and interference microscope. Once we have such devices in hand, endless variations may be constructed, such as polarizing interference microscopes, multiple beam interference, phase modulated interference, and so on. There are also acoustic microscope, field emission microscope, and phase contrast microscope. There may still be invisible differences in refractive index in various parts of its structure. The phase contrast microscope converts those into visible differences of intensity in the image of the specimen. Because of instruments manufactured according to completely different principles of physics, we observe that similar specimens have very similar structures. We believe this because we understand clearly most knowledge of physics associated with the manufacture of instruments – it is by means of these instruments as we have ascertained what we already know. In addition, the gratifying intersection of the microscope and the biochemistry gives people confidence, since biochemistry has confirmed that the structure which the scientists specify with a microscope may be determined according to the chemical properties.

That is an argument for realism at least. If we do not want to explain what we through microscopes by considerations like coincidence and miracles, the structure obtained through different microscopes with various properties has to be ascertained as the component of the specimen itself.

Another important philosophical implication of scientific instruments is given by Heidegger, who argues scientific instruments are of *Gestell* for scientists in practice. The *Gestell* as a way of practice decides what a scientist can see, discover, and make up. This kind of discussion has certainly transcended the conclusion from PSE, which frequently adopts analytical method to discuss scientific experiments and instruments.

10.2.6.3 *The ontological and epistemological classification of scientific instruments*

Rom Harré thinks that previous studies not only ignores scientific instruments on the whole, but also interprets scientific instruments wholly based on the misunderstanding that all of the instruments are homogenous. In this way, we lose the ability to analyze scientific instruments in laboratory further. He claims to classify scientific instruments and tools carefully. He points out that,

> In some cases in which an apparatus is serving as a working model of some natural system, the changes brought about by experimental manipulations must be interpreted as analogous to states of the natural system being modelled. In some cases in which an instrument is causally affected by some natural process, the changes in the instrument are effects of the relevant state of the material world. These effects must be interpreted in terms of the causal relation presumed to hold between the process in nature and the state of the instrument. It will be convenient to use the word "instrument" for that species of equipment which registers an effect of some state of the material environment, such as a thermometer, and the word "apparatus" for that species of equipment which is a model of some naturally occurring structure or process, such as the use of a calorimeter to study the effect of salt on the freezing point of water or in vitro fertilization.
>
> (Harré, R., see Radder, 2003, p. 20)

Harré classifies scientific instruments into two types according to the philosophical significance of them: material Models as domesticated versions of natural systems and Apparatus-world complexes (Harré, R., see Radder, 2003, pp. 26–9).

Such a classification is really meaningful. For example, what is the image and meaning of scientific instruments as material models of domesticated versions of nature? Harré cites some cases as 'the drosophila in laboratory', 'Atwood's machine', etc. And Harré (Radder, 2003, p. 27) says as follows:

> Example 1: a drosophila colony is domesticated version of an orchard replete with a breeding population of fruit flies that display variation by selection. If certain conditions on manipulability of laboratory colony are met, it makes possible the experimental, laboratory study of inheritance.
>
> Example 2: an Atwood's machine is a domesticated version of a cliff that a stone falls down or leaning tower from which objects can be dropped. The machine consists of a graduated vertical column with various movable attachments, allowing for the releasing of standard weights from different heights, and the determination of locations of the falling masses at different times.
>
> Example 3: a tokamak is a domesticated version of a star. A powerful magnetic field confines hydrogen atoms in a small volume, fusion to helium being ignited by an external energy source. The process set going is a domesticated version of stellar fusion.

The key point includes both the relationship between the model of domesticated version and the world, and the relationship among models, laboratory research, and the material world, which is also very important. Harré thinks that (Harré, R., see: Radder, 2003, p. 27):

> Domesticated versions of material setups and processes that occur in feral form in nature, versions that we know as experimental apparatus and procedures, are relative to their feral ancestors, simpler, more regular, and more manipulable. The drosophila colony in the laboratory is a simpler biosystem, with more regular life patterns and more manipulability than the swarms of flies in the apple orchard.

Therefore, the metaphor of domestication permits strong back inference to 'wild', since the same kind of material systems and phenomena occur in the wild and domestication. Thus, an apparatus of domesticating this sort is a piece of nature in the laboratory.[5] However, the richness of back inference will depend on the simplicity we can extract from the model. It is another issue how relations of similarity and difference are weighted according to the interests of the researcher in performing experimental manipulations. The most important thing is that, there is no ontological disparity between apparatus and natural settings in domestication. The choice of apparatus and procedure guarantees this consistency, since the apparatus is a version of naturally occurring phenomena and material settings in which it occurs.

Another kind of scientific apparatus is associated with nature / world complex. For example, Wilson Cloud Chamber, the Stern-Gerlach apparatus, and many other well-known pieces of laboratory equipment are all of that type. Harré thinks that, the apparatus/world complex is that which scientists, engineers, gardeners, and even cooks bring about something like Bohrian artifacts. Manipulating them brings into phenomena that did not exist in the wild or natural world. In general, there is no material structure in nature like the apparatus. For instance, ice cream does not occur in nature, only in kitchens with freezers (Harré, R., see: Radder, 2003, p. 28).

This leads to two important distinctions. Harré uses two case metaphors to discuss their differences (Harré, R., see Radder, 2003, p. 31).

> As far as I know, there is no setup similar to Davy's equipment anywhere in the universe. Free metallic sodium exists only by virtue of the apparatus/world complex Davy built. Humphrey Davy used electrolysis on molten common salt in a crucible to bring metallic sodium to light. Think of how much is presupposed in describing this experiment as the 'discovery of sodium' or as the 'extraction of sodium'. There is no metallic sodium in the universe to the best of my belief.... Sodium-as-a-metal is a Bohrian phenomenon.
>
> This experiment contrasts sharply with Faraday's use of a tube of rarefied gas to study discharge phenomena. A similar setup to the apparatus existed 'in nature', in the electron wind in the rarefied upper atmosphere. Therefore,

The philosophy of scientific experimentation 185

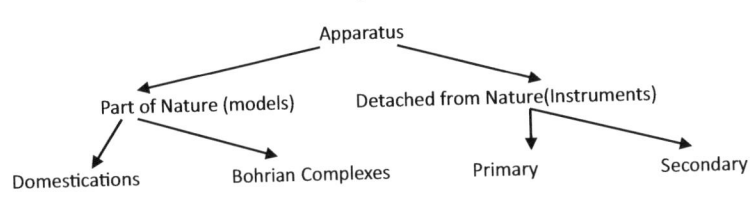

```
                        Apparatus
             ┌─────────────┴─────────────┐
    Part of Nature (models)      Detached from Nature (Instruments)
       ┌─────┴─────┐                ┌─────┴─────┐
  Domestications  Bohrian Complexes  Primary    Secondary
```

Figure 10.1 Types of apparatus (Radder, 2003, p. 33).

we can understand the glow in the laboratory tube as an analogue, in a domesticated version of the upper atmosphere of the aurora borealis.

Just as the cow and the aurora can serve as a metaphor for the relation between apparatus and the world, the homely image of a loaf of bread can serve as a metaphor for the Bohrian apparatus/world complex. A loaf is brought into existence from wheat and other ingredients by the use of material structures that do not exist in the wild, such as flour mills and ovens. Loaves do not appear spontaneously in nature.

Obviously, an effect taking place in laboratories or scientific instruments is a domesticated version of nature. Similarly, there are other phenomena taking place which is not produced by nature itself, but with the aid of the interaction between instruments and the world together. Here, instrument is no longer a natural model, or a transparent window of observing the world.

In principle, an actual phenomenon produced in an apparatus is a representation of a potentiality in the world. This point attribute phenomenon to potential but natural propensity no matter whether it happens in natural environment or in causality with apparatus. According to Harré, it still treats apparatus as transparent windows of world, but it ignores the contribution of apparatus to the form and qualities of phenomenon. Reflecting on this issue will let us understand the Bohr interpretation deeper. The Bohrian phenomena are neither properties of the apparatus nor properties of the world that are elicited by apparatus. They are properties of a novel kind of entity, an indissoluble union of apparatus and world, namely the apparatus/world complex.

There is a question on 'In what form does metallic sodium exist before the electrolysis begins?' Nature, in conjunction with Davy's apparatus, affords metallic sodium, just like it in conjunction with Wilson's apparatus affords tracks and thereby electrons as particles. Therefore, the question 'In what form do electrons as particles exist before the cloud chamber is activated?' is equally illegitimated in an analogous way.

Finally Harré gives the classification of the full set of scientific instruments as follows (see Figure 10.1: Harré, R., see Radder, 2003, p. 33).

Thus, the detailed classification of scientific instruments allows us to go into the in-depth discussion of ontological and epistemological issues behind the instrument and tools. It is necessary to pay attention to scientific instruments, since it is problematic that a theory-dominated approach ignores them.

186 *The philosophy of scientific experimentation*

10.3 A theory of working knowledge: objective knowledge and practice

Scientific instruments are all material things. The instruments and facilities which are used in scientific research are often mediated between the research and its object. What is the epistemological meaning of them for philosophy of science?

From PSE, Baird classifies scientific instruments and facilities. And he puts forward that these things contain concrete knowledge very different from the knowledge of representation.

10.3.1 Model knowledge and practice

Model is mainly material also, such as hydraulic dam model in engineering design, Lei's building model in ancient Chinese architecture design, and Watson and Crick's DNA model. According to Baird,

> such things work like theories, but they use part of materials of the world, not words, to represent. For this reason, they provide a different entry for our cognitive apparatus. Conceptual manipulation provides an entry; material manipulation – 'hand-eye manipulation' – provides a second entry.
> (Baird, see Radder, 2003, p. 45)

Watson and Crick's DNA model provides a fine example for model knowledge. They combine the metal plates and rods together to represent the structure of DNA (Figure 10.2).

Watson and Crick's model of DNA provides a two-part argument for understanding material models as knowledge (Baird, see Radder, 2003, p. 45).

> There is a negative part: it makes little sense to think of the Watson – Crick Model in theory terms. They did not use the model as a pedagogic device. They did not simply extract information from it. The model was not part of some intervention in nature. It was not a part of an experiment.
>
> There also is a positive part: Watson and Crick's model performs theoretical functions with contrived bits of the material world instead of words. Their model has the standard theoretical virtues. It has been used to make explanations and predictions. It was confirmed by evidence-X ray and other- and it could have been refuted by evidence, for example, if DNA had been found with obvious different quantities of adenine and thymine. Though made of metal, not words, there can be little doubt that Watson and Crick's model of DNA is knowledge.

Baird discusses on a three-part account for how material models represent the world. This is the 'DDI account', which claims that model representation includes denotation, demonstration, and interpretation.

First, models denote some part of the word, which is similar to Harré's notion of domestication version. In order to properly understand a model,

Figure 10.2 Watson and Crick's model of DNA.

one must know the denotation of every parts of the model. Thus, the sticks in Watson and Crick's model denote bond lengths, not rigid metallic connections. The plats denote 'base groups' – adenine, thymine, and cytosine – that is, groupings of atoms that can be treated as units in their own right.

Models also need demonstrations, which can be manually carried out. It is a magic moment when Jim Watson discovered base-pair bonding in DNA. He plays around with cardboard cutout versions of the bases and their bonding distances. Through these manual manipulations he finds that the hydrogen of one of the two purine-bases (adenine) bonds with one of the two pyrimidine-bases (thymine), while the other purine (guanine) hydrogen bonds with the other pyrimidine (cytosine). So, A bonds with T and G with C. The bonding distances guarantee that bases could bond on the interior of the DNA chain while a two-strand backbone twisted in a double helix without distortion.

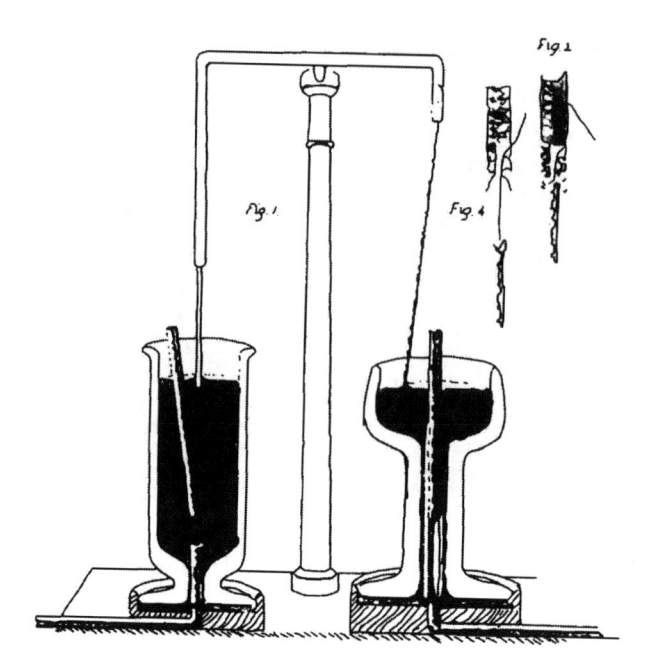

Figure 10.3 Two versions of Michael Faraday's electromagnetic motor (Baird, see Radder, 2003, p. 48).

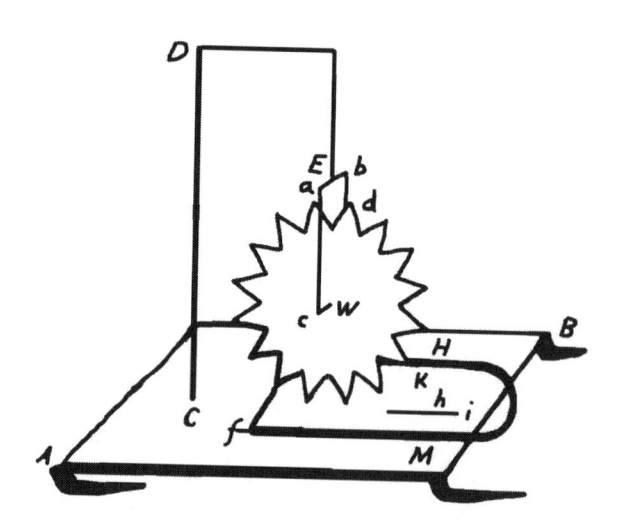

Figure 10.4 Barlow's Star, Peter Barlow's variant of Faraday's motor (Radder, 2003, p. 48).

Models are interpreted finally. One immediate result of the Watson-Click model is the requirement that there be the same amounts of adenine as thymine and of guanine as cytosine. Chemical analysis distinguishes this out. The most significant interpretation is the possibility it presents for genetic replication.

Therefore, the model can also be operated in practices. For example, the demonstration of the model mentioned previously is operated manually in practices.

10.3.2 Working knowledge and practice

According to Baird, the working knowledge is 'a different kind of interaction with the material and a different kind of knowledge borne in materials' (Baird, see Radder, 2003, p. 47). This knowledge is not only stored in the materials, but also related with operational actions. The focus should be on the connection between knowledge and action. Baird points out that Faraday's 1821 electromagnetic motor provides a good example (see Figure 10.3). When Faraday firstly made the motor, there was a considerable disagreement on the explanation it exhibited. Yet, no one contested what the apparatus did: it exhibited a kind of rotary motion as a consequence of a suitable combination of electric and magnetic elements. It also exhibited a phenomenon for which, at that time, there was no adequate theoretical language yet. In addition, other inventors created the other forms of motor model quickly, inspired by Faraday's model (see Figure 10.4). That in itself provides a case that thing itself can carry working knowledge without any theoretical support.

Despite Faraday and other scientists at that time lack an agreement on theoretical language to describe, they could reliably create, re-create, and manipulate this phenomenon. The motor itself was actually their 'working knowledge'. Indeed, Faraday's motor worked reliably even though there was no theoretical explanation; and other researchers could learn from the motor just as Davenport could learned from Henry's electromagnet. It sufficiently demonstrates that the thing itself contains the 'knowledge' telling people how to operate it. This is working knowledge.

10.3.3 Measuring knowledge and practice

Baird argues that measurements, combining both model knowledge and working knowledge, are a synthesis of these two kinds of knowledge. Measurements provide a standard or a sample through comparison and measurement in reflecting the difference between other types of practice and knowledge. For example, a mercury thermometer is a part of the world, for its linear expansion of mercury with temperature is a nature phenomenon, of which scientists make use. It is also a tool to measure other object, obtaining the temperature in contacts with objects. Thus, it contains the knowledge of measurements and of the measured objects. It is a model as well as an operation. According to Baird, mercury thermometer is built with the working knowledge. As a result from the builder's selection of possibilities, some choice of signals are generated which makes transformations to the signals in order to render 'a measurement'. The instrument operator builds a representation of possibilities into a material form of the instrument. Therefore we have a scale on the thermometer's glass tube that displays the possibilities that we embrace with the thermometer. Here is a representation of the structure of temperature, which is model knowledge. When both kinds of knowledge are integrated, the instrument suggests that it extracts information from nature, and we have encapsulated model and working knowledge into a material form of the instrument (Baird, see Radder, 2003, p. 50).

10.4 The contribution and deficiency of the new experimentalism

10.4.1 Three aspects of outstanding contributions of the new experimentalism

First, new experimentalism breaks up the theory-dominated tradition in philosophy of science. For a long time, philosophers of science usually think about theories while ignoring the role of scientific experiment. Experiment is usually considered to be dependent on theory or paradigm, which is impossible to maintain its own life. In those related researches in the new experimentalism, we reconsider the nature and the role of the experiment which partially restore the view of positivism, and which let the philosophy of science back down to earth.

Second, new experimentalism reconsiders the relationship between experiment and theory. People like Hacking, Heidelberg, and Radder have more detailed discussions in that aspect. The most well-known view is from Hacking, namely, 'experimentation has a life of its own'. His view contains four propositions: (1) The material practices of experiment contains not only observation and data, but also the implementation of experiment, techniques used in experiment, and the understanding about the scientific experiment; (2) The experiment work is not simply to examine or explain some theory. Rather it is an answer to the purposes, opportunities, and constraints from the experiment itself; (3) The philosophical significance of experiment does not depend on their theoretical explanations; (4) During the experiment, new artificial phenomena are often generated or associated with, and the occurrence of these phenomena not only illustrates of universal laws of the nature, but also reflects the basic stance of the new experimentalism, namely, considering the experiment as practice and intervention on the nature. Experiments interact with the world and therefore change each other. Experimentation creates new phenomena, in which its philosophical significance cannot be fully expressed by its representational role for theory. This view puts tremendous shocks on the claim of the theory-ladenness of observation held by most post-positivists in philosophy of science.

In addition to Hacking's view, there are many discussions on this issue in new experimentalism. In 1994, Mayo published an article in *Philosophy of Science*.[6] In this article, he specifically discussed three significances of Hacking's claims about 'experimentation has a life of its own'. In his book *The Negligence to the Experiment*, Franklin analyzed the wonderful realizing methods of the experiment in Particle Physics from 1950s to 1960s, and he was trying to prove that the theory relying on laboratory is possible. In his 'Bubble Chambers and Experimental Workplace',[7] Galison discussed in high-energy physics experiments how various experimental instruments hugely impact the experiment practice and the laboratory culture, and he analyzed the important role played by experiment in theoretical knowledge.

Third, the viewpoint of new experimentalism on scientific realism is helpful to construct practical realism. Although new experimentalists do not completely resolve the debates on scientific realism, they acknowledge the diversity of experimental and scientific activities, which gives scientific realism a good approach from the perspective of experimental intervention. For example, Hacking points out that we cannot resolve the debates between realism and anti-realism from the representing level. Only by departing the representing level and by understanding the world from experimentation, will we find out the solid foundation of the realism. Experiments are not ideas or theories, but something we are doing with no concept and intervening into objects, and then influencing and transforming them with specific instruments. If experiments intervene, influence and transform the objects to achieve the expected results, the whole process will be a reason for its objective reality. A theoretical entity is proposed only when it is understood in its causal relationship with other objects, in building some kind of instrument and in getting prospective results according to these causal relationships. Therefore, only when it is possible to operate on the entity, and do experiments on its objects, is the reality of the entity credible. That is to say, the experimental study provides strong arguments for scientific realism in a sense of practices.

10.4.2 The shortage of new experimentalism

New experimentalism is very necessary to argue against theory-dominated philosophy of science. However, there are still many problems in that position itself.[8]

On the one hand, the new experimentalism only represents a trend of experimentation study in philosophy. Its literature distributes in a too dispersed way, which mainly focuses on historical narratives and descriptions, while its philosophical analysis and explanation seems inadequate. In other words, new experimentalism mainly draws on methods and case studies from history and sociology. For example, Gooding analyzed Faraday's magnetic effect of current experiments and Hertz's electromagnetic effects, while Hacking analyzed the amplification of the microscope. Both of them illustrated that experimental observation itself intervenes in the reality without reference to any high-level theory. Therefore, in the theory-dominated philosophy of science, such studies do not seem to rise to the level of theory. The observation and experimentation are supposed to be practice, which is a tool for serving theories without a trace in the text. For that reason, in the viewpoint of representationalism, experimentation has nothing worthwhile to analyze, since observations and experimentations have been 'proved' to be theory-loaded.

On the other hand, new experimentalism also has inconsistencies in itself. Some problems, including the relationship between theory and experimentation and the role and status of the scientific practice, might illustrate theoretical inconsistency in new experimentalism. The most prominent defects are the four aspects as follows:

First, not only experiment but also theory has its own life. When Galileo put his telescope to the sky, he did not want to test a theory about the satellite of Jupiter. But we can not deny that theory will continue to guide experimental study and give directions to discovery. For example, no one can deny that the prediction of Einstein's general theory of relativity promoted the Eddington's study about solar eclipse. In fact, there is indeed a complex dialectical relationship between experimentation and theory. The new experimentalism does not suggest how to exclude the role of theory from science, which is the first question encountered by the new experimentation. In fact, if the new experimentalism goes to the other extreme, it is even more problematic. But if it does not push into the extreme, the trouble encountered is to explain the relationship between experimentation and theory in more details.

Second, the new experimentalism seeks to prove that experimentation has a more fundamental status. From this standpoint, it tries to prove that experimentation and theory can be divided, though it can't prove that. It was proved that there are experiments that do not depend on high-level theoretical guidance in the view from new experimentalism, from which we can conclude that the existence of experiment is independent from theory. However, it does not prove all experiments and theories are independent from each other. Logical positivism argues that the experience is the standard for testing theory, which is broken up by the claim of theory-ladenness of observation. Similarly, trying to prove that the experimentation/observation is independent from theory in order to rebuild empirical study might be also not successful.

Third, although new experimentalism argues that the experimentation is more foundational, its basic function is still in representing theory. The new experimentalism brings the actual experiment practice into the scientific explanation, but it still confines to the representationalist framework of traditional philosophy of science. It will not be able to replace the framework in traditional philosophy of science with experimental concept. The representation as a result of practice is more important than the practice itself. The practice is to represent the results as a text, only when it becomes a kind of statement or report of observation, qualified as a particular narrative. Therefore, in the new experimentalism, experimentation and observation cannot avoid completely the influence of the representationism.

Finally, its partly returning to the logical positivism is not accepted by many philosophers of science, which affects their understanding of the new experimentalism. As new experimentalists, Hacking, Gooding, and Mayo made great efforts to re-examine some of experiments in the history of science, and found the role of experiment in intervention, cross-review, errors control, and elimination, which suggests that the experimentation has a fundamental role in selecting and improving theory, and which represents science as a progressing trend in a whole. This approach is more like returning to the accumulative view of scientific progress claimed by logical positivists, which is believed to have many problems and has been overthrown by the revolutionary concept of historicism. Unless the new accumulative concept from the new experimentalism gets more evidences and sufficient reasons, the argumentation of the new experimentalism will not be taken seriously.

In the next chapter we turn to discuss the philosophy of new empiricism which also involves deep connections with PSP. Rouse has repeatedly mentioned the new empiricism in his discussion, including the works from Hacking, Hayes, Cartwright, etc. In the next chapter, we mainly discuss Cartwright's viewpoint as a representative for the new empiricism.

Notes

1 The differences between observation activities and observation statements are extremely important. Latour seems to have recognized this. In addition, such a distinction not only resurrects the positivist explanation of relationship between the observational statement and the theoretical statement in representations, but also indicates that the description and explanation do not apply to observe activities. However, the observational statement is a subsequently explanation corresponding to the theoretical one.
2 In a lot of translation works in the history of science, we only find one book related to the scientific instruments, that is: Thomas Crapper's *A Brief History of Science: As Seen through the Development of Scientific Instruments* (translated by Runsheng Zhu, The China Youth Press, 2004). Compared with other works of history of science, this book discusses more about scientific instruments in different era; there are more atlases of scientific instruments in it; it clearly puts forward that scientific instruments are the decisive powers of the development of science. However, the position of this book is still in the view point that the instrument is a service for discovering theory, rather than paying more attention on how to expand the scientific practice in using scientific instruments.
3 G. Bergman, 'Outline of an empiricist philosophy of physics', *American Journal of Physics*, 11 (1943), pp. 248–58, 335–42.
4 G. Maxwell, 'The ontological status of theoretical entities', *Minnesota Studies in the Philosophy of Science*, 3 (1962), pp. 3–27.
5 Notify that we usually distinguish between artificial nature and natural nature, in fact, artificial nature also links with the natural nature through the domesticated version.
6 D. Mayo, 'Neo-experimentalism, topical hypotheses, and learning from error', *Philosophy of Science*, 1 (1994), pp. 270–9.
7 Included in: P. Achinstein and O. Hannaway (eds.), *Observation, Experiment and Hypothesis in Modern Physical Science*, Cambridge, Mass.: MIT Press, 1985, pp. 309–73.
8 The traditional philosophy of science mentioned here refers to the historicism and the theory-dominated philosophy of science position before.

11 New empiricism

A close relative of the philosophy of scientific practice

In this chapter, we discuss the role and the significance of new empiricism in PSP based on Cartwright's research. Rouse, an advocate of PSP, cites Cartwright many times and discusses many ideas of new empiricism. This suggests that new empiricism is an important philosophical resource for the rise of PSP. We cannot ignore the new empiricism approach in philosophy of science if we want to study PSP.

11.1 Plural local realism: the dappled world

Cartwright clearly advocates a view of plural realism. However, Rouse does not clearly support a certain kind of realism. It appears as contrary to PSP which resolves the dispute between realism and anti-realism and which claims that intervention to the world is the interpretation of the world. But in fact, they are consistent. And, the plural and local realism that Cartwright proposes even gives PSP a plausible realistic conception.

Why local realism is necessary? Why this realism is plural? And why it is a local realism?

First, PSP understands scientific knowledge as local. This locality of knowledge reflects a limitation of the practical space of practitioners, and of temporality in time. Though practice itself can extend and continue endlessly in time, every specific practice is local. The standardization of knowledge is moving from a locality to another via the motion and expansion of laboratory, which involves transformation, changing and development. In epistemology, nature of knowledge is local and it also need a corresponding local ontological commitment. In other words, local realism is a natural and inevitable requirement of view of local knowledge in PSP.

Second, the practice of the practitioner is not only always local – either in the laboratory or various scientific practices in general – but also an intervention in the surrounding world. The world is around the practitioner from the start. It must not be non-temporality or non-locality. Thus, it implicitly shows that there is a world in which the practitioner intervenes. Even Rouse thinks that the contradiction between realism and anti-realism has been solved through the intervention of a practical hermeneutic. So, realism is required. It is not a holistic realism since world is a whole system, rather it is a local realism that the world and actors are integration and associated with local practices of practitioners. So, local realism is an inherent requirement of intervention of practice.

Third, why this realism is plural? Either practices in living world or more unified scientific practices are carried out by practitioners through specific situation. This kind of practice will differ from man to man and vary by region and condition. The situation, condition, time and objective of practices which every practitioner will face are different. And unity of them is the exception or caused by the setting of 'ceteris paribus'. We can get the situation which same law is true through making different practitioner under the same situation as far as possible by 'ceteris paribus' setting. Since the world is dappled, the classification of objects is varied, and the laws of nature are patched, it is very reasonable to recognize that the agents and their practices involve serious difference everywhere. So, for this world, plural realism is more real. Pluralism is an inherent requirement of PSP. And this is the most novelty of new empiricism.

Forth, though a unified world with hierarchical levels is more desirable, but it seems to be less credible as a view of world in the sense of metaphysics, and a dappled world is more real and credible. Cartwright believes this and says:

> metaphysical nomological pluralism is the doctrine that nature is governed in different domains by different systems of laws not necessarily related to each other in any systematic or uniform way; by a patchwork of laws. Nomological pluralism opposes any kind of fundamentalism.
>
> (Cartwright, 1999, p. 31)

So, by reference to Neurath, Cartwright opposed to the unified view of world that the world is a big, unified hierarchical system (Figure 11.1, left) and advocated the belief of dappled world (Figure 11.1, right). In fact, not only in new empiricism, there is a tendency in the philosophy of system that the way we see the world can be systematic, but it is not mean that the world itself

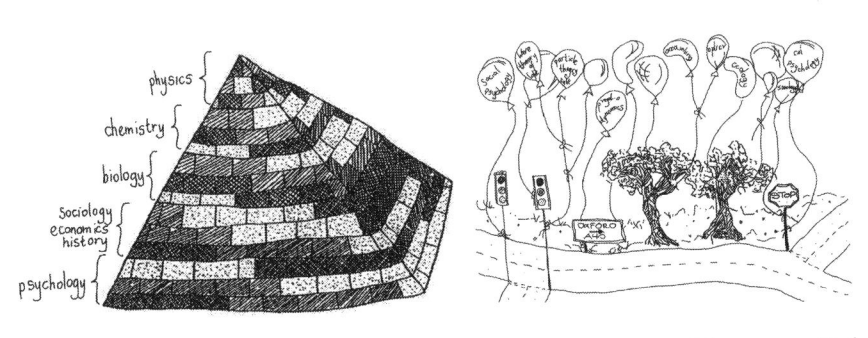

Figure 11.1 Two views of science: left, unified hierarchical view; right, non-unified dappled view.

Source: From Cartwright, 1999, pp. 7–8.

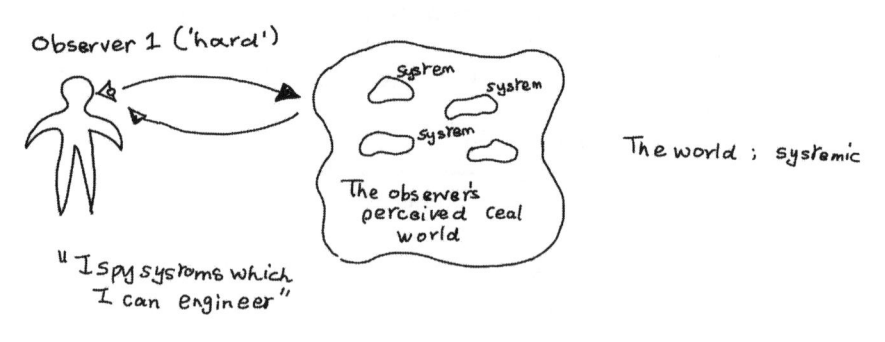

Figure 11.2 The hard systems stances which Checkland against.

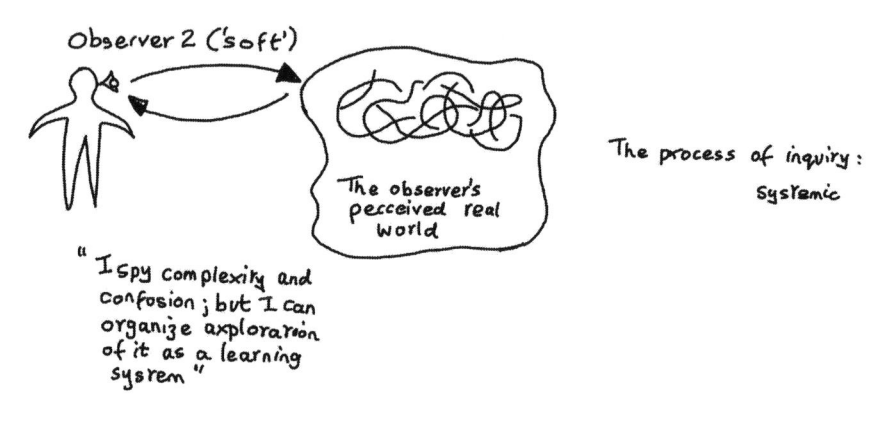

Figure 11.3 The soft systems stances which Checkland support.

is systematic. For instance, in 'soft systems methodology: a thirty year retrospective', the founder of the soft systems methodology Checkland change his idea of which thinks the world as a system, but thinks that system just a way of seeing the world and a method of organizing our perspective. So, we need not take the world as a system (Figures 11.2 and 11.3; Checkland, 2000, S11, S58).

11.2 Ceteris Paribus: when law is true

People have a fear or misunderstanding of admitting that the nature of knowledge is local, as if admitting it will deny the truth. In fact, it is indeed a misunderstanding. As Cartwright says, the pernicious effects of the belief in the single universal rule of law and the single scientific system are spread across the sciences. Indeed, a good many physicists right now are in revolt (Cartwright, 1999, p. 16). This is because locality of knowledge does not deny the existence of truth but admit the locality of truth. They think there could be specific truth in the context of locality. So, why do people have such a misunderstanding? It is because the opposition to the local view of knowledge is the view of truth related to universality that is the universalism of truth. Local view of knowledge opposes to those who viewed the truth universally and without any condition.

For people who hold the view of universalism of truth, this view of local knowledge has two problems: first, why do some laws at least appear to be universal on the expression? For example, the expression of a law; second, the application of a law seems to be universal. People hold the view of universalism of truth will say that truth is both concrete and universal, both relative and absolute, and there are both relative truth and absolute truth.

Let's see how Cartwright refuted this universalism of truth. Then on the basis of Cartwright, we further elaborate such a view of local truth and laws.

Cartwright takes a most simple example: Newton's second law.

> In order to make my claims as clear as possible, I shall consider the simplest and most well-known example that of Newton's second law ($F = ma$) and its application to falling bodies. Most of us, brought up within the fundamentalist canon, read this with a universal quantifier in front: for anybody in any situation, the acceleration it undergoes will be equal to the force exerted on it in that situation divided by its inertial mass (the author blackening). I want instead to read it, as indeed I believe we should read *all* nomologicals, as a *ceteris paribus* law.
>
> (Cartwright, 1999, p. 25)

Then, she proposes that we may write Newton's second law as: for anybody in any situation, if nothing interferes, its acceleration will equal the force exerted on it divided by its mass (Cartwright, 1999, p. 26). This transformation for the formulation of Newton's second law by Cartwright is very important (see Table 11.1).

The difference between them is whether there are conditions and how to set the conditions. They do not exist in the former: say, fundamentalism and universalism. And in any case, the acceleration of an object will be equal to the force exerted on it divided by its inertial mass. In the latter, the local notion of knowledge, it is not impossible for the object to be governed by Newton's Second Law. But this is on the situation of nothing interferes. Later is known a *ceteris paribus* law.

Conditions and law of *ceteris paribus* are not only the expressions of conditions hypothesis sentence or counterfactual conditionals. It in fact gives

Table 11.1 Different expressions of Newton's second law

expression of Newton's Second Law by Universalism and fundamentalism	expression of Newton's Second Law by view of local knowledge with '*ceteris paribus*'
for anybody in any situation, the acceleration it undergoes will be equal to the force exerted on it in that situation divided by its inertial mass	for anybody in any situation, *if nothing interferes,* its acceleration will equal to the force exerted on it divided by its mass

a scientific method of researching. It offers the condition of transforming natural nature artificially: we can make *ceteris paribus* so that laws could suit for those situations. Therefore, we can expand and extend the scope of law's application and make it seem as universalization. Therefore, according to Cartwright, instead of being always universal, laws of nature depend upon the ceteris paribus conditions concerning to the laboratories, which means that those laws are always valid in their local conditions.

Thus, Cartwright set a nomological machine.

What is a nomological machine? According to Cartwright, it is a fixed (enough) arrangement of components, or factors, with stable (enough) capacities that in the right sort of stable (enough) environment will, with repeated operation, give rise to the kind of regular behaviour that we represent in our scientific laws (Cartwright, 1999, p. 50).

Nomological machine have three roles: first, the construction of regularity, combination and application of principle; second, construction of specific situation; third, shielding conditions. With a nomological machine, we have conditions of *ceteris paribus* which will make laws be true. On the conditions of *ceteris paribus*, laws could be applied on the setting conditions. And the meaning and function of science is to change the world by the way of *ceteris paribus* and to make the world more and more suitable for the situation of industrialization. Next, let's follow Cartwright to discuss an example of nomological machine from physics.

This is an example of planetary motion. Cartwright points out that in the case of the orbiting planet, the constituents of the nomological machine are: sun is characterized as a point-mass of magnitude M, and the planet, a point-mass of magnitude m, orbiting at a distance r and connected to the former by a constant attractive force directed towards it (Cartwright, 1999, p. 50). Based on the nomological constituents, what is the relationship between these two point-mass in a *ceteris paribus* sense? Cartwright points out that Newton's achievement was to establish the magnitude of the force required to keep a planet in an elliptical orbit:

$$F = -G^{mM}/_{r^2}$$

In particular, Cartwright points out that the shielding condition is crucial here. To ensure the elliptical orbit, the two bodies must interact in the absence of any additional massive body and of any other factors that can disturb the motion (Cartwright, 1999, p. 51). Cartwright said:

> The example of the planetary motions is important for me since it has been used by philosophers and physicists alike in support of the view that holds more "basic" regularities as first and fundamental in accounting for observed regularities (i.e., in explanation, it is laws "all the way down"). This view emphasizes the unifying power of the appeal to Newton's

laws with respect to Kepler's. I do not deny the unifying power of the principles of physics. But I do deny that these principles can generally be reconstructed as regularity laws.

(Cartwright, 1999, p. 52)

The discovery of Neptune is considered to be a victory of the law of gravitation. Scientists suspected that there may be another planet besides Uranus. It was suggested by the deviation that the orbit of Uranus displayed with respect to the predictions that can be made from Newtonian principles. So the discovery of Neptune is often considered to be a victory of law of gravitation and consequently a victory of Fundamentalism. But, if Newton's Law is universal, it will suit for everywhere without shielding conditions. But, the irregularity indicates that Newton's law is not available everywhere. So, scientists suspected that there was another planet which had impact on Uranus and this thus leaded to the discovery of Neptune. The discovery of Neptune in turn illustrates that Newton's law of gravitation is not applicable to three-body problem. So, the example of Neptune, in Cartwright's opinion, is a failure of Newton's law of gravitation. Of course, scientists in the conditions of Universalism do not realize that this is a failure of nomological machine and it needs shielding conditions. It shows that Newton's law of gravitation is not applicable to three-body problem.

Cartwright thinks that models function as blueprints for nomological machines. Cartwright describes conditions under which models are true for law and the locality of law by taking an example of Coulomb's law as blueprints of nomological machines. For example, we can set a specific situation:

As shown in the Figure 11.4, two electrons e_1 and e_2 are released from rest into a cylinder. The cylinder is open from one side only, and it is open to a uniform magnetic field which pointed towards the negative z-axis. How will the two electrons move? Obviously, motions of electrons are determined by the Coulomb repulsion in the cylinder. But when e_1 enters the magnetic field, it is not only governed by Coulomb's law. So, the insulation cylinder plays a role of shielding conditions.

Electron e_1 is governed by other law when it enters magnetic field. So, we solve problem by the patchwork of laws in different specific situations.

First situation is in insulation cylinder, the force between the two electrons which the distance between the two electrons is r_1 is equal to:

$$f = \frac{1}{4n} \frac{e_1 e_2}{r_1^2} = m_e a_a$$

At this time e_2 will be locked inside the cylinder, e_1 will enter the magnetic field B with a certain velocity v_1. The magnetic field on e_1 will move it in a circular motion (as in the figure) with a force equal to:

200 *New empiricism*

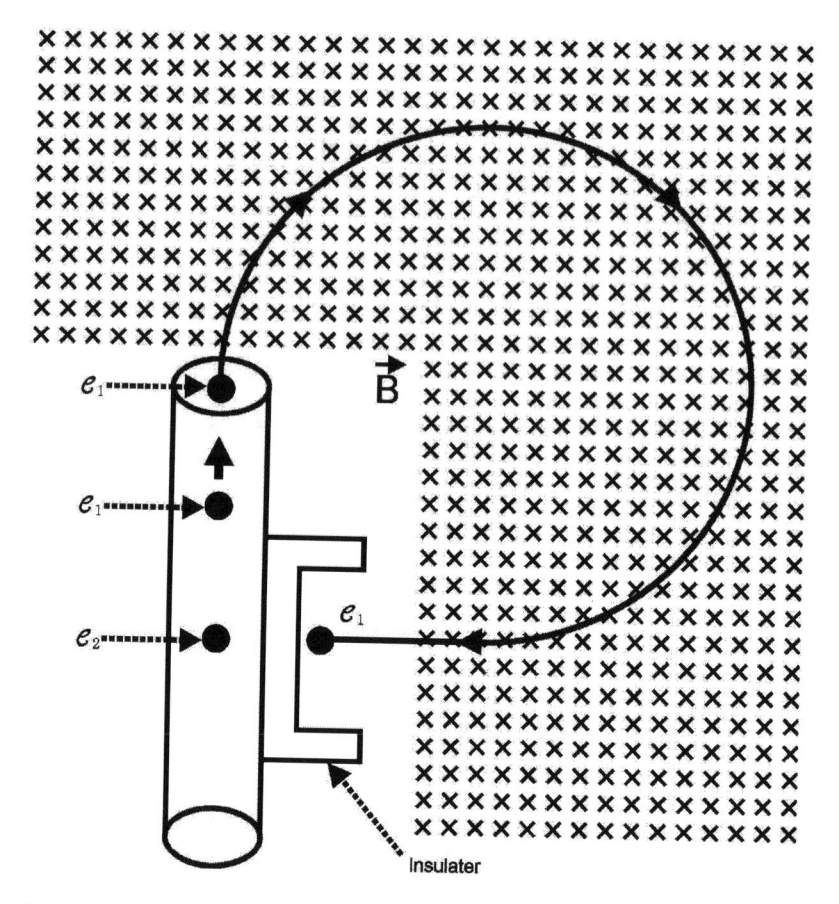

Figure 11.4 Specific application of Coulomb's law with magnetic field (quoted from Cartwright, 1999, Figure 3-1b).

$F = e_1 v_1 O \times B$

This will take the electron e_1 into an insulated chamber attached to the cylinder. The dimensions of the cylinder and the chamber can be set so that the distance between the final positions of e_1 and e_2 is less than r_1. Then content of Coulomb's law is varying. So, the specific content of law will vary with the specific situation. And this shows that the law and its application are situational and local. As Cartwright puts, 'I argue against laws that are unconditional and unrestricted in scope. Laws need nomological machines to generate them and hold only on condition that the machines run properly' (Cartwright, 1999, p. 59). It is clear that there is no universal cover of law (Cartwright, 1999, p. 6).

11.3 Knowledge of nature and the law

Originally, scientific knowledge is about the understanding of occurrent properties. People often think that law knowledge is a basic component and core of scientific knowledge because scientific knowledge has often been expressed in the form of law knowledge. Cartwright challenges this view. She says, our most wide-ranging scientific knowledge is not 'knowledge of laws' but 'knowledge of natures of things' (Cartwright, 1999, p. 4).

In chapter 4 of *The Dappled World*, Cartwright rejects the conventional categories of British empiricism and turns instead to a more ancient one which is the view of Aristotle. Cartwright thinks: a concept like Aristotle's notion of 'nature' is far more suitable than the concepts of 'law', 'regularity', and 'occurrent property' to describe the kind of knowledge we have in modern science: knowledge that provides us the understanding and the power to change the regularities around us and produce the laws we want (Cartwright, 1999, p. 78).

We all know that Aristotle's physics appears to be wrong for Department of Physics today. But, why Cartwright turns to support this more ancient view? Because it is not enough to understand everything of the world only by science. We are inclined to ask, 'How can there be motions not governed by Newton's laws?' The answer is: there are causes of motion not included in Newton's theory.... The wind is cold and gusty; the bill is green and white and crumpled. These properties are independent of the mass of the bill, the mass of the earth, the distance between them (Cartwright, 2006, p. 32).

Cartwright inclines to think that knowledge of nature is more basic than knowledge of laws. And early Aristotle's physics was mainly to discuss natures of things, not motion of things.

Cartwright points out that we could not replace 'natures' by 'laws of natures'. For our basic knowledge – knowledge of capacities – is typically about natures and what they produce (Cartwright, 1999, p. 80).

And Cartwright also points out that modern explanation similarly relies on natures. She will argue, though modern natures are like Bacon's and unlike those of the Scholastics in that they are attributed to structures and qualities we can independently identify. Third, having made the empiricist turn, we no longer identify natures with essences (Cartwright, 1999, p. 80).

Of course, natures as Cartwright says differ from Aristotle's in three important ways. First, as in Cartwright's example of an atom in an excited state, Cartwright assigns natures not to substances but rather to collections or configurations of properties, or to structures. Second, like the early empiricists and the mechanical philosophers of the scientific revolution, modern physics supposes that the 'springs of motion' are hidden behind the phenomena and that what appears on the surface is a result of the complex interaction of natures. Cartwright thinks that we no longer expect that the natures that are fundamental for physics will exhibit themselves directly in the regular or typical behaviour of observable phenomena. It takes the highly controlled environment of an experiment to reveal them (Cartwright, 1999, p. 81).

But, Newton's laws including law of gravitation are not about occurrent properties of things. Cartwright points out that Newton's 'law of gravitation' is not a statement of a regular association between some occurrent properties – say masses, distances and motions (Cartwright, 2006, p. 52). And what is 'force'? 'Force' can never be out of specific context of particles and motions. So, it is an abstract term that describes the capacity of one body to move another towards it, a capacity that can be used in different settings to produce a variety of different kinds of motions.

Let's follow Cartwright and discuss the capacities of Coulomb's law. Cartwright thinks that Coulomb's law tells not what force charged particles experience but rather what it is in their nature, *qua* charged, to experience. 'Natures' are closely related to 'powers' and 'capacities'. To say it is in their nature to experience a force of $q_1q_2/4\pi\varepsilon_0r^2$ is to say at least that they would experience this force if only the right conditions occur for the power to exercise itself 'on its own', if they have very small masses so that gravitational effects are negligible. But it is to say more: it is to say that their 'tendency' to experience it persists even when the conditions are not right; for instance, when gravity becomes important. *Qua* massive, they tend to experience a different force, Gm_1m_2/r_2. What particles that are both massive and charged experience will depend in part upon what tendency they have *qua* charged and what *qua* massive (Cartwright, 1999, p. 82).

It is to mark this fact, the fact that charge always 'contributes' the same force, that Cartwright uses the Aristotelian notion of *nature*. But, as Cartwright remarks in referring to Bacon, these modern natures differ from Aristotle's in one very central respect. Although it is in the nature of charge to be subject to a force of $q_1q_2/4\pi\varepsilon_0r^2$, this nature does not in any proper Aristotelian way reveal the essence of charge. Cartwright points out what charge is depending on a lot of factors independent of Coulomb's law (Cartwright, 1999, pp. 82–3).

Of course, Cartwright thinks that it differs from the more ordinary 'capacities' we refer to in everyday life in three ways. And those ways are important to the kind of understanding that exact science can provide of how a nomological machine operates (Cartwright, 1999, p. 53).

First, the capacity is associated with a specific feature – charge – which can be ascribed to a body for a variety of reasons independent of its display of the capacity described in the related law – here Coulomb's law. This is a part of what constitutes a scientific understanding of the capacity. But description of capacities in everyday life may not be that. For example, Cartwright says that she is irritable.

The second way that Coulomb's capacity differs from everyday ones is that it has an exact functional form and a precise strength, which are recorded in its own special law. And capacities in everyday life may be not.

Third, description of combined effect of capacities has explicit rules. We know some very explicit rules for how the Coulomb capacity will combine with others described by different force laws to affect the motions of charged particles. What happens when a number of different forces are exerted together on the same object? We use those rules to calculate a 'total' force [not the resultant force, the concept of resultant force let us think that there is an independent force in addition to each component force (Cartwright, 1999, p. 54).

11.4 The concrete and the abstract

Just like there is not an abstract job without a specific job. All abstract laws and knowledge are in the context of specific application of knowledge and laws. It is important for understanding locality of knowledge.

Cartwright takes an instance of writing 'the dappled world':

> Writing this chapter is what my working right now consists in; being located at a distance r from another charge q_2 is what it consists in for a particle of charge q_1, to be subject to the Coulomb force $q_1 q_2 / 4\pi\varepsilon_0 r^2$ in the usual cases when that force function applies.
> (Cartwright, 1999, p. 39)

Further, she says:

> What did I do this morning? I worked. More specifically, I washed the dishes, then I wrote a grant proposal, and just before lunch I negotiated with the dean for a new position in our department.... *Working* is a more abstract description of the same activities I have already described when I say that I washed dishes, wrote a proposal, and bargained with the dean.
> (Cartwright, 1999, p. 40)

And

> the abstract-concrete relation is not the same as the traditional relation between genus and species. The species is defined in terms of the genus plus differentia. But in our examples, the more concrete cases, like washing dishes or travelling very fast, have senses of their own, independent (or nearly independent) of the abstractions they fall under.
> (Cartwright, 1999, p. 40)

There is not an independent and abstract job separated from washing dishes and writing proposal. And how could the concept of Physics be abstract?

Considering $F = ma$

According to Cartwright's account, force is to be regarded as an abstract concept. It exists only in the more specific forms to which it is led back via models of the kind listed in specific condition. As Cartwright says,

I did not wash the dishes, apply for a grant, talk with the dean, and then work as well (the author blackening). These three activities constituted my working on the occasion. Similarly, the world does not contain two-body systems with a distance r between them – or blocks and springs or blocks and rough surfaces – plus forces as well.

(Cartwright, 1999, p. 45)

Cartwright continues to criticize this kind of common sense bias of philosophy and science:

We can contrast this view with another which I think is more common, a view in which you can hive the force off from the situation and conceive it to exist altogether apart from the situation. Apparently, this is never thought to be possible since the situation is clearly physically responsible for the force; but it is taken to be logically possible. The simplest version of this idea of physical responsibility is that the situation produces the force (though it may not be by a standard causal process since perhaps the two come into existence together). The force comes into existence and it carries on, on its own; it is there in the situation in addition to the circumstances that produced it. This account makes *force* function like a highly concrete concept on the same level as the two bodies and their separation, or the block and the spring.

(Cartwright, 1999, p. 45)

Force – and various other abstract terms from physics as well – is not a concrete term in the way that a colour predicate is. It is, rather, abstract, on the model of working, or being weaker than-, and to say that it is abstract is to point out that it always piggy backs on more concrete descriptions.... And being abstract, it can only exist in particular mechanical models.

(Cartwright, 1999, pp. 45–6)

11.5 Truth and social construction

First, Cartwright borrows the social constructivists' perspective to criticize Universalism and Foundationalism in natural science. Second, there is a little bit different between Cartwright and social constructionist. Those two points above are clear in Cartwright's argument.

Cartwright points out social constructivists argue against taking the laws of physics as mirrors of nature. Do scientists have special lenses that allow them to see through to the structure of nature? They have not. Their language is evolved from the natural language and not the one in which the Book of Nature is written. The concepts and structures that they use to describe the world must be derived from the ideas and concepts that they find around them. We can improve these concepts, refine the structure, but always the source must be the books of human authors and not the original Book of Nature. So, even the science is bound to be a thoroughly human and social construction, not a replica of the very laws that God wrote (Cartwright, 1999, p. 46).

Cartwright asks a question for lay people: what then of the startling successes of science in remaking the world around us? Don't these successes argue that the laws on which the enterprise is based must be true? She borrows social constructivists' words to point out that the successes are rather severely restricted to just the domain she mentioned – the world as we have made it, not the world as we have found it. But, Cartwright is somewhat different from social constructivists that she admits a few notable exceptions, such as the planetary systems. And other parts of our most beautiful and exact applications of the laws of physics are all within the entirely artificial and precisely constrained environment of the modern laboratory (Cartwright, 1999, p. 46). Contrary to social constructivists, she thinks that this artificiality can be true. The artificiality just shows that conditions are strictly under control. If here adding artificial counterpart of new experimentalist nature domesticated of fruit flies, even the real scientific laws can transit to its natural version.

Of course, social constructivists also point out that: even when the physicists do come to grips with the larger world. First, they do not take laws which they have established in the laboratory and try to apply them outside. Rather, they extend the whole laboratory to the external world. Second, they construct small constrained environments totally under their control. They then wrap them in very thick coats so that nothing can disturb the order within; and it is these closed capsules that science inserts, one inside another, into the world at large to bring about the remarkable effects by which we are all so impressed (Cartwright, 1999, pp. 46–7).

The difference here again: the conclusion Cartwright inclines to draw from this is that, for the most part, the laws of physics are true only of what we make. The social constructivists tend to be scornful of the 'true' part (Cartwright, 1999, p. 47).

Newton's law can be true of exactly those systems that it treats successfully; for we have seen how we can take it to be true of any situation that can be simulated by one of the models where the force puts on a concrete dress. That does not mean that we have to assume that Newton has discovered a fundamental structure that governs all of nature (Cartwright, 1999, p. 47).

This is a question that bears, not on the truth of the laws, but rather on their universality (Cartwright, 1999, p. 48). It is quite obvious that Cartwright opposes Universalism of which the law is universal, not Realism. It is because, the Newtonian models of finite numbers of point masses, rigid rods, and springs, in general of unextendible, unbendable, stiff things can never simulate very much of the soft, continuous, elastic, and friction-full world around us (Cartwright, 1999, p. 48). So, it is a different question to ask, 'Do Newton's laws govern all of matter?' from 'Are Newton's laws true?' (Cartwright, 1999, p. 48). In other words, we are not against that Newton's laws are true under *ceteris paribus* conditions, but merely against its universalism and foundationalism of reduction.

Cartwright, with above understandings, thinks that view of locality could be a solution. 'Once we recognize the concept of "force" as an abstract concept, we can take different views about how much of the world can be simulated by the models that give a concrete context to the concept' (Cartwright, 1999, p. 48).

It is thus clear that Cartwright holds a view of pluralism which takes an intermediate position between Realism and Social Constructivism. This view accepts scientific laws as true in its situation which is often artificial and highly circumscribed. But it is against universal application and generally true of scientific laws. I think it is a typical view of new empiricism and closest to the view of PSP.

11.6 To what degree the new empiricism supplements PSP

In my opinion, new empiricist philosophy of science has supplemented and surpassed PSP and make implications of PSP to be more complete and reasonable in following aspects:

First, the metaphysical premises of plural empirical realism have been established. New empiricism shatters integral and systematic view of the world and advocates a dappled view of world. This lays the foundation of realism for local science. Rouse's arguments which connect practitioners and objects through introducing practices avoid the realistic problem, or to say, solve realistic problem incompletely (because realism is its implicit position). This makes his local view of knowledge lack foundation. And Cartwright publicly admits that her view has the realistic position, but it is a plural, local and dappled realism which is different from integral realism and systematical realism. And this is logically consistent with local view of knowledge. Obviously, this view provides local view of knowledge a stronger foundation.

Second, in the argument about how local knowledge extends to like-universal knowledge, the new empiricism makes local knowledge extend to like-universal knowledge by setting the condition of *ceteris paribus* to make laboratory moving from one place to another, and by admitting that law could be true in the condition of *ceteris paribus* and then could be moved from one place to another so can be extended into dominant law. This argument is more plausible than Rouse's which appeals to element of power because it is the discussion from the inside of knowledge and practice. Of course, Rouse's articulation on the relation between knowledge and power is not unreasonable. In my opinion, internal factors such as Cartwright's *ceteris paribus* and shielding play some more important role, and power is also effective. Both of them are at work and even inseparable. So, new empiricist philosophy of science is a good complement to PSP. And this is precisely why the new empiricism has surpassed PSP.

Third, the articulation of shielding conditions and ceteris paribus of law by the new empiricist philosophy of science makes us realize that: on the one hand, the natural situations in which science is applied are narrow; while, on the other hand, human beings have transformed 'natural' nature into artificial nature through technology and artificial conditions in the laboratory. At this point we can see how strong the practice is and how broad the scientific application is. Give me a laboratory, and I will change this world! It is indeed true. So, we should not only use power of science and technology, but also watch out for them, not only the abuse of science and technology, but also their evolution by self-organization.

Fourth, the argumentations of the nature of knowledge in new empiricist philosophy of science make people to realize that science is incomplete, the natural phenomena which we face with is complex the world is dappled, and laws are patchwork. There is a lot of knowledge of natures, but not knowledge of laws. So, the science and practices in different regions are different. This strongly supports the local view of knowledge.

Fifth, new empiricist does not say publicly and clearly that knowledge and power are inherent interconnected. But, in fact new empiricist had proved that science transform natural world into artificial laboratory by setting of *ceteris paribus* and rule the world by moving laboratory around. So, the one who falls behind in this transformation by science will lose power of controlling nature and society.

Of course, new empiricist philosophy of science has its problem. It pays little attention to external problem of science like relationship between knowledge and power. In fact, the justification of knowledge and the 'external' power are intertwined. Rouse demonstrates extension of local knowledge and similarity of universal knowledge by appeal to power operation of knowledge. And Cartwright demonstrates extension of knowledge and issue of knowledge's truth from the internal perspective of science such as shielding and *'ceteris paribus'* law. In fact, these two aspects are benefit by associating with each other.

Next, we will discuss application of PSP. It does not mean that the discussion about inherent problems and contents of PSP finishes, but that we have studied main contents and problems in-depth. We have also used this PSP to study Chinese knowledge and practices (including new experimentalism's and new empiricist philosophy of science), and we will show the role and effect of this application research in application chapters.

The research is endless. In fact, our research in application chapters will in turn support or criticize this PSP.

12 The starting point of scientific research
Opportunity, question, or observation?

Following the basic logic and the inference of the Philosophy of Scientific Practice (PSP), this chapter first argues that scientific research begins with opportunity, and then discusses upon that view and the other two views which supports question and observation. And discuss the significant differences between this view and the other two views and their significance. Second, this chapter gives some historical and practical cases to supports the view that 'scientific research begins with opportunity'. Finally, it discusses upon how this view could cover and integrate possible logical positions and practical positions of the other two views, thus promoting a more coherent, historical, and diversified view for PSP.

12.1 Background

In philosophy of science, the logical starting point of scientific research was a very controversial issue, and the different understanding to this is also an important boundary stone that distinguishes the different views in philosophy of science. Since Karl Popper puts forward that scientific research does not begin with observation but with question, logical positivism who hold scientific research begins with observation seems to have been unattended. And Popper's view was quite prevailing.[1] However, due to the rising of PSP, as well as SSK and feminism, a new concept on the starting point of scientific research is quietly formed: scientific research begins with opportunity. Based on PSP, this chapter makes comments on that new position that 'scientific research begins with opportunity' by means of case study and discusses upon the differences among the three views in philosophy of science and upon the rationality and significance of the new position.

First, we argue for the view that scientific research begins with opportunity, and then discuss upon the significant differences between this view and the other two views.

Second, we give historical and contemporary cases which support this view. I will give three cases to falsify that 'scientific research begins with question'. Because according to Popper, though we only need to find one counterexample to falsify this universal affirmative judgment, three different examples will increase the strength of our argument on this issue. The first case, which is derived from the work of Rouse and Pinch et al, is that scientists have historically studied the formation of solar neutrinos and thus forms a chance. The second case comes from the complexity research of SFI, whose works upon economics and artificial life are also caused by current society and research institute that are eager to seek opportunity to attract funding. The third case is a doctoral student's work on the micro-insect flight status in Tsinghua University. It shows that the starting point of this study is closely related to research opportunities which are provided by research resources, for instance, instruments and equipment etc. And the research opportunities restrict and modify the objectives and projects again and again.

Finally, we discuss upon how that view covers and integrates the possible logical position and practice of the other two views. And thus, we promote a more coherent, historical and pluralistic view of the starting point of scientific research based on the PSP.

12.2 The significant difference between starting from opportunity and from observation or question

Does scientific research start from opportunity, question, or observation? Making distinctions between these questions are not insignificant.

It is known that the view of 'science starting from observation' has been always supporting the accumulated view of science. And it is the cornerstone of the dichotomy between observation and theory. Since the idea of 'observance permeated by theory' or 'experiment loaded with theory' is put forward, observation and theory cannot be divided thoroughly. Therefore, 'scientific research begins with observation' is equivalent to 'scientific research begins with the observation permeated by theory', and further is equivalent to 'scientific research begins with theory'. Therefore, the view that 'science begins with observation' has been greatly challenged.

Since Popper put forward that 'scientific research begins with question', this view was regarded as an important concept supporting critical rationalism. However, after the concept of scientific revolution hold by falsification was challenged, the idea that science begins with question escaped from the challenge. It is still believed that scientific research begins with question. In fact, since people have not found a more appropriate view of the starting point of scientific research to replace the view that science begins with a question – and this view is better than the view that science begins with observation to fit in the representational view of science – starting with a question has not been questioned by the philosophy of science, which is dominated by representationalism.

In fact, Popper's argument that scientific research begins with question is not conclusive. His argument is based on the intuition that, in his view, if observation was a starting point for scientific research, it cannot answer the interrogation that 'what do we observe on earth'. He still regards the question as an important part of the mode of knowledge accumulation. However, just by the question that science will study, it is impossible to imply that question must be the starting point of scientific research. Popper does not give an argument to this query. Of course, in the view of theoretical representationalism, if the proposition of observance permeated by theory is not only set up but universal, once the basic status of observation has been shaken, there is no need to defend that scientific research begins with observation.

However, from the point that seeing science as an activity, what can be the direct cause of the present scientific research? Must it be a question? A lot of facts prove that this is not the case. In the logic of the activity, neither is the incompatibility that should not exist in the scientific theory, nor is the conflict between theoretical expectation and experimental result.

First, from the perspective of PSP, theory is not a seamless network of faith, otherwise there will be no conflict; and conflict is often seen as normal things by scientists, rather than as intolerable things by our theorists of philosophy of science. Theory is not our 'world picture', but a wide range of presentation and operation. Therefore, even if there is a conflict between experiment and theory, it may only conflict with theoretical representation. Scientists do not see this conflict as a big thing in practice. In the history, as scientists had shown in the case of finding 'ether', scientists had continued seeking what they believe as the traditional 'transmission media' for electromagnetic waves. But before the transmission media had been found, and the Michelson-Morley experiment gave 'zero' results, scientists were also very calm, and calmly accepted the results. Therefore, when incompatibility cannot be solved, it was not completely intolerable for scientist, contrary to the opinion hold by traditional philosophy of science.

Secondly, not all the problems that have emerged as questions can constitute a scientific research issue, especially those potential problems and the problems that have not been aware of, which mean 'nothing' for scientists. And even if the problem came into the field of the study vision of scientists, if there was no chance, ability and resources to support it, it would not necessarily become the issue which scientists currently study. In other words, if there was no opportunity, scientists would not be aware of the question. If there was no opportunity, even if scientists were aware of the question, they would not carry out real research.

In the view of PSP, it has more significance for philosophy of science to make the distinction between opportunities and questions.

Representationalism holds that in the accepted theory it is necessary to determine how the question is made of, including the exploration of implied incompatibility, the construction of new experiments to verify the connotation of theory, the assessment of the theory that are not clear in the field. These are the theoretical tasks that need to be distinguished from practically determining which issues are worth using existing resources for research. In the view of representationalism, the former assessment is a normative assessment that is worthy of study. The latter assessment is a practical assessment, which although is subject to a local context, and it is not worthy studying because the scientific view only considers the normative question. The practical question is dominated by the theoretical vision.

However, in PSP this distinction is wrong. The concept of opportunism breaks this difference. What constitutes a research opportunity cannot be differentiated with the thinking about the existing local resources and the need. There is no abstract research opportunity that differentiates the specific situation from which the opportunity is created, and thus there is no question which is independent from the context in which they are addressed. And not all theoretically identifiable questions can constitute research opportunities. If there was no one to study them, whether due to a lack of resources, interest or collaborators, or because there seemed to be no solution, then these questions would not appear in our current research (Rouse, 1987, p. 88), not even in our field of concern.

This evaluative concern for the research opportunity is the circumspective concern in the sense of Heidegger. In this kind of circumspective concern, people are of great concern to the current resources and knowledge which is present at hand. For example, what are the existing available achievements? What extent can we make innovations based on such achievements? What are the available tools and techniques, and what extent of progress can our conditions make because of these? Moreover, our assessment is still practical and interventional. Coupled with the requirements of the system, we often think about the possible scientific research progress from our own local situation and the impact of competitive environment. Thinking about the routine scientific research and groundbreaking research around us, researchers doesn't just like this to evaluate and make progress of research from their local situation and the skills and resources they have grasped. The textbook of scientific methodology discusses four basic principles – rationality, innovativeness, necessity, and feasibility – which relate to how to acquire the principles of scientific questioning (i.e. acquire normativity). The last principle is the feasibility, which is talking about the assessment of opportunity, resource restriction. Thus, while we are talking about that scientific research begins with question, we do not realize it with problem. We admit that we must evaluate the opportunities that constitute a problem at first, that is, a feasibility assessment, but we do not set it in the basic position. In fact, we have destructed the Popper thesis on the basis, but we ourselves do not know it.

Now, even if we do not strongly have to criticize the view that scientific research starts from question, we can also build a weak version of the view that 'scientific research starts from opportunity'. Of course, even if the concept of opportunity we want to establish is a comprehensive or integrated starting point of scientific research, it is not mean to hold observation, question and opportunity at same time. And it does not mean that they together constitute the starting point of scientific research. Our view is that when we say scientific research starting from opportunity, to a certain extent, it integrates or contains the significance of observation and question for research.

212 *Opportunity, question, or observation?*

```
  opportunity ◄── practice1 ──► question
        ▲  ╲                        ▲
        │   ╲                       │
  practice3  ▼  observation  ►  practice2
```

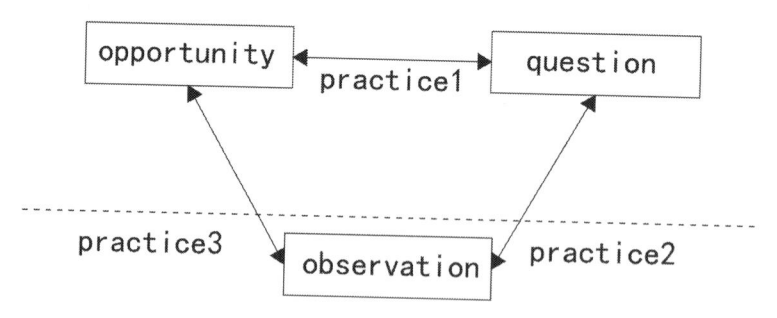

Figure 12.1 The relationship between opportunity, question, and observation.

In fact, we believe that only the opportunity truly reflects the actual scientific research. It is the resource-based opportunistic circumspection that makes some questions go into the horizons of scientists, mature enough to become a scientist's current research question, and the opportunity also makes observation become meaningful. Science is indeed to study the question, not the opportunity. But the opportunity gives question and selects the question and sharpens the question. Likewise, the opportunity makes observation to be an in-depth observation and a meaningful observation. Of course, besides the opportunity, we also need resource. 'Opportunity + Resource' creates the research approach and forms the research process. We must study what is not decided by the question, but decided by the existing conditions, which is the opportunistic circumspection based on the resource. The question only constitutes the space that we may want to study, but the opportunity and resource restrict the possibility of space, giving the actual research approach, determining the actual starting point of the research. Observation, question and opportunity together form a chain of the starting point of scientific research, forming a practically hermeneutic cycle of scientific research. Let us construct such a hermeneutic chain (see Figure 12.1).

Through the opportunistic circumspection, based on evaluating the resources controlled by our own and our peer, we can find the appropriate research project or question through practice (1). We can practice (2) more specifically through question and observe the new differences and promote the original research. Based on the development of original research, we look for the new opportunities for research through the social consultation and communication between scientists in laboratory (practice 3).

At this point, SSK scholars have given good research support. In *Laboratory Life*, by examining the scientists' conversation in laboratory, Latour and Woolgar boiled the conversation down to four types:

1 The use of 'known facts': the discussion is concerned with how long this phenomenon has been known; the role of communication is to disseminate information, which is helpful to rediscover the practice, the paper and past ideas related to the concerns of the current issues.

Opportunity, question, or observation? 213

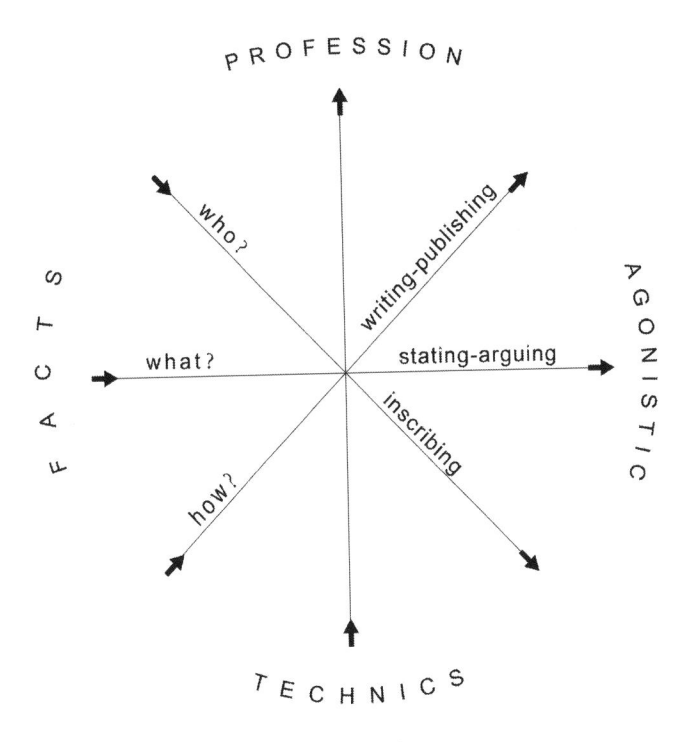

Figure 12.2 Laboratory conversation topics.

2   Seek the right approach and assess its reliability: in this attention, notice the investment amount acquired by the research team and avoid the prospect that study the artefacts.
3   The project evaluation aiming to theoretical problem also involves many problems, such as the future of the discipline and the determination to the direction of the laboratory.
4   The competitive evaluation aiming at other researchers. And these four types' discussions are often transferred from one theme to another them, moving from one interest centre to another interest centre.

Figure 12.2 expresses the view of Latour and Woolgar. (The main concerns of the laboratory conversation are: the facts that have been constructed, the individual creators of these facts, the series of claims in the process of fact making, and at last, the practical subject and the inscription devices allowed for operation.) Most of them involve the resources that the investigator may control and the circumspection for possible research opportunities (Latour & Woolgar, 1986, Figure 4.1, p. 167). Notice that the entrance of the arrow is a thinner line and is labelled as 'who', 'what', and 'how' respectively, and the opportunistic circumspections to them lead to a thicker arrow

214 *Opportunity, question, or observation?*

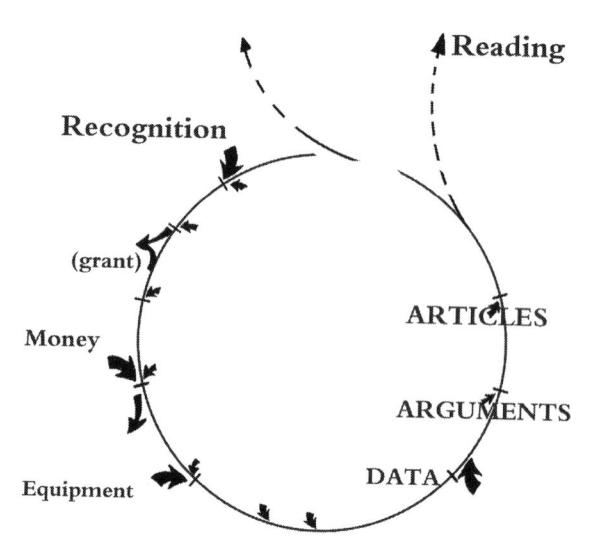

Figure 12.3 The capital required by researchers' study.

exit: 'inscribing', 'stating-arguing', and 'writing-publishing', which are more apparent outcomes of laboratory communication. Similarly, Figure 12.3 also reflects the view of Latour and Woolgar. That is researchers focusing more on research capital (resource), starting their research from the resources that may be controlled by them and the circumspection for opportunities, and often from one resource to another. These resources include recognition, grant, money input, equipment input, the output of large amounts of data, arguments and articles, as well as the formal reading, so the opportunity and resource has obtained a cycle which is like hermeneutics.

In fact, the people who are earlier than Rouse to concern about the view that research begins with opportunity are Pickering and Cetina. They all have this budding point of view. Pickering pointed out that the scientists' research strategy is decided by the creative exploring opportunity, which is made by the individuals in scientists who have different resources and contexts (Pickering, 1984, p. 11). Every scientist has his own unique resources in the face of choosing and handling questions. This includes both the aspect of material equipment and the aspect of vocational training and professional skills, which will determine what measures and what scientists will take to carry out scientific research in the face of the problems. Scientific research is always to make the best use of scientist's own resources. Just as Cetina's metaphor says,

> The cooper is an opportunist. They understand their important opportunities in specific places and take advantage of these opportunities to complete their plans. At the same time, they recognize what is feasible

and adjust and develop their plans accordingly. When they act, they constantly produce and reproduce the kind of useful production to make them successful in line with their purpose of temporary decision ... talking about opportunism in research does not mean that scientists are unsystematically, irrational, or professionally oriented in their practice ... opportunism represents a process rather than an individual characteristic.
(Cetina, 2001, p. 65)

The work done by scientists in the laboratory is like that of the coopers, where the decisive role is the resources, the skills and the opportunities, and therefore it is the opportunity rather than the question that plays a decisive role in scientific research, especially at the starting point of scientific research.

Pickering also points out that though this opportunism cannot explain all the problems in practice, it is in terms of how the new tradition grow. In the Pickering's study of disputing case, the best case that can show his 'opportunism in research' model is the newly discovered J/ψ particle in November 1974 and the arousing controversial interpretation on whether it is the 'color' or 'Taste' (Wang Yanfeng, 2005, p. 102). Pickering believes that the differences between the cultural resources occupied by the new and old physicists' faction determine the differing views on the above issues. He points out that Mary Gaillard of the European Research Center and Alvaro De Rújula of Harvard University have different academic experiences, and they can only develop because of their own major in the face of a transforming process from one research tradition to another, to find different opportunities to seize the new resources for them, and to become an active supporters of charm quarks and normative field theory. It shows that when scientists in the face of choice, on one hand, they will be inevitably impacted by the original academic background and other cultural resources; on the other, the impact of this original academic background is not entirely decisive. In the face of new context, they will also strive to make their own research resources fit the existing research. In the new context they constantly adjust their motives, their research objectives and problem context, which is a research choice dominated by opportunistic view. They will adjust continuously in the study, to make best use of their resources. In the study scientists make a war to nature taking a way of opportunism. Philosophers of science also need to make appropriate explanations and interpretations to this to help us understand the behaviours of scientists. The concept of opportunity provides a better interpretation and understanding.

12.3 The cases of 'scientific research begins with opportunity': the solar neutrino experiment, the complexity study, and insect flight measurement

The clear view of 'scientific research beginning with opportunity, rather than question' comes from J. Rouse. In interpreting the history of the exploration of the sun neutrino experiment discussed by Pinch, Rouse found that what had been the subject of research in this historical case of science depended on the existing resources and the chances of how to take advantage of them. The case is like this:

> At the time of the first proposed experiments to detect solar neutrinos (the mid-1950s), it was generally believed that the solar neutrino flux was too small to be detected with current technology. The initial impetus for such experiments came from an experimenter who had developed the equipment and techniques and had trained technicians to detect neutrino interactions. Having devoted considerable resources to develop a neutrino detector, he began to try to detect neutrinos, and he was trying to enlist some theorists to provide an expected value for the solar flux that he could use as a target. Changes in the accepted view about stellar nuclear reactions suggested to several theoretical astrophysicists in 1958 that the solar neutrino flux might be just sufficient to be detectable. By 1962, this possibility, based upon the availability of a sensitive neutrino detector, led several such scientists to collaborate on the complicated calculation of just what the expected value for the solar neutrino flux might be. What is important to recognize is that there was no intrinsic theoretical interest in solar neutrinos.
>
> (Rouse, 1987, p. 87)

Citing Pinch's study, Rouse suggests that 'indeed, those scientists whose expertise was required to produce models of the sun's inner structure and nuclear reactions "could see little point in working out the Sun in the detail required and were more interested in the late stages of stellar evolution"' (Rouse, 1987, p. 87). The measurement to the solar neutrino flux which due to the advances in research instruments was not the goal of scientists, at least not the initial goal of scientists, and scientists had no theoretical interest to detect solar neutrinos in the beginning. After seeing these, Rouse sensitively pointed out that the initial reason for the project to study solar neutrino was not a question, but an opportunity.[2]

By learning from the immature scientific research of complexity science, I find that the complexity study begins with economics and combines with the artificial life research of computer programming. It does not begin with physics, which also has strong opportunism characteristics. According to Wardrop's biography *Complex*, writing for the Santa Fe Institute, we believe that the process of establishing the research direction of complexity study in Santa Fe Institute is clearly a practical constructive evolutionary process of the complexity study, which get the chance through discussion, consultation, mutual competition and cooperation, and the integration of exterior resources.

Was the research direction of the Santa Fe Institute clear from the very beginning? No. It had a process of opportunistic exploration. When Corwin, the founder of the Santa Fe, was trying to establish such a research institution, he was just disappointed that traditional research institutions such as laboratories and schools of universities were unable to engage in interdisciplinary research and integrated studies. He was hoping that one social institution can accommodate such interdisciplinary and integrated liberal study. In the beginning, there was only a hazy idea. His ideal research institution was a new type of independent research institution. The optimal program was that this institution both had advantages of the two worlds: it can not only have the broadness of university but also the ability of Los Alamos to blend different disciplines. Most importantly, this institution must be a place to attract the best scientists – those who really knew what they were talking about in their own research field. This organization was able to provide them with far more general content than usual. This institution should be a place where senior scholars can explore their own immature ideas without being ridiculed by their colleagues, and the best young scientists can work with world-class masters to make their own work productive. In short, this institution should be a rare place to train scientists since the World War II. If this idea was only epistemological and had no sociological process, it couldn't come true. Corwin talked about this idea with his two friends, which made him put this idea to Los Alamos' weekly Chinese food seminar. It turned out that those senior researchers liked his idea. So far, this idea only emerged from the water at this point and at this place. Moreover, at this moment, some senior scholars had begun to feel hazily or deeply that it was time to study the complexity system from their research field. For example, Carothers argues that complexity theory should be the next driving force of science in the 21st century (of course, at that time, they did not have the concept of complexity theory). Some academics (such as senior researcher and astronomer Stirling Colgate) supported this motion by the consideration of raising the level of a university, and some (such as senior researcher Nick Metropolis) supported this because they felt it was important to develop computers. Scientists' ideas were constantly colliding, conflicting and negotiating in the open and private situations, and the direction of discipline integration continued to sway with the competition between scientists' ideas.

The most critical and most fundamental question was: what should the research institution study? Different scholars thought differently. Some scholars tried to make this research institution a research centre studying for large or even giant computers. Some scholars believed that the research tasks may be much broader and more targeted than this. The subject of the study was from the vague 'new way of thinking' to begin to clear gradually through the debate of scientists. Several seminars had further inspired everyone to move towards a clearer view. First the view was the integration of disciplines, then was 'chaos science', and then was 'emergent' science, and finally it was determined to the 'complexity science'.

After the theme of the study was identified, how to intervene in the complexity research practice can be considered as a real beginning of the study. This was still a question. It was directly related to the starting point of scientific research. Was it a question of complexity? Despite having the theme of complexity study, the study was still unable to start and develop. Perhaps having both a dramatic opportunistic feature and a path dependence, such study for complexity or complex system was impossible to conduct without substantial financial support. However, when the institute was looking for support here and there, some large companies also had a headache for economic forecasts. The economic system was undoubtedly a complex system, and the system was often run far beyond the predictive ability of classical economists. It was not because that the ability of these economists had problems, but because that their economics was based on the simplicity of the over simplified model. So even with the support of giant computer, those models cannot get the economic results of the emergence. One of the chief executives of a giant company (Citibank president) wanted to get different economic inspiration. And the Santa Fe Institute was also looking for sponsors, wanting to make something different from the past to stimulate the research of the sponsors – this factor that was seemed as opportunity by the traditional philosophy of science seemed to have played a decisive role here – This kind of research did not require concrete results, if had some different new ideas. It had no time limit. If the Santa Fe Institute could start the study and make progress year by year, it would have been enough. After the two aspects of the need were combined, both sides felt deeply challenged. This economic need stimulated the work of Santa Fe Institute and promoted the work of the Santa Fe Institute, letting the dialogue between economists and physicists became the third important symposium since the establishment of the Santa Fe Institute (September 1987), in addition to the first two symposia focusing on the development of new disciplines integration. It was the opportunity of the contact with the Citibank president that led to this seminar and pushed Arthur's 'increasing returns' new non-linear, complex economics ideas into the front, also pushed the idea of 'adaptability creates complexity' into foreground, which was contained in the report of 'the global economy as adaptive process' made by John Holland, who studied the concept of complex adaptability for 25 years. The symposium sparked a heated debate, and the two sides were swaggering. But apart from the fact that physicists and economists were incompatible with each other and did the different work with each other, they found that there were a lot of common talk between them and had some consensus. The path dependence, locking, non-linearity, self-reinforcing mechanisms, QWERTY keyboards, possible inefficiencies, the origins of Silicon Valley, and so on in Arthur's 'increasing returns' and especially the Holland's complex adaptive systems were also gained the consensus of physicists and economists.

In view of the need of economics, the complex adaptive system raised by Holland's report has much like economic paradigms, since many examples and analogies used in it are economics. But this does not cover his main view on complex adaptive systems. That is: first, every system like this is a network composed by many 'actors' that act in parallel. Each of the actors will find that they are in a systematic environment formed by the interaction of themselves and other actors. Each of the actors acts and changes action continuously based on the movements of other actors. In this system, there is no centralized control. Second, a complex adaptive system has a multi-level organization, each level of the actor playing a role of building bricks for the higher level of the actors. And complex adaptive system can learn lessons, which often improve and re-arrange their work of building blocks. Third, all complex adaptive systems will expect future. This expectation is based on the models of assumptions that the inner world knows about the outside world. Fourth, the complex adaptation systems will always have many habitats, and each of such habitats can be used by an actor who can adapts himself to it and develops in it. Moreover, while each of the actors fill a habitat, it will open more habitats, which opens up more living space. Therefore, since it is always in constant development and constant changing, this system will never be able to achieve a balanced state, the matter that Holland went to the institute to work also changed the original 'complexity system' study of the Santa Fe Institute, which turned it into a 'complex adaptive system' study. Because Holland made Murray Gell-Mann and others suddenly realize that their research plan had a big omission: what on earth are the emergence structure doing? How do they respond and adapt to their own environment? Holland didn't be discovered and became the plan of the Santa Fe Institute until he was 57-years old, after he had studied 25 years on the concept of adaptation.

What the above example tells us?

First, the needs of economic circles and economic strength are fit in with the desire that the institution need to be founded urgently. It determined that the Santa Fe Institute cannot carry out complexity research from basic theory, but only from the economic complexity first. Second, the external needs of the research project made two people (Arthur and Holland), who engaged in non-mainstream economic research and computer research gain an excellent opportunity to play their talents: one of them is a non-mainstream economist; the other is a computer expert, while studying complex adaptive issues in long-term. Third, their research, in turn, accommodates the needs of the current economic community, and the reputation of the Santa Fe Institute has been flourished. And more importantly, the Institute has received more and more funding from it, thus gaining more good research opportunities. From the above ideas and the process of activities, we see that in fact the scientific knowledge generated by specific research have these local characteristics. None of the knowledge can be separated from the specific occasions and specific people. They all need incentive opportunities. But for Holland, perhaps the complexity study in Santa Fe is not a complex adaptation system, but only a complexity system.

Knowledge is not only like a stream, but also has the convergence effect of opportunism through practice, social consultation. The most important factor needed for this convergence is the excellent local researchers and research opportunities, which provide interesting and important questions that can be grasped.

Since the complexity study has a vivid biography, that is, Wardrop's *Complexity*, thus leaving us a practical research paradigm which have the significance of SSK and the significance of PSP. Through the study of the concept of complexity, the study of the thought and the study of the history of the characters, we recognize the local characteristics of the knowledge and the characteristics that science begins with the opportunity, and specify the path dependence and practical dependence on opportunities and resources.

We also have an actual representative case of scientific research beginning with the opportunity rather than the question. Since this case is an ordinary doctoral thesis research, it has more extensive and universal significance. In this case, the research objectives have changed successively, and the reality condition restricts the research. Contrary to all expectations, the new experimental method has also been created through practice. It is the existing resource of Tsinghua University and the research opportunity which researchers can grasp play a decisive role. There are a variety of scientific and technological researches and engineering frontier researches in the department of precision instruments in Tsinghua University. In 2005, I encountered a doctoral student who studied the optics engineering project, and his doctoral thesis studied the measurement of insect flight. The initial goal of this question was to do bionic research for tiny aircraft research. Usually we are surprised at the complex flying skills of insects when we see their flight. So how does the insect fly? This question does not constitute a scientific problem. Because even if it is a question, we cannot study it when we don't have the measure tools and instruments. Why we interest in flying insects? The reason is that we want to use insects to study tiny aircraft. However, we only know in our daily experience that tiny insects can fly, hover, and suddenly turn, but we don't know the dynamics of these movements. Those cannot form research questions before the suitable instruments are available. So initially, the study is just based on some vague ideas. In fact, the initial purpose of the researcher is ambitious. He not only wants to acquire insect flight data systematically, but also wants to propose a set of tiny aircraft theory based on the data. However, this goal has not been achieved after a practical attempt, but he has reached the purpose of establishing a new measurement system for the initial insect flight (which of course is quite remarkable). The main reason is that we don't have any theory to the flight of insects, such as bees flying, except experience intuition, and the most important reason is that we even don't know how to measure the free flight of insects. The opportunities that constitute the study to the flight of insect in the space depend on many factors. Whether have a better observation instrument that track flight is one of the most important bottleneck factors. Therefore, how to build a set of equipment for observation and accurate measurement of insect free flight has become an important prerequisite for the study of insect flight. At last, the author of the thesis completes a hybrid system which composed by two subsystems: the measurement system of magnetic sensor coil and image tracking system. Through the accomplishment of this

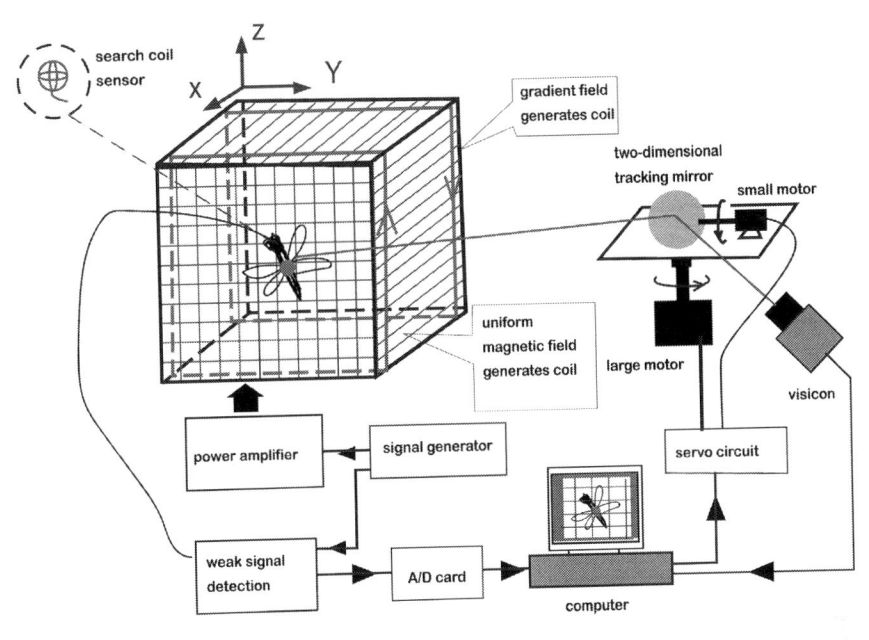

Figure 12.4 The structure chart of the hybrid tracking system of a magnetic field and image.

system, the researcher tracks the bears flying (hover, direct flight and turn) posture preliminary and obtains several related data. The researcher also carries out experiment on the bumblebee successfully through various attempts, that is measuring the flapping and torsion angles of the bumblebee through bonding the micro sensor coil on the wings of bumblebee (see Figure 12.4; Cai Zhijian, 2005). Finally, due to the success of this method, that is, the successful practice of the method of bonding coil, an opportunistic work that just gives it a try becomes an important part of this research. So, in this study, the opportunism study of which Popper's trial and error practice constitutes the core of the research work because of the success of it.

There are still some cases that the opportunity determines the success of scientific research. For example, Rouse re-examines the construction process of the DNA model by the view of the discovery of the opportunity. It indicates that several different local academic resources, the study of research vision cause the difference of the sensitivity of discovering the opportunity and the identity of meaning. These thus result in the case of discovering DNA model (Rouse, 1987, pp. 89–90).

The above cases at least falsify the opinion that scientific research begins with the question if it is a universal affirmative proposition. The above cases also at least explain the existence of the situation that scientific research begins with the opportunity, which is better than the other opinion.

Opportunities can fuse other factors, such as questions and observations, and make scientific research be able to re-downward-to-earth.

12.4 The significance of 'scientific research begins with opportunity'

The significance of this view for scientific research is that: scientific research is a fighting for the unknown world, and a collective social competition and cooperation process. The scientific resources controlled by each of the scientific research institutions and organizations are or different (not only the amount but also the quality). There are also differences on research experience, ability and practice for each of the researchers working there. Therefore, seeing from the process of scientific research, we are sure that practice has the locality, temporality, and opportunity. The researchers should not only understand and evaluate them, but also find a variety of possible opportunities based on this locality sufficiently. Opportunities create a colourful feature of the study and provide a practical research approach. Researchers can distinguish the degree of scientific ability and research cultivation through the grasp of opportunistic circumspection.

The significance of this view for the philosophy of science is that: it provides a critical reflection on the view as 'scientific research beginning with question' of fundamentalism and representationalism. Although 'scientific research begins with question' is better than 'scientific research begins with observation', that replacement is still repeating the fault on the level of logic positivism. The view that scientific research begins with the question still recognizes scientific activity as serving for proposing a super theory, which is basically consistent with the philosophy of science which think of theory as some objective with priority. In fact, from the point of view of scientific research, with questions to the accumulated view of scientific development, or critical rationalist view of science, or Kuhn's view of paradigm shift, there is a shadow of absolutism which reflect the neat characteristic of modernism. And the view of 'scientific research beginning with the opportunity' puts the development of scientific research in a local position, and promotes the development of knowledge opportunistically from a variety of different laboratory activities. It is not only closer to the real scientific research process, but also embodies the implication of postmodern thought because it promotes diversity and decentralization. The view of opportunity makes the process of scientific research more reasonable and more realistic, and thus makes scientific research more understandable.

In short, it provides a new understanding of the diversity of scientific research and the diversity of science and culture.

From the next chapter, we will apply the basic corollaries and the results obtained by using the basic perspectives of PSP to discuss the most important issues of scientific research, which is the relationship between observation, experiment and theory. This is one of the meaningful studies of the New-experimentalism and an important basic work for the PSP.

Notes

1 At least in the traditional philosophy of science few people doubt upon this point until now, which even becomes the standard view in Chinese philosophy of science and natural dialectics' textbook.
2 For the sake of clarity, we quote the words of Rouse as follows: 'the original reason for the project was not a problem but an opportunity' (Rouse, 1987, p. 87).

13 A new solution for an old problem
The relationships of observation, experiment, and theory

Observation, experiment and theory, which one has some priority over the other two? Or none of them enjoy privilege? In traditional philosophy of science, observation and experiment are equal to each other, and both contribute propositions of observation statements to establishment and development of theory. For new experimentalism, there is still not instinct different between observation and experiment. For example, Ian Hacking's argument, 'experiment has a life of its own', merely indicates that observation and experiment are more basic than theory, remaining a position of traditional representationalism. However, since the rising of PSP, experiment has been emphasized as direct scientific practice rather than indirect one like observation. Standing on practical philosophy, we will re-examine the three elements relationship from the perspective of history and practice, placing them in their practical processes, and discuss problems like the role of observation at the research start point, intended / unintended observation, the role of experiment among candidate theories, unique role of experiment, probing experiment without accepted theory, and so on. We attempt to show that only intertwine of life between experiment and theory can be the truth of scientific practice.

13.1 'Theory-ladenness of observation'

The proposition 'theory-ladenness of observation' proposed by Hanson, which gave the fatal blow to the logical positivism and played a pioneer role in historicism, has broadly become the new received view in philosophy of science. In China, it is treated as standard viewpoint covering philosophy of science by most of textbooks for postgraduates. However, by the 1980s and 1990s, the rise of new experimentalism (Hacking, 1983; Franklin, 1986, 1990; Galison, 1987; Gooding, 1990; Mayo, 1996) in philosophy of science challenged this proposition again. 'Experiment has a life of its own' and 'experiment has multiple lives of its own' have been proposed. However, perhaps there is great space left for logical positivism, these slogans were put forward so suddenly that somewhat radical apparently. Meanwhile, as people were not ready to deal with the problems concerning these themes, challenges from new experimentalism were temporarily shelved. By the 20th century, due to the laboratory practice research of SSK, the hermeneutics of scientific practice, and the new experimentalism, scientific experiments and practices gained attention in the trend of practical return leaded by PSP, and the proposition 'theory-ladenness of observation' get to be re-examined and corrected.

It is important to re-examine the judgment 'theory-ladenness of observation/experiment' as a universal affirmative proposition. Its importance rests with not only the criticism on the viewpoints of logical positivism, but also the construction and expansion for those of the PSP.

This proposition once was a strong opponent of positivism and became the backbone of the important theoretical perspectives of the later historicist philosophy of science. Once accepted, the relation between observation/experiment and theory no longer exists. However, if they integrate into a triune mixture, problems about the progress, comparison, and truth of scientific theory will become perplexing because of the lack of necessary and reasonable foundation. Regaining the status of observation/experiment, especially the latter, as the requisite and appropriate foundation of theory and providing it with adequate argumentation of the cases of history of science, are of great significance for rectifying the extremely theory-dominated viewpoints of philosophy of science, resetting the viewpoints of experiment and theory as well as the complex dialectical relation between scientific accumulation and mutation.

13.2 A strategy for new experimentalist criticism and the implication of an 'experiment has its own life'

The new experimentalist philosophy of science does not intend to return to the stand of positivism but seeks to find a relatively reliable basis for scientific research from experiment to criticize the viewpoints of the extreme theory-ladenness.

Let us discriminate the perspectives of the so-called theory-ladenness of observations or of experiments first. It is a perspective as such: observation of x is shaped by prior knowledge of x. The languages or symbols we use to express what we know also affect observation, but without them, there could not be what we recognize as knowledge. Nevertheless, as the later development of philosophy of science, this viewpoint of 'theory-ladenness' is magnified and enhanced. The consequence of this magnification is that the previous consciousness which all observations relied on is 'theories'; the use of ordinary language also becomes the evidence of 'theory-ladenness'. The enhancement results in such a consequence which observation/experiment definitely depends on or is loaded with strict theory. This means that the ambiguous proposition of 'theory-ladenness of observations or of experiments' has different versions: a weak version and a strong one. The former just claims that there must be some concepts and instruments about nature prior to the leading-in of observation/experiment, while the latter believes that observation or experiment is preceded, influenced or loaded by more systematic or more systemized theories (actually, the degree herein is still different).

The first critical strategy of new experimentalist towards the 'theory-ladenness' is to oppose to the strong version instead of the weak one. However, as pointed out by Hacking, the weak version of the 'theory-ladenness' has enlarged the theory. We cannot go without language-load or some of our vague concepts and faiths which have been attained during our growth, but they are absolutely irrelevant with the observation/experiment. We need to apply instruments and equipment previous to specific observation/experiment, but this does not imply that theories, especially those relevant and specific ones, must be loaded. Our definition of theory is primarily very strict in the traditional philosophy of science, so why not preserves the strictness but arbitrarily extends 'theory' into something of spirit, thought, and beliefs when it comes to the relation between observation and theory? Therefore, in the sense of the weak version, everything that is loaded with previous experience lacks its meaning, since there is no difference between whether it is spoken out or not.

The second basic strategy is to illustrate that there exist theory-free experiments and observations and even meaningless observations through massive case studies on experiment and observation, in order to demonstrate that observation and experiment have 'its own life' independent from large-scale theory (Hacking, 1983, p. xiii). This success of this strategy is that the proposition of 'theory-ladenness of observation/experiment' is a universal judgment. Then, according to Popper, we can falsify the proposition only by finding a counter-example, which is the very practice of new experimentalists. In the actually process of proof, new experimentalists ask whether every observation is loaded with theory. It's actually unnecessary. Hacking, the herald of new experimentalism, found many examples of observation and experiment that prior to theory, proving that experiments and observations that precede theories do exist and some observations are completely anterior to theories. For example, in 1827, a botanist Robert Brown reported the random motion of pollen suspending in water which is the phenomenon later called 'Brownian movement'. The Brownian movement was discovered 60 years ago and even considered to be the life movement of living pollen. For a long time, Brown made this painstaking observation but had no explanation for it. Researches did not simultaneously happen until the first decade of the 20th Century. Experimenter such as J. Perrin and theorists such as Einstein have showed that the reason for pollen's random motion is that the pollen is flicked by the molecules around it. This typical case indicates that Brown and his predecessors who were just curious about the phenomenon did not know the pollen molecules' movement which was later named 'Brownian movement' or its reason, and their thought about it, if any, is merely pointless speculation.

There are many other similar cases. For instance, Galileo was not directed by Newton's Theory of Gravity when he pointed his telescope at the sky;

Humphry Davy accidentally found that green alga in river and lake, when exposed to the sunlight, could generate a gas which was inflammable in the later experiment, but did not know the property of this gas, let alone the related theory; many observations of C. Herschel in astronomy were nothing but observations, which is also the case with W. Herschel's observations and experiments in radiant heat. Gooding (1990, pp. 135–62) analyzes Faraday's whole course of discovering the electromagnetic interaction and manifests the reconstruction of Faraday's case by the means of symbols and drawings to indicate that experiments need not to be loaded with accurate theories. These show that some observations need not to be loaded with theories or accurate and strict theories.

The third strategy for criticizing 'theory-ladenness' is to prove that the doctrine of 'theory-ladenness' by no means implies that all reports of observation must be loaded with scientific theories, even though it is an important and airtight proposition about ordinary language (Hacking, 1983, p. 172). New experimentalism emphasizes opposing the viewpoints of the strong version. For example, by reinterpreting this invention, it demonstrates that Faraday's motor effect does not need exhaustive theoretical knowledge. Therefore, the opinion that all the experiments in the sense of the strong version are 'theory-ladenness' is wrong. New experimentalists prove that there exist some experiments or observations that are anterior to theory and loaded with no theory. Certainly, this way of cases explanation just explains the existence of some XS at odds with the proposition of 'theory-ladenness' rather than the non-existence of experiment or observation of 'theory-ladenness'.

The fourth strategy is to illustrate that some experiments need not resort to high-level theories, otherwise, they will cause or evoke new high-level or wide-range theories. Galileo, when pointed his telescope at the sky, did not grasp the theories about the moons of Jupiter or the test that the theories about the moons of Jupiter should be experimented or observed. On the contrary, his new discoveries provided later development of theory with support of phenomenon and experiences. Buchwald made a detailed research on Hertz's experiment of cathode ray. Now, we generally take Hertz's experiment as a mistake because it was conducted in Hertz's context (namely, with the wrong theory-ladenness). Actually, Hertz could not see the electric field's deviating effect on the cathode ray due to a deficient arrangement of instruments. By lowering the pressure in the tube and arranging electrodes more properly, Thomson found the effect of the electric field. Therefore, experiment does not correspond to theory one to one; it is dispensable to turn to a higher level of theory for solve problems; and new effect can be found even by adjusting instruments. New discoveries can be made just via the exhaustive knowledge on the effects of instruments. According to Deborah Mayo (1996, pp. 294–5, 442), experiment knowledge can be reliably ascertained through a series of technologies eliminating mistakes and error statistics instead of the support of high-level theories. This denotes that the data acquired at experiment level is partly independent from the strict high-level theory relevant to the interpretation of the data's meaning. This also demonstrates that experiments using instruments have their own life of different varieties.

13.3 New experimentalist achievements and shortcomings in understanding the relation between observation and theory

We have witnessed that traditional philosophy of science, such as logical positivism, has been in a dilemma because of rigidly drawing a clear boundary between observation/experiment and theory, and embarking on scientific rational reconstruction with experience as criterion. We have also seen that the historicist philosophy of science has not only blurred the difference between observation/experiment and theory, but also brought the origin of relativism, which is the consequent of breaking the boundary between observation/experiment and theory, as well as expounding with the viewpoints of 'theory-ladenness of observation/experiment'. New experimentalism refutes the doctrine of the 'theory-ladenness of observation/experiment' by successfully pointing out the existence of some theory-free observations and experiments through empirical observation and experimental cases. It also applies successfully reduction to absurdity during the proof process and casts doubts on the explanation on this universal proposition.

Nevertheless, criticism of new experimentalist on the proposition of' theory-ladenness' is right in the middle position. It approves the division between observation/experiment and theory but disapproves complete division. Thus new experimentalism seemingly situates in an embarrassing state. It cannot assimilate the advantages of logical positivism and those of historicism while expelling all their drawbacks. This is the first problem of new experimentalist.

Its second problem is related to the first one. If the relation between experiment/observation and theory must be tested one by one and understood in specific contexts, we can by no means have a general standard understanding upon experiment/observation and theory. New experimentalism must be fundamentally against essentialism and belongs to contextualism. Its research approach needs to largely interfere with case studies on experiments and observations to indicate that some of observations/experiments have no 'theory-ladenness' while some do. This argument is certainly tiresome and cannot provide a universal opinion. However, at least, it can tell us to be extremely cautious and to carry out contextual test when research the relation between observation, experiment and theory.

In the criticism of new experimentalist upon the viewpoints of 'theory-ladenness', we see its effort to avoid from retreating to positivism while expressing the view of progressive accumulation and development of scientific knowledge. Even though this view of progress is based on the accumulation of experiment knowledge and different from the previous viewpoints of positivism, the progressive and accumulated view of science still needs a large number of explanations and interpretations. In this regard, new experimentalism does not give a conclusive answer.

Now, new experimentalist philosophy of science facilitates us to reset scientific experiment as the foundation for scientific development, and to gain a new knowledge about the concept of scientific practice. The laboratory culture study in new experimentalism and SSK converges with the hermeneutics of scientific practice, just like the contemporary opposition to logical positivism. It is still an important research approach of PSP, despite its many shortcomings.

13.4 The evolutionary relation between scientific experiment and theory

We are now going to put forward our explanation on the relation between 'experiment/observation and theory' on the basis of new experimentalism. We think that all the problems of the traditional philosophy of science may result from that it originates from a theory-dominated view of science. This is fundamentally incorrect, while if it is reversed, the view will become practice-dominated. Then many problems of traditional philosophy of science may vanish.

First, if we regard science as a kind of practice and a complicated cognized activity based on experiment/observation, we will have intention which related to research and confused by some unsolved phenomena from the perspective of unknown phenomenon of a certain concealed realm. Being similar to a lion which roars only when 'its tail is wrenched', as said by Bacon, nature also talks, so we will find that we can know its character and behaviour in its practice of 'speech act'. Hence, it occasionally happens that we get to know one thing because we are researching another thing. For example, Röntgen accidentally found X-ray when studied cathode ray. Therefore, observation sometimes precedes at the starting point of research. Even though we can know that problem is the obvious starting point of researches, the emergence of this explicit problem is relevant to long-term negligence or observation on other matters.

Second, we neither totally split observation/experiment and theory, nor mix them even according to the stance of new experimentalism. Even though our stance seems to be contradictory and embarrassing, the conflict and embarrassment on the surface will immediately disappear if we treat the relation between observation/experiment and theory as an evolving and active process. This kind of stance can be free from essentialism, change during the process, and adopt the viewpoints of pluralism of evolution process, since pluralism can be reconcilable for the process. For this reason, we hold that there is a flowing boundary between observation/experiment and theory. The boundary, neither identical nor unified, should be analyzed specifically according to different and concrete observations/experiments and theories as well as its specific circumstances.

Therefore, we think that the extreme theory-dominated philosophy of science has lost the expounding function of observation and experiment which is the most important part of science. Just as Chalmers (1999, p. 206) said, 'new experimentalism makes philosophy of science come down to earth again and proceed on a valuable road. It well rectifies the theory-dominated research approach'. In fact, observation, experiment, and theory are different in a strict sense. Even from the perspectives of linguistics and semantics, their different names have already implied that they are different things in human cognition.

There are complicated and diverse relations between observation/experiment and theory. Relation between theory and experiment varies from different development stages and sciences. Only by adopting evolutionary, procedural and multifarious viewpoints, can the true relation between observation/experiment and theory be reflected correctively and comprehensively. The relation between experiment which has its own independent evolution and theory which has its own life is tangling together and co-changing. This is our tentative conclusion.

We propose that the Chinese circle of philosophy of science should draw on the research findings of the hermeneutics of scientific practice and the research approach, ideas, and achievements of SSK, in addition to new experimentalist research, in order to carry out a new systematization of the history of science and technology on the one hand, and to probe into local experiments and laboratories for case studies on observation/experiment on the other hand. It is also recommended to align with natural scientists and engineering technologists. Critical research should be conducted to make rational and appropriate description so as to take experiment as science's proper basis.

From the hermeneutic prospective, there are also issues about hermeneutic cycle of experiment in the research on the purports of experiment practice, which we will dedicatedly discuss in the next chapter.

14 New studies on replicability of scientific experiments

Experimental replicability is considered to be self-evident and indispensable in the traditional notion of science. The principle of replicability is one of the important criteria of scientific fact and the universality of science. That principle is analyzed in depth by both SSK and New experimentalism who pointed out shortcomings of the traditional views on this issue. However, due to the difference of research methods and epistemology between SSK's and New experimentalism, their conclusions are not the same. From the point of PSP, this chapter discusses and compares both approaches of research on experimental replicability and then puts forward some comments.

14.1 Traditional views and challenges

Many issues in scientific practice are dominated by experiment rather than theoretical studies. There are many criteria of theoretical evaluation in those studies. For example, in *The Essential Tension,* Kuhn (1977) listed five criteria: accuracy, consistency, scope, simplicity, and fruitfulness. Moreover, besides criterion of evaluation for theoretical studies, there are also evaluative principles for experiments. And replicability is one of the most important principles.

In general, experimental replicability means that both processes and results of experiment are repeatable, and that a successful experiment does not vary with different time, space, and subjects. Replicability is one of the criteria of a successful experiment. It is accepted as a general principle by both scientists and philosophers of science. Only when certain events recur in accordance with regularities, as is the case of repeatable experiments, can our observations be ascertained successfully. We do not take our own observations quite seriously, or accept them as scientific facts, until we have repeated and tested them.... Indeed, the scientifically significant *physical effect* may be defined as that which can be regularly reproduced by anyone who carries out the experiment in and appropriate way.

According to the traditional view, experiment is a mere handmaid of theory (Gooding, 1990, p. 209), for its primary function is to provide support or falsification for verification of theory. Scientists often appeal to the replicability of experiment in order to defend the validity of their claims for new discoveries to get peer recognition. Replicability of experiment is the most important standard especially when the results of experiment is in doubt and questionable. Replicability indicates the factivity and reliability of scientific knowledge, of which the generalization in the Mertonian sense of normativity is unshakable.

Within the development of natural science, principle of replicability has faced with serious challenges in such as Complexity and Postmodern Science (Griffin, 1988). In Complexity science, events are sensitive to their initial states which is very hard to repeat initial states accurately in the experiment. And David Griffin (1988, p. 27) argues, 'science requires repeatable experiential demonstration, it does not require one particular type of demonstration, such as the laboratory experiment'. In his opinion, traditionally experimental methods reflect assumptions of Materialism and Reductivism to which Postmodern Science opposed. Especially, philosophers of scientific experiment (Hacking, 1983; Gooding, 1990; Radder, 1992, 1996) and sociologists of scientific knowledge (Shapin & Schaffer, 1985) take very different views of replicability.

Is the principle of replicability valid? Does it threaten the certainty and generality of scientific knowledge, if it does not function like a criterion for scientific fact? How is experimental replicability characterized? Based on PSP, we will discuss different ideas on experimental replicability in both philosophy of scientific experiment and sociology of scientific knowledge. Then an argument will be given about how to characterize the principle of experimental replicability.

14.2 The SSK notion of experimental replicability

The issue of experimental replicability is an important part in SSK laboratory studies. For instance, H.M. Collins's *Changing Order*, Steven Shapin's *Leviathan and the Air-pump: Hobbes, Boyle, and the Experimental Life*, and Andrew Pickering's *Constructing Quarks: a Sociological History of Particle Physics* are all outstanding studies. *Changing Order* is a representative of Collins's view of replicability. In this section, our discussion is mainly based on this book, while the others are also mentioned when necessary.

Scientific discovery has been originally some personal knowledge of scientists. It is the replicability that makes it a new component of public. Collins (1985, p. 18) firstly describes what is the questionable replicability: 'The discovery, if it is to be a discovery, must participate a new set of public rules – a new set of ways of "going on in the same way"'. In the traditional view, scientists will often defend the validity of their claims by the replicability of observations and experiments. In common understanding upon science, replicability or replicability is a 'touchstone', a Supreme Court for scientific system. Scientific discoveries must be strictly tested to be recognized by all.

The next question would be to what extent a scientific discovery or experiment is considered to be repeated? Collins thinks that scientists tend to judge

Replicability of scientific experiments 233

the adequacy of an experiment through its result. (Collins claims this in the case study of TEA Laser.) Hence, repetition of experimental results is considered to be the key of the whole replication. Collins proposes an 'empirical' sorting schema in *Changing Order*. He points out that replication of experiment is often thought of as determined by completing each step of this schema. The schema is:

> Level one: eliminate all activities not to do with the subject of 'r'.
> Level two: eliminate all activities that are not scientific.
> Level three: eliminate all activities where the identity of the experimenter is inappropriate.
> Level four: eliminate all activities that are not experiments.
> Level five: eliminate all experiments that are not competent copies of the original.
> Level six: divide the remainder into those which are positive and those which are negative.
> Level seven: decide whether 'r' has been replicated.

However, we could find many problems of replicability if we step by step by following this 'empirical' schema. We will find it at each level unable to provide clear demarcation which allow us to move on. Collins (1985, pp. 39–46) reflects all of those steps and his arguments are as follows:

1 We have different views on a lot of questions in science. Thus, it is hard to determine demarcation of subject 'r'. The parapsychology is a good example.
2 If we assume there is a demarcation in first step, it is still hard to eliminate pseudo-science. And, so long as science is developing, there are bound to be activities which were once thought to be 'science' but no longer thought to be so – alchemy comes to mind – and there will be others once thought as pseudo-science but subsequently as real science – acupuncture is an example.
3 It is hard to tell whether pseudo-scientists were qualified as a replicator even they replicated original experiment. It is because we need to develop a set of criteria concerning appropriate background, training, and personal qualities. And this will associate with developing nature of science and its criteria.
4 What is an experiment? It is also not very clear. In actual experimental work, experiments get hardly success for the first time. Thus, experimenters could usually recognize that most of things he or she does in practice might be trial or even erroneous. Those activities comprise not merely successful experiments, but also 'preliminary run' of them. Lots of reports of failed experiments remain unpublished, which is known as the 'file drawer problem'. Collins thinks that, since most experiments are too delicate to fail in most of time, they are not proper experiments until

an appropriate level of skill has been attained by agents in experiments. This means that no one ought to report a negative result until they had demonstrated their skill by producing a positive result. It seems that we could not know whether one carries out an experiment without ascertaining one's actual capacity of experiment.

5   Nearly all efforts to rationally theorize the experimental replication, or to put forward a statistical algorithm for aggregating negative and positive results, are doomed to fail. For example, In Rosenthal's psychological experiment upon expectancy effect, including no less than 345 results, roughly two thirds of them are negative or null, while the other one third are positive. Nevertheless, it turns out that the positive results outweigh the negative ones so that the hypothesis is clearly upheld. And that calculation seems perfectly reasonable, since it is assumed that the variation among experiments is entirely accounted for by random fluctuations in unknown background variables. However, it might suggest that, since experiments are such delicate things, it would be plausible to believe that most of the 345 results are defective even including the positive ones. The statistical algorithm cannot distinguish between well-done and failed experiments, which means it is useless for explaining successful replication of experiment.

6   There remains a number of problems in deciding whether experiments produce positive or negative results.

- If we suppose there is no paranormal effect, critics would tend to argue that any experiment which seems to demonstrate such an effect must be defective by that very fact. However, it might be replied that all critics are not qualified to charge others experimental results, unless they do have abilities to produce positive results, since nearly all negative results would involve no authority for judgement. These feedback loops complicate the status of results as success or failure for experiments.
- It concerns the nature of the 'positive result'. There is an interesting problem in parapsychology that the abilities of subjects decline remarkably after sometime. The subjects might get bored, tired, or distracted. However, it can be argued that since the 'decline effect' is such a regular feature of experimental work, it demonstrates indirectly that there is really something going on. If the entire phenomenon are artifacts, there would be nothing to decline!
- What level of statistical significance is to be accounted as a positive result varies in different disciplines. Sociologists consider a result as positive if its likelihood arising purely by chance is less than 5%. But in physics most of results are subject to some higher requirement, or even do not recognize any necessity for statistical analysis.

Finally, we don't know how to distinguish the results of experiment as positive from negative. Thus, such efforts of validating replicability of experiment are also problematic upon confirming results of experiment.

Therefore, Collins concludes that the replicability of experiment is not self-evident, since it is difficult to be guaranteed by experimental fact. Whether experimental phenomena could be considered as repeatable is a result of social consultation involving a few members of scientific community. In those scientific practices, called 'open-system' by him, there is an order in the establishment and maintenance as a collective consensus of scientists, rather than the evidence of the experiment. 'Replicability, the vanguard of common sense theories of science, turns out to be as much a philosophical and sociological puzzle as the problem of induction rather than a simple and straightforward test of certain knowledge' (Collins, 1985, p. 19).

In addition, Collins's (1985, p. 83) study of experimental replicability finds an interesting phenomenon: 'experimenter's regress'. Whether a primary experiment is established depends on the results of the experiment r is true, and whether r is true needs to be tested by means of appropriate instruments. But the ability of experimenter and the suitability of instrument need to be measured by whether the result of r is true, which is rightly the thing we don't know yet. And therefore, whether r is true depends on whether we believe that r is true ... which falls into an infinite regression cycle. Besides Collins, Shapin employs the concept of experimenter's regress in his arguments on Hobbes and Boyle about whether the air pump can produce facts. And he argues that there are some problems in the replicability of experiment. 'In replication, there is no unambiguous set of rules that allows the experimenter to copy the practice in question' (Shapin & Schaffer, 1985, p. 226). For Collins, experimenter's regress is the core of social constructional interpretation of science (Radder, 1992, p. 66). But it is just a paradox for people who advocate the certainty of science by replicable experiment. Both Collins and Shapin claim that the only way to break down that paradox is to introduce a mechanism of social consultation. Collins stresses the consultation as core interests of scientists, and Shapin highlights the role of social convention in scientists' decisions.

ability of experiment, appropriate of instruments

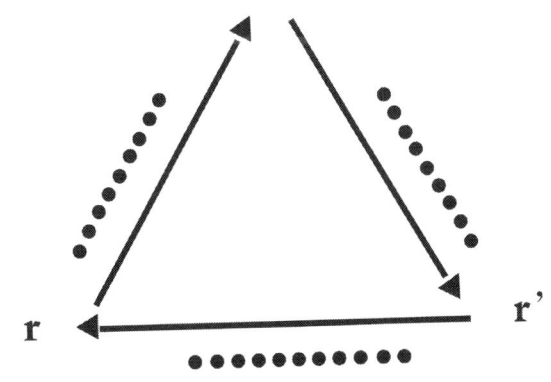

r ← → r'

Figure 14.1 Experimenter's regress.

14.3 The new experimentalist view of replicability

New experimentalism is a new school in philosophy of science since 1980s. New experimentalist philosophers not only start to search for a relatively reliable foundation for science, but also focus on descriptions and explanations of science, who think of scientific experiment as more basic than theory. They are hoping to get a more reasonable explanation of scientific development. In a book review of Franklin's *The Neglect of Experiment*, Robert Ackerman (1989) clearly puts forward the notion of 'New experimentalism', which usually refers to Ian Hacking, Hans Radder, Allan Franklin, Peter Galison, David Gooding, Deborah Mayo, and other philosophers.

Since it concerns experimental explanation, New experimentalism must conceptualize experimental replicability. Many of new experimentalists agree with Hacking (1983, p. 231): 'Roughly speaking, no one ever repeats an experiment'. Yet this is not the whole story if we care about more details. New experimentalists' view of experimental replicability is different from SSK. They think of replicability as much problematic in scientific practices, though it also plays a very important role in the development of science. The following work has been done in New experimentalism as an interpretation of experimental replicability:

14.3.1 'The concept of experimental replicability' reconsidered

From the representationalist perspective, experiment is just a general concept. But from the perspective of PSP, experiment is actually a very complex process of practice. Therefore, what is the experimental replicability is also necessary to make some clarification. For instance, we must answer the question about 'what is replicated' in experiment. Radder (1992, pp. 64–5, 1995, pp. 64–7), a new experimentalist, claims that there are three types of answers here.

(1) Replication of material realization in the experiment. Material realization is an important aspect in Radder's philosophy of experiment. He thinks that two basic aspects of natural science are material realization and theoretical description (or interpretation) of an experiment. The so-called material realization 'refers to the features of action and production. That is to say, it reflects the fact that science concerns not only thinking, reasoning and theorizing but also doing, manipulating and producing' (Radder, 1992, p. 64). Material realization is a process of operation. And the theoretical aspect of science can be realized in the material level through that operational process.

This type of replicability requires people to perform the same actions under instructions of theoretical interpretation, and to generate the same situation of experiment according to descriptions of material realization. If we set p as theoretical description or interpretation, and q as experimental result. Thus, in the replicability of material realization, we may have $p \to q, p' \to q', p' \to q$ or $p \to q'$. In other words, people with different theoretical interpretation may all have the same type of replicability. Replicability of material realization need the same process and material phenomena of experiment, rather than the same theoretical interpretation. A layman might also reproduce experiments under correct instructions. For example, Shapin finds that Boyle in fact rarely repeated his experiment. Some people (including other experimenters and layman) repeats material realization of air-pump experiments without getting the correct theoretical results. Boyle thought that the candle in the air-pump went out because of air burning out, but for Hobbes it was merely a gas flow.

(2) Replicability of an experiment under a fixed theoretical interpretation $p \to q$. For example, in the experiment of liquid boiling, if q present 'liquid boiling at $y°C$', it is very difficult to repeat the experiment under the $p \to q$. But if we add a range of error 'z' and let q represent 'liquid boiling at $(y \pm z)$ °C', the experiment can be repeated under this new theoretical interpretation. Therefore, we must repeat experiment accurately according to intended theoretical interpretation.

(3) Replicability of the result of an experiment. This means that we can get the same results by different experimental processes. At this time there may be $p \to q, p' \to q$ or $p'' \to q$ which p, p' and p'' are different theories. For example, in the experiment of liquid boiling, we can use different thermometer, such as mercury or gas thermometer. A successful reproduction of a result under a fixed interpretation entails that people agree on the correctness or truth of the interpretation or result in question.

Moreover, we can ask 'Who reproduces the result?' And there will be four ranges: replicability by any scientist or any human being, in past, present or future; replicability by contemporary scientists; replicability by the original experimenter; and replicability by the lay performers of the experiment. So, in answering the questions 'replicability of what?' and 'replicability by whom?', we obtain a richer notion of experimental reproduction (see Table 14.1).

Radder points out that the boxes 2, 3, 4, 6, 7, 10, and 11 can be exemplified by cases from the recent historical and sociological literature on experimentation. Boxes 5 and 9 are most probably empty. After all, the claimed range

Table 14.1 Types and ranges of reproducibility (Radder, 1992, p. 66)

Of what? By whom?	Reproducibility of the material realization	Reproducibility of the theoretical interpretation	Reproducibility of the result of the experiment
By any scientist or any human being, in past, present or future	1	5	9
By contemporary scientists	2	6	10
By the original experimenter	3	7	11
By the lay performers of the experiment	4	8	12

of those types of replicability would presuppose an unrealistic continuity of belief in the correctness of the theoretical interpretation, as well as an unrealistic stability of the material and social conditions required for the material realization of experiment. For the same reason, box 1 is also probably empty. Boxes 8 and 12 are empty by definition, since the laymen in question are presumed not to be theoretically informed.

14.3.2 Replicability in experimental practice

Experimental practice is a very complex process. Empirical research would reveal those problems involved in the principle of replicability. And according to above classification of replicability and experimental practice, New experimentalists mainly consider replicability of experimental practice in following aspects:

1 Restriction from tacit skills is necessary for experimental practice in replicating experiment. In experimental practice, scientists and laymen usually need original experimenters' instructions to reproduce an experiment (especially material realization of experiment). It is because 'the required craftwork might be so extraordinary that only one or a few people are able to agent it. In such a case successful intervention in nature would require a very individual "feeling"' (Radder, 1995, p. 70). This 'feeling' is actually a tacit skill of experiment which is not clearly represented or coded by formal language. SSK holds also a similar view. For example, in an analysis of construction of TEA laser, Collins mentions that no single scientist or scientists group could construct a laser just according to the official published papers without directly contacts with scientists who actually build laser. Replication of some actions, such as preparation of the devices and objects, process of interaction, and strict control over the end of experiment, is all restricted by tacit skills of experimental practice.

2 Fallibility and constant corrections of experimental practice influence experimental replication. Gooding (1992, p. 68) takes the emergence of unexpected events as an important feature of experimental testing, who argues that experimental practice is a complex dynamic process. Hacking (1992, p. 57) illustrates the co-evolution between theories and laboratory equipment as 'a contingent fact about people, our scientific organizations, and nature'. In experiment, theories and instruments are developing each other, and there is no eternal stability. Assuming a fixed theory, the replication of an experiment often faces with experimental errors and new phenomena which might never been explained by the assumed theory. Thus, we need to modify the original theory in order to match the experimental objects and instruments.

3 Limitation of replication rises from specialization and locality of experimental practice. Gooding (1990, p. 190) argues that 'philosophers have tended to assume that, because the laws of physics are universal and

timeless, the results of experiment have these same qualities. In fact, most of the time, the opposite is true. Experimental practice is particular and local'. Different types of research involve different characteristics. By studying Faraday's electromagnetic induction experiment and Morpurgo's search for quarks, Gooding (1990, p. 211) shows that 'fine structure of experiment shows why it is hard to prove that one experiment or observation replicates, or is identical to, another'.

14.3.3 Replicability as a non-local norm

New experimentalists claim that the replicability of experiment will encounter many problems in experimental practice, which however does not weaken its significance for science, since the principle of replicability often plays a role as a non-local norm in scientific practice.

Non-local norm is intelligible, though the replicability of experiment is originated in specific situations. However, in scientific practice we can produce stable process (including material realization, theoretical design, and stability of experimental results) through standardization and decontextualization. Thus, local knowledge will be expanded outside laboratories, which requires non-local norms as a result from a standard expansion of the original locality.

Hacking (1992, pp. 41–2) points out the role of standardization in extending local experiment:

> in the case of experimental techniques, a great many of them fade away, and only the most gifted experimenter can duplicate what textbooks causally say was done. New instruments make obsolete the skills needed to build old instruments; replication requires perverse antiquarianism ... the Zeeman effect and the anomalous Zeeman effect are now what they were when they were discovered, and it is the practice of teaching and naming that makes things seem so constant.

Gooding (1990, p. 191) emphasizes the process of demoralization which Bruno Latour also holds. He argues that things which originally refer to individuals and places (some models) are thrown away from the conversation of scientists, thus the facts and rules have to be independent from situation. For example, a bit of 'Cavendish physics' becomes part of the physics practiced in every laboratories. Honorific titles (such as 'Stokes' law', the 'Hall effect') do not function as modalities.

Radder (1992, p. 71) carries out a more detailed analysis at this point, who thinks that function of non-local norm of replicability can be realized through a lot of stabilized program, such as stabilization through reproducing the material realization, stabilization through explicit and theoretical design, stabilization through replication, and stabilization through standardization. According to him, 'replicability also functions as a nonlocal norm, the application of which leads to experiments or experimental results that are stable against a number of variations in their local contexts. Therefore, experimental natural science cannot be adequately understood as a mere sum of local cultures' (Radder, 1996, p. 10). Naturally, this statement is just another version of Nancy Cartwright's view that modern science can make laboratory move around.

14.4 Comparing the views of new experimentalism and SSK

In the above discussion, New experimentalist and SSK's notions of replicability both focus on the specific cases in history of science. And through microscopic empirical research, both suggest that in the experimental practice, the replicability of experiment is not self-evident in the way that traditional philosophy of science claims. They all think that experiment necessitates a kind of practice of skills. And tacit skills (New experimentalism) or tacit knowledge (SSK) puts impacts on replicability of experiment. In addition, both approaches actually have been deeply influenced by 'Duhem-Quine thesis', which contains two complementary propositions: for one thing, theory is not fully determined by experience (no confirmed hypothesis can usually be saved by auxiliary assumptions). For another, empirical data have theoretical-laden (experimenter cannot know the world independently from a theoretical picture of world). However, there is also some difference between New experimentalism and SSK that is worthwhile for notice:

1. By clarifying the replicability of experiment into three types with four ranges, New experimentalism suggests that experiment is replicable in a certain sense especially like material realization. SSK focuses on the experimental results to illustrate why experiment is not replicable, and therefore its discussion which merely includes the reproduction brought out by other scientists (boxes 9 and 10 in Table 14.1) is incomplete.
2. SSK combines micro-analysis of cases with macro-explanation of sociology. SSK is trying to suggest that, since empirical judgments cannot rely on observation and experimental facts, a sociological explanation is necessary. The argument of experimenter's regress requires us to explain experiment in pure sociological terms: since replicability is believed to confirm the reliability of experimental result, scientists already have consistent standards to enable their evaluation of experiments. Therefore, SSK tends to employ social conceptions like collective belief, convention, and core interests to explain experiment, while New experimentalism mostly takes the strategy of combining micro-case study and philosophy. Through in-depth analysis of specific cases, New experimentalism suggests the complex structure and dynamic process of experimental development. And in the normative aspect of philosophy, it defends the reasonable and particular role of replicability to a certain extent.
3. Difference among methodology leads to difference upon ontology. SSK holds the under determination of experience, which supports the

Table 14.2 Differences on the notion of replicability between SSK and the new experimentalism

Subjects	Sociology of scientific knowledge	New experimentalism
Types of experimental replicability	material realization; fixed theoretical interpretation; experimental results	experimental results
Methodology	micro case +sociology	micro case +philosophy
Ontology	natural / social Dualism	natural and social mixture
Epistemology	Relativism	Naturalism

involvement of social relations in scientific practices. That seems still to presuppose a dualism between the natural and the social. Assuming that dichotomy, it needs to introduce social relation to ensure replicability, which considers social relation as the only real relation. And New experimentalism argues that people (social) interacts and co-evolves with nature in the experiment without final cause.

The relativistic implications of the experimenter's regress follow only if we accept that social interaction is always sufficient to explain what observers think they see. We know that sometimes it is. This is why the issues of power and authority are so important to science studies. Because social interactions are not always sufficient to explain experience, we need a more realistic view of empirical access.

(Gooding, 1990, pp. 212–13)

4   The relativism in SSK sense is different from the naturalism held by New experimentalists. From the analysis of experimenter's regress and SSK's notion of replicability, we can characterize the sense of relativism on the whole. In fact, SSK argues that traditional view of science can be eliminated by social constructivism. And New experimentalists not only describe the type and scope of replicability in the experimental practice, but also point out the possibility of the non-local norms. Thus, New experimentalism is a typically naturalistic approach. That normative naturalism seems likely to bring us back to a old-fashioned notion of scientific objectivity through a more detailed and clear understanding of scientific practice (Table 14.2). Table 14.2 has suggested the difference in the studies of replicability between from New Experimentalism and from SSK.

14.5 Conclusion

SSK argues that experimental replicability is not merely related to the experiment itself, but more importantly explained by social facts. Though the complexity of replicability is recognized, New experimentalism still insists that the principle of replicability plays a very important role in science. Through the standardization and decontextualization in constructing non-local norms, operational process, phenomenon and result of experiments can be stably generated, in the sense of which the experiments have already been replicated. In practice, the replicability of experiment is not as powerful as considered in traditional philosophy of science, but the blames and doubts against it will not threaten the certainty and universality of science. The regression of the experimenter may be inevitable, but it is not necessary to resort to social facts.

In a word, experimental replicability is not self-evident for the reasons as follows:

First, it depends on a hypothesis that laws of nature are universal and incorrigible. Experiment can generate the same phenomena in any space and time only based on that assumption. And it is a metaphysical assumption since we can never exhaust all confirmation.

Second, it is also difficult to replicate an experiment completely in experimental practices. What we have said that 'experiment is replicated' is merely in some sense of experiment. And the replicability of experiment is too complicated to be explained according to particular contexts, and therefore there is no general rule for the replicability of experiment.

Third, both experimental practice and knowledge are local, and it might appear, but actually it does not need, to guarantee the certainty and universality of science in terms of the replicability of non-local norms. New experimentalism argues that the important role of replicability can be explained through the standardization and decontextualization which produce stability of experiment. We believe that the standardization and decontextualization cannot completely eliminate the local characteristics of knowledge. Joseph Rouse also points out that, in the process of standardization, experiment loses all of the inner relationships with any special person, task, or situation. Thus, experiment is no longer a thing of *Zuhandenheit*, readiness-to-hand, in Heidegger's word, but something functionally averaged. The result of averaging is to make the process of experiment more universal, but also less sensitive or discriminating. The information loss caused by standardization and generalization may be similar to the increase of entropy in the ordered system (Rouse, 1987, pp. 113–17). Therefore, in a strict sense, knowledge is determined because it is local, specific, and certain. It is undeniable that scientific knowledge can be extended out of its local situation in order to get a universal appearance. Rouse argues that one must not only refine and adapt the procedures, strategies, and equipment, but also partially reconstruct situation within which the know-how is to be applied.

> The removal of traces of their contextual origin from scientific results reflects a pragmatic choice within a larger field of practical involvements, a balancing of gains and losses, rather than a move from local practical involvement to a universal "theoretical" stance.
>
> (Rouse, 1987, p. 118)

15 Local knowledge (I)
Traditional Chinese medicine (TCM)

It requires a normative view to study Chinese medicine practice. We can study Chinese medicine from different perspectives, such as Chinese medicine itself, Western medicine, or others. From the main perspective of PSP and Chinese medicine itself, we can specify its properties, characteristics, historical evolution, and the relationship between theory and practice, to grasp its essence. Objectively and historically, the history of Chinese medicine can be divided into three main stages: 'period before systemization: knowledge accumulation through practice', 'period of systemization and autonomy', and 'period of differentiation and development since the eastward transmission of western sciences'. Chinese medical practices in different stages involve their own characteristics. Yin-Yang and Wu-Xing (five elements: metal, wood, water, fire, and earth) are philosophical concepts and metaphysical ideas of Chinese medicine. It had not only academic network connection role, but also ideological dominance role. There are three kinds of relationship among Chinese medicine theory and practice: stimulation, obstruction, and discrete.

In addition, we will have a new understanding of scientific status of Chinese medicine and Western medicine from the perspective of local knowledge in PSP.

15.1 The basic characteristics of Chinese medicine

Someone has ever criticized Chinese medicine, especially on its embodied metaphysics. And it does not discuss the scientific mechanism of physical nature and human body, no scientific experiments nor repeatable. One thousand doctors have a thousand treatments for the same disease. It has no scientific basis and regular pattern. Its theory is relatively outdated, and so on.

Regardless of the scientific mechanism, scientific experiments, and experimental repeatability of Chinese medicine, we can discuss the nature of Chinese medicine first. Once the nature of Chinese medicine is made clear, those questions above-mentioned may be solved. As researchers in philosophy of science thinking about the basic attributes and characteristics of Chinese medicine, we need help from the discussion upon distinguished

Chinese medicine doctors, scholars on Chinese medicine, and interpretation of Chinese medicine classics to look at these issues.

15.1.1 Chinese medicine as knowledge: unifying the metaphysics and the physics

According to the nature of the object world, ancient China has a historical notion stated as 'the metaphysical is called Dao, physical is called Qi'. Some Chinese medicine doctors, for instance, Lihong Liu think that Chinese medicine is 'the knowledge unifying the Dao and Qi'. For example, in the theory of Five Zang (Five internal organs, i.e., heart, liver, spleen, lung and kidney) the Chinese characters for Zang has a component representing flesh except for that of the heart. Flesh represents the physical nature, while without flesh indicate the metaphysical nature of the heart. According to Liu, the understanding of nature of Five Zang gives out the way to understand the whole Chinese medicine, and further Chinese traditional culture. For example, in the theory of Wu-Xing, Gold, Wood, Soil, and Water are tangible materials, but Fire is intangible, which seems to be a metaphysical thing. Chinese medicine not only refers to liver, spleen, lung, kidney, but also heart, which is also a representative of Dao. Therefore, Chinese medicine is a kind of knowledge which involves both physical and metaphysical. In addition, status is different between physical and metaphysical thing. On one hand, it is the Dao and the Qi, unifying shape and spirit, or shape and Qi. On the other hand, Chinese medicine places extreme emphasis on Dao, Spirit and Qi. Hence, it's a kind of knowledge in which Dao dominates Qi, Spirit nominates shape, and metaphysical thing dominates physical thing. It can be confirmed in two paragraphs from *Huang Di Nei Jing*:

> Heart is in the prime location among organs of living, intelligence and wisdom will thus be produced.... Twelve officials coordinate with each other, meridians interlink to each other, otherwise, disasters will be approaching. Heart dominates five organs; twelve officials have their duties. According to such way, people can be health and longevity, free from crisis. Govern the country for the same reason, it will be booming prosperity.

Knowledge as unification of the metaphysics and physics had been one aspect of Chinese medicine being criticized. Critics think that modern science is characterized by empirical research. Traditional philosophy of science such as logical positivism rejects metaphysics in terms of meaning criteria. But pace the range of western science and medicine, could it not be considered a science? Owning to knowledge of unifying the metaphysics and physics is indeed beyond the category range of the classical western academic system, Chinese medicine is excluded from science; however, owning to the modern western academic system, which starts with a new direction and discipline of unifying metaphysics and physics, perhaps Chinese medicine can be recognized as science. For instance, ethnobotany, originating in the United Stated in Late 19th century, is increasingly becoming a new discipline, merging with nature science, social science, and humanities. (Further discussion in the other part of Chapter 15.) Additionally, contemporary Philosophy of Science such as *PSP* insists that explanation is also present in the natural sciences. In other words, hermeneutics and its significance seem to apply not only to literature, history and other humanities, but also shown to exist in the natural sciences. This clears up the strict boundaries of science and the humanities. For Chinese medicine, knowledge of unifying the metaphysics and physics is no longer a reason for criticism.

The distinguishing of Chinese medicine practitioners explains the nature of Chinese medicine to some extent. As a result, Chinese medicine belongs to the Technique class in Chinese academic classification. The concept of academic and technique tradition is united in humanistic tradition with the same effect on medicine. There exists distinguishing of doctor ratings in Chinese medicine. However, this level is not the level of academic power, but the level of doctor itself. Then, how to evaluate it in Chinese medicine doctors? In *Huang Di Nei Jing*, doctors are divided into two levels, Shang Gong and Xia Gong. The so-called Shang Gong is a superior doctor, while Xia Gong is a doctor of craftsmen level The distinction would be that 'craftsmen doctor cure patient's by body, but superior doctor by spirit' (Zhang Zhicong Variorum, 2002, *Ling Shu*, p. 2). And 'so superior doctor prevent patient from disease before he will be sick, drug treatment after patient fall ill, it's like someone dig the well for water when he's thirsty, it's too late' (Zhang Zhicong Variorum, 2002, *Su Wen*, p. 13). Criticisms on Chinese medicine are from the level of having been already sick, namely physical level. Comparing Chinese medicine with Western medicine, many people insist comparing them from already sick level, instead of not being sick. It's certainly unfair. Of course, someone would ask, 'Is it sick before falling ill?' Western medicine also gradually recognizes its importance of curing disease before being ill. Nip heart disease in the bud rather than curing heart disease. It's better to prevent than to cure.

Where is true level of Chinese medicine? Where is essence of Chinese medicine? Many people think that the problem of clinical efficacy is caused by its outdated theory. According to Liang Shuming, a famous modern Chinese scholar, Chinese culture is a kind of precocious culture; therefore Chinese medicine theory has a precocious nature. Those who think theory is behind clinical may not really comprehend the classical theory. It is indeed a big problem. It requires strong understanding for people to grasp and innovate because of its precocious nature. To have the so-called elites becomes the basic and key point of the innovation of Chinese medicine and culture. However, there were such a few elites, which was believed to have led to Chinese medicine leaving behind. It becomes the most important and urgent problem in the development of Chinese medicine.

Therefore, from some Chinese medicine scholars' view, the problem of Chinese medicine is to inherit and educate. In the field of Chinese medicine, the biggest problem is how to pass on and carry on. There is a fearsome phenomenon that the education of Chinese medicine classic is becoming weaker and weaker. Most Chinese medicine colleges have taken classic courses as merely elective courses.

15.1.2 Chinese medicine knowledge is strongly practical

Above all, Chinese medicine treatment involves strong individual practice. Individuality of Chinese medicine is reflected in the diagnosis and treatment of specific activities every time. Its vivid portrayal is that 'The same disease has different treatments and different diseases have the same treatment'. There are three elements of individuality: doctors' individuality, patient individuality, and individuality of environment of diagnosis and treatment. These three aspects are forming mechanism of individuality of diagnosis and treatment. Doctor individuality plays a leading role among them. Chinese medicine knowledge system is just like a tall building, which is composed of many bricks. These 'bricks' are Chinese medicine doctors who may be left no name. The quality of bricks is different, so there are different location and role in the tall building. It's necessary for the personality of doctors, which is an important part in Chinese medicine. Even some aspects were taken as living soul. Doctor's personality, endowments, life experience, and environment will all have a major impact on doctor's academic direction selection, theoretical knowledge, technical level, achievement height, and so on. Of course, most of the doctors without leaving their names and writings are cornerstone of the building. Because of their unsung working, glories are created in the past, and hope are expected in the future for Chinese medicine.

Secondly, diagnostic methods of Chinese medicine are strongly practical. The famous 'Wang-Wén-Wèn-Qie' method, or seeing, smelling, listening and feeling, Four Diagnostic Methods of Chinese medicine is strongly practical in Chinese medicine. In ancient times, Chinese medicine created fruitful observations and verification methods during a long practice of medicine clinics, in the absence of scientific instruments and detection technology of human body.

The 'Wang-Wén-Wèn-Qie' is four methods of diagnosis and treatment of diseases gradually taking shape during a long practice of medicine clinics.

'Wang', or seeing, is that doctor observes the spirit, shape, and fettle of patient's whole or local body with sense of sight. 'Wén', or smelling, is that doctor distinguishes voice and smell of patient with sense of hearing and smell. 'Wèn', or listening, is that doctor knows about the occurrence and development process of the disease, current symptoms, and other conditions associated with the disease through querying patients and their families. 'Qie', or feeling, includes feeling and pressing the pulse, which is the method of pressing the patient's pulse and touching skin, hand, abdomen, four limbs, and other part of the body.

Chinese medicine believes that the running, inducing, and transmitting of spirit and blood can transfer disease. Jing and Luo, or main and collateral channels is the passageway of reflecting pathological changes. It functions like connecting organs and four limbs, and communicating between inside and outside. It's just like telephone network which links the human body into a unified whole closely. Therefore, the change of a part can influence the whole body through Jing and Luo. Indication of Visceral disease can be found on the surface of body, which is stated like 'the disease of entrails is bound to take shape outside'. However, Chinese medicine can speculate the change of entrails through observing of the outside of the body. It's the basis and evidence of diagnosis and treatment in Chinese medicine.

15.1.3 The locality of Chinese medicine

The locality of Chinese medicine is obvious. First, the disease spectrum is clearly localized: medicine is applied science for disease prevention and treatment. Disease is the object of research and disposal, as well as the main trend of the target and basic promotional power. Because of the vast territory, a diverse landscape, and different living and productive conditions, Chinese people's bodies have a range of regional differences, and thus different susceptibilities to disease. Thus, the disease spectrum is regional.

Since different regions have different disease spectrums, doctors and scholars pay attention to the demand for prevention and treatment naturally in their local region. They carry on targeted academic research and clinical experience accumulated. Therefore, they become the experts in understanding and disposal of these diseases in the field of Chinese medicine. In practice, they gradually form a corresponding and full of local features in the academic style and techniques. And gradually the differences among the regions will be shown. Chinese medicine practitioners participated disputes about professional interests and reputation. As a result, the exchange between doctors and patients has become similar between disease and knowledge. In this atmosphere, the doctors often do only two things: one is actively consolidating their dominant position in the understanding and disposal aspects of the disease in their local region; and then they try to expand their influence to other regions. They try to give the public a feeling that their knowledge and experience does not seem to be limited to the disposal of the disease in their own region; but has universal significance and can be extended to other areas of similar or more similar regions.

Secondly, natural and social environment have strong effects on the development of Chinese medicine. This kind of difference mainly reflects in the economic conditions. Doctor's method of diagnosis and treatment is quite different under different economic condition. People always pay more attention to health in rich area. They usually prevent themselves from disease and keep healthy. Once someone feels uncomfortable, he treats himself immediately. The method of diagnosis and treatment is accordingly complex, exquisite, and expensive. While people pay less attention to health in poor area. Even though someone falls ill, he insists working. The method of diagnosis is simple, extensive and less-expensive. If social and economic conditions improve generally, medical knowledge of rich area becomes popular. While social and economic conditions are not good, medical knowledge of poor area spread consequently. The proverb 'the poor makes people forget life, the rich makes people cherish life' not only indicates real influence which economic factor had on the view of life, but also reflects the influence which economic factor on medicine itself.

Thirdly, various schools come out one after the other. The debates among schools relate to the different regions. The third local factor is interpersonal relationship about politics and power. It mainly shows in competing for the disclosure power among different schools of Chinese medicine. Essentially, the research for the disclosure power is the research for reputation and benefit.

There are two concerns about this problem. First, since Chinese medicine theory has flexibility, there are multiple doctrines at the same time for the same object. Who wins is the result of negotiation. Winning factors include current practical utility, accumulating authority in history, and the ability of relying on other social forces (religion, community, customs, regime, economic rights, and so on). Second, since transport developed slowly in ancient Chinese society, this leads to geographic isolation to some extent. Thus, different Chinese medicine schools reflect certain geographic relevance. During Song Dynasty, Chinese medicine has a good foundation and accumulated a rich experience. All of this provides the basis condition of improving theory and studying new problems. At that time, wars were going on, and people's lives were not stable, so disease broke out here and there. Since social needs is very urgent, innovational thought came out in Song Dynasty. All these jointly promoted the development of Chinese medicine and effected on academic orientation of medicine field directly. The view of 'ancient treatment can not cure disease nowadays all the time' became doctor's prevailing opinion, breeding development of new medical theories and innovation of clinical practice.

The locality of Chinese medicine presents as the strong relevance between locality and individuality. In short, different places, periods and doctors correspond to distinguished features in Chinese medicine. Through the above comprehensive survey, it is included by the fundamental basis for anthropology, sociology, and politics. The 'disease spectrum' represents the anthropology (cultural) factors. The people living in different place have different physical characteristics, adding environmental conditions, and living habits, result in differentiated susceptibility to diseases. And the 'natural and social environment' represents the natural and social (economic) factors in common. The different natural conditions and social factors have direct effects on selecting different medical model. 'Various views and influence of medical schools' represent various local academic powers and practical relations (political) factors in the formation of Chinese medicine. The personality and endowments of individual as anthropology (cultural) factor, living experience as sociological (economic) factors, environment and opportunities as powerful (political) factors, have more or less effluences on Chinese medicine clinics. These complex interactions among the dynamics factors of nature, society and culture describe each specific moment character in the development of Chinese medicine knowledge. Chinese medicine knowledge is therefore just like other kinds of knowledge as well.

15.2 The historical evolution of Chinese medicine

Compared to Western medicine, Chinese medicine is characterized as 'empirical medicine' by those people studying modern science, who think of theoretical foundation of Chinese medicine as weak while practicality as strong. It's lack of scientific norms followed by Western medicine. Some even think Chinese medicine as witchcraft rather than science. They think there are too much metaphysical ingredients. How to treat the methodology of Chinese medicine? Is it rational about theoretical system of Chinese medicine? If there is rational theory system, what's the relationship between theory and practice? To answer these questions, we need to review the history of the evolution of Chinese medicine, to grasp its development law, and then to examine scientific practices in different historical periods. Based on that, we can understand its character of theory and practice. In the end, we can explain the relationship among ideology, theory, and practice when combine with philosophical basis and core theories of Chinese medicine, such as 'Yin-Yang', 'Wu-Xing', and 'main and collateral channels' and 'viscera theory'.

Generally speaking, the Chinese medicines are divided into three stages: 'period before systemization', 'period of systemization and autonomy', and 'period of differentiation and development since the eastward transmission of western sciences', according to degree of theoretical and autonomy. The practice of Chinese medicine represents various historical appearances in different stages.

15.2.1 Period before systemization: accumulation of knowledge through practice

Huang Di Nei Jing was finished in the Eastern Han dynasty, which is a sign of the formation of the early theoretical system. Its appearance is a major event in the history of Chinese medicine, and itself a classic of traditional Chinese medicine. Since then, Chinese medicine transferred from empirical medicine to relatively standardized theoretical medicine. Prior to *Huang Di Nei Jing*, the general methodology of Chinese medicine practice is indeed 'empirical'. The doctors generally grasp a set of individual characteristics of clinical cognitive technology through examining pathogenesis of patients and summarizing corresponding diagnosis and treatments. By the accumulation of knowledge, practical model of Chinese medicine gradually formed almost entirely through experience rather than theory.

15.2.2 Period of systemization and autonomy

Huang Di Nei Jing was not written by one individual at a relative short time, while it was finished by many hands in a fairly long historical period, from Warring States Period, Qin dynasty and even to Han dynasty. It took shape merging with the relevant aspects of medical practice and knowledge by efforts of many anonymous scholars. Its biggest feature is emergence of the 'main and collateral channels', 'Viscera', 'Qi and blood' and another important theoretical category. It unified the culture of 'Yin-Yang' and 'Wu-Xing' which prevailed in Han dynasty. Hence, Chinese medicine entered into the era of theoretical era based on 'Yin-Yang and Wu-Xing'. It has been developing up to now, while not having fundamental changes of original framework, only some non-structural adjustments.

15.2.3 Differentiation and development since the eastward transmission of western science

During the introduction of Western medicine into China since 19th century, Chinese medicine has been seriously divided into three schools, instead of being replaced totally, which are the school of preserving tradition, the school of westernization, and the school of integration. The proponents of preserving tradition insist on the theory of Chinese medicine, such as the theory of Yin-Yang and Wu-Xing, to argue for an autonomous development of Chinese medicine in the future. The proponents of westernization, however, claim to discard the theory of Yin-Yang and Wu-Xing in Chinese medicine, and to accept the new guidance from the western medical sciences, with merely preserving some clinical methods that are already justified as valid in practices. Besides, the school of integration, which is the most popular choice supported especially by the government since 1949, argues for the possibility of an organic integration between Western and Chinese medicine, to construct a totally new mode of medication beyond both, and to reconcile completely the conflict between the two systems of medicine.

15.3 The influence of Yin-Yang and Wu-Xing theory on Chinese medicine

'Yin-Yang' and 'Wu-Xing' theory is the philosophical foundation of Chinese medicine. It rules and unifies the huge theoretical system of Chinese medicine. Undoubtedly, to understanding the relationship between the theory and practice of traditional Chinese medicine, and based on which to grasp the essence of theoretical elements of traditional Chinese medicine, it's of great significance to understand problems as how it turns out to be organizer of Chinese medicine theoretical system, how its influence changes in the face of Western medicine challenges, and its possible role in the further development.

15.3.1 The influence of ancient Chinese culture on the formation of the core concept of traditional Chinese medicine: 'Yin-Yang' and 'Wu-Xing'

15.3.1.1 Quest for theory

In the period before systemization, medical practice exists independently and does not depend upon theory at all. After all it is only limited to the local environment of the practice, reflecting the not high medical level and many problem. It's only a primitive way of medical practice. With the constant enrichment of clinical practice and the gradual accumulation of clinical experiences, the original model of Chinese model is no longer satisfying the needs of Chinese medicine. To improve practical ability of Chinese medicine and to form theory demands in the eternal, this demand creates the presuppositions concerning Yin-Yang and Wu-Xing.

15.3.1.2 Adaptation of Philosophy in Pre-Qin and Qin-Han Period

Different schools of Pre-Qin had involved in the discussion of 'Yin-Yang' and 'Wu-Xing' at different degrees or from different perspectives, especially Taoist and Naturalists with many achievements. Until Han dynasty, 'Yin-Yang' and 'Wu-Xing' not only expanded the influence further but also continued integration into a universal view of the world and methodology. Meanwhile, some specific medical theories which has direct and close relationships with medical practice were gradually developing.... In this societal context, on the one hand Chinese medicine is ready to take in philosophical thought, while on the other hand the philosophical thought gained strong expansion capability. Thus, the theory of Yin-Yang and Wu-Xing and the practice of Chinese medicine successfully integrated into a Chinese medicine theoretical system of distinct characteristics of the era and style of Chinese culture.

15.3.1.3 Background: characteristics of Chinese Academic Research

According to some historians of Chinese thought, the Chinese academy is specialized, but not in the same way as modern western academic disciplines. One of the most prominent phenomena is that principal investigator (person) and region is taken as the criteria of specialization rather than the research object (object). Its research objects mainly focus on the scope of the ancient classics without taking nature as an object directly. Chinese medicine is dominated and unified by Yin-Yang and Wu-Xing theory, owning to this

15.3.1.4 The factor of Power

Yin-Yang and Wu-Xing theory is typically a holistic system. Together with the political ideology of Confucianism, it occupies the legal status and conciliates with other thoughts in traditional Chinese patriarchal system. It becomes as the theoretical basis of traditional Chinese ideology of society to provide reasonable and legitimate theoretical basis for mature 'Grand Unification' political model in Han Dynasty. Therefore, it's inevitably influence to lower levels and plays a significant role for the theorization of Chinese medicine. Isomorphic nature of knowledge and power has enabled the Yin-Yang and Wu-Xing theory to not only had 'soft' academic networking, but also 'rigid' ideological dominance upon Chinese medicine.

15.3.2 The challenges of 'Yin-Yang and Wu-Xing' in modern medicine

'Yin-Yang and Wu-Xing' have long been regarded as an important ideological foundation of medicine indisputably. However, it encountered increasingly strong challenges after western academic rising and becoming significant knowledge in China. There were two kinds of presentatives in the field of Chinese medicine, the moderate and the radical. The moderate gradually reduced or got rid of theoretical guidance of 'Yin-Yang and Wu-Xing' in practice. While the radical thought of it as 'pseudo-science' and completely took a negative attitude towards it. To understands the reasons and systematically analysis on all of this, is the only way to understand Chinese medicine in the contemporary context.

15.3.2.1 In a changing time

There are many reasons of the rise of western medicine; however two main aspects shall be considered. One of which is that western medicine carries out its institutions and modes backed by political and economic superiority of western society. The development of modern Western world has indicated its economic foundation and institutions as superior to traditional Chinese society. The other is that some aspects of western medicine better met the needs of that time and made up for the deficiencies in Chinese medical diagnosis and treatment. Since entering the modern era, wars and migration have occurred frequently. In vigorous development activities, mostly the world's ecological and social environment was in changing. The result was many war wounds, including the outbreak of infectious diseases, the lack of nutritional, and other health disasters has mass impact. Therefore, western medicine, which paid attention to medical standards, disease classification, and unified treatment could get rid of these issues by means of organization effectively. While Chinese medicine tended to individual treatment, and established the basis on the main clinic practice of individual doctors, had weaker respond Western medicine established the bridgehead for the development of medicine through exerting owns specialty and the new practice, and then expanded the scope of their territories gradually. Chinese medicine lost control in these key areas, and then makd their territory shrinking. The shrinking of territory will inevitably lead to the reduction of practice opportunities and doubts for theoretical basis of Chinese medicine.

15.3.2.2 Pressure from the new rising discourse power

With the continuing success of Western science, multifaceted impact beyond its actual scope began to appear. The significant result is the identity for the 'Western-Centrism' and 'scientism'. Although from the appearance different historical periods have different performances of reacting national spirit and resistance to Western-centrism. However, the 'Western-centrism' and 'scientism' are still popular in public. Economic and political backward made the Chinese people and academics at that time learn advanced Western countries sincerely. Therefore, it spread to almost all aspects of social life. Since Chinese medicine does not have the style of Western science, it was regarded as heretics by some scholars. Such pressures from inside and outside academic circle caused huge negative impacts on the survival and development of Chinese medicine jointly. The concept of 'Yin-Yang and Wu-Xing' subjects to the most intense criticism as a typical non-scientific superstition.

15.3.2.3 Chinese Medicine's slow development and failure to respond

In addition to the external factors previously mentioned, we cannot avoid those problems from Chinese medicine itself. After the mid-Qing Dynasty, it appeared the tendency of emphasizing theory and ignoring practice in the interior of Chinese medicine. The typical event is that the imperial hospital cancelled acupuncture in 1822. With respect to internal medicine, acupuncture has a stronger sense of practicality and fewer theoretical ideas. Some acupuncture activities are even difficult to find any rationale and fancy theory basis. Internal science is not like this. 'Theory', 'method', 'prescription' and 'drug' are clearly and logically. It has philosophical and political systems of isomorphic in theory. Thus, it's favoured by the regime. Since then Chinese medicine not only can't effectively respond to the challenges of Western science, but also no longer has a significant history of innovation to academic influence.

15.3.3 The rationality of 'Yin-Yang and Wu-Xing' theory

Nowadays science stands at a turning point of time again. Reductivist approach of modern western science began to face increasing challenges after its rapid growth in several hundred years. Besides the advantage of western medicine, its defect is prominent increasingly and becoming more and more clear. These are increasingly apparent problems of western medicine, which do not pay close attention to the whole and overall evolution, and the role for environment influencing bodies and epidemic diseases. In this social environment, Chinese medicine seems to get on the point of turning to a new life.

15.3.3.1 Knowledge contribution to complexity science

In life, social and cognitive sciences, Complexity Science has been unfolding. Chinese medicine has become one of Complexity Science cases which have drawn widespread attentions. 'Yin-Yang and Wu-Xing' is also regarded as reference model of constructing the complex system by dynamics theory. It is contemplated that Chinese medicine contributes to the development of Complexity Science meanwhile also having a new birth. The theory of 'Yin-Yang and Wu-Xing' provides referable knowledge foundation of complex system dynamics. It can be also fully explained by Complexity Science and gain modern scientific significance.

15.3.3.2 Epistemological contribution to scientific methodology

The theory of 'Yin-Yang and Wu-Xing' is one kind of epistemological analogy. More and more scientific practices demonstrate deductive and inductive methodology which has universal validity under reductivist science view cannot satisfy the need of dealing with the complex issues of nature and society in reductivist science. However, Yin-Yang and Wu-Xing is on behalf of analogist method which can be successful in when dealing the same. Therefore, it seems that the theory of 'Yin-Yang and Wu-Xing' is likely to play an important and constructive role in the formation of the new scientific epistemology.

15.3.3.3 Conceptual contribution to culture diversity

Although different cultures mode has different characters, people still hope to build the general form for comprehensive description beyond various modes. 'Yin-Yang and Wu-Xing' seems to be a candidate. However, especially we need to make a new interpretation for 'Yin-Yang and Wu-Xing' which is consistent with the needs of the era while not fully retaining its inherent connotation. One of the most valuable thought is multiple complementary and inter-dependent. 'Yin-Yang' represents multiple complementary, while 'Wu-Xing' represents inter-dependence. If we integrate 'Yin-Yang' and 'Wu-Xing' together, it can lead to the ideas of one universal natural and one world. These thoughts just embody the true spirit of calling for peace and protecting the environment as nowadays required.

We can easily find from the practice of Chinese medicine that the practice of knowledge and action is a continuous historical process. It relates to intrinsic characteristics and the external environment closely. It can exhibit various postures with the change of time and place. It can exist naturally and had nothing to do with the theory, which can integrate with theory and interact with each other, and which also can be united superficially with heterogeneous theory and maintains its adaptability in the adverse changes of environment. Thus, the practice of knowledge and action in Chinese medicine itself is extremely important. It's inappropriate to evaluate one kind of knowledge theory as right or wrong in an over-simplistic way. It's more desirable for comprehensive judgment in the light of reflecting the influence of scope, time, intensity, relationship with the practice, and the ability to adapt to changes and other factors. The theory of Yin-Yang and Wu-Xing has its ups and downs in the development of Chinese medicine, which provides a vivid case.

15.4 The debate around the scientific status of Chinese medicine

Chinese medicine is considered not as a 'science', owing to its different nature and form from the scientific requirements of western medicine, such as experimental replicability, searching for structural evidence (e.g., the main and collateral channels), and difference of empirical forms. Of course, there are problems on conceptual clarification. If the notion of science is just western knowledge, which has form and nature of modern science, Chinese medicine is not such a kind of science without any doubt. However, we think scientific debate about Chinese medicine should be investigated in a broader view. For example, since it is effective at saving lives, it is as true, reliable, and efficient as Western medicine referring to practice and medical pathology.

It's a long-standing debate whether Chinese medicine is science. For instance, some people think Chinese medicine is an ancient oriental science up to now. The main and collateral channels doctrine and acupuncture techniques have been hailed as great as the four great inventions of ancient China, and of world's major scientific achievements of significance. However, other people think the main and collateral channels doctrine in Chinese medicine does not exist, while its 'fabricated' statement, owning to unable to find the exact anatomical evidence and to verify it similarly as western medicine. These give rise to debate upon whether Chinese medicine is science.

This debate became more acute because of several celebrities joining in. Many celebrities have denied Chinese medicine as science in modern China. For example, Yan Fu thought Chinese medicine as lack of actual observation and logical reasoning. He attributed Chinese medicine as Feng Shui, Astrology, Fortune-telling, and so on. Chen Duxiu believed the Chinese medicine has neither structure of human body's organs nor analysis engaging in drug potency. It is only attached to the theory of Wu-Xing. Liang Shuming asserted that it's not medical theory as good as craftmanship in Chinese medicine. Ten doctors have ten kinds of prescription with hardly any in common. Thus, neither treatment nor drug has objective criteria. At first, it's a purely academic dispute, while it was expanded into the social-political and cultural events. A paper *'Farewell to Traditional Chinese Medicine and Herbology'* made the dispute more and more intensifying, that whether Chinese medicine is science or not. Zhang Gongyao proposed 'farewell' Chinese medicine in the name of science. He thought Chinese medicine is lack of empirical and logical basis. They all thought Chinese medicine was not a 'real' science. Because its empirical and theoretical statement are lack of strong support of clear and reliable enough principles and causal relations. Chinese medicine is even considered as hodgepodge, which contained philosophy, metaphysics, folk medicine, and witchcraft.

Throughout all these arguments, apart from the kind of people with ulterior motives of political culture attempt, they still commenced around the issue of 'the standard of Scientific Demarcation'. Their core focus still closely related to scientific concepts and its semantic interpretation. Scientific demarcation is to make a clear distinction between science and non-science. It certainly leads to the problem that if scientific standard is unitary or plural. From the view of modern philosophy of science, the standard of Scientific Demarcation has become more and more pointless. In the contemporary perspective, the demarcation criteria are gradually losing its core significance. Nowadays scientific demarcation criteria are not unitary but plural. Meanwhile, there are situation that the standard resolves. After the emergence of PSP, scientific demarcation criterion is also a fundamental change. Following with the viewpoint of local knowledge, the standard of science cannot be unitary, according to which western knowledge is science and other ethnic knowledge is not science. Those who try to put Chinese medicine as non-science is to measure Chinese medicine with western scientific standards, standing on and judging from the perspective of Western medicine. They agree that a measure of scientific theory is to see if it is determined to establish clear and reliable relationships or causality principle. The reason of Chinese medicine as nonscientific is that its empirical and theoretical statements have not satisfy such a standard. Therefore, Chinese medicine is labelled 'pseudo-science' openly.

In fact, Chinese medicine is not western medicine, which is not in a form of modern medicine in the Western sense. It's just like a chicken is not a duck, a duck is not a chicken. There is no problem. However, if it's not science in Western modern sense, can we conclude that it's pseudo-science? There should be more arguments.

Second, before Western medicine spreads to China, Chinese medicine has been responsible for the treatment of people with diseases. And its main and collateral channels and acupuncture have been widely used in daily practice. These are also facts. Could it be that people have always been dealing with nature and treat their bodies relying on pseudo-science? Is it always so effective? It should be said that Chinese medicine has a similar quality and function as Western scientific medicine over thousands of years in medical practice. Otherwise, Chinese people have already died out by relying on witchcraft or metaphysics every day. Chinese medicine is the best proof for thousands of years in history of China. Chinese medicine has such qualifications of dealing with the nature and bodies of Chinese people. Namely, the theory and practice of traditional Chinese medicine have rationality and legitimacy in China. Traditional Chinese medicine is a kind of practical natural knowledge and activity system, which has profound theoretical basis and practical clinical experience. For thousands of years, Chinese medical practice and theory employed a completely different way to deal with human regulatory systems and life processes. It has been always effective in guiding clinical subjects, especially in the clinical practice of acupuncture. Owing to a huge role in diagnosis and treatment in clinical practice, it has caused widespread attraction in the medical profession at home and abroad. If we are still in accordance with the western knowledge and colonial values, Chinese medicine would be difficult to get a scientific status, even it can cure diseases in practice.

15.5 A local view of the characteristics of Chinese and western medicine

All human knowledge has the characteristics of local knowledge, according to the view of 'local knowledge'. In other words, there is no purely universal knowledge, no matter in the West or the East, in ancient or modern. This idea opens a new sight of re-examining the scientific debates between Chinese and western medicine which are bound to deal with the healthy question in different ways, according to the specific situations, values or positions. It can be said that not only Chinese medicine but also Western medicine is the product of local knowledge. Therefore, there is no legitimacy in assessing one kind of local knowledge by means of another kind. It's not to say that local knowledge is incommensurable, but they can acquire understanding of each other's knowledge to some extent through communication and dialogue. The premise of acquiring local knowledge is to change the traditional mindset and values. The view of local knowledge embodies a transformation, related to changes of the understanding of the essence of knowledge.

Kant argues that scientific knowledge is not analytic but synthetic. The basis of synthetic proposition needs empirical foundation beyond logic. And then people recognize that knowledge depends upon our creative participation rather than finished things in anytime and anyplace. It relates to specific contexts. The subject of knowledge is a cultural community under certain historical conditions and environments. Whether Chinese or western, the aim of stressing local knowledge for both medicines is to declare that knowledge arose from a series of changes in specific historical conditions or situations. That is the consolidated result of local character. In the Chinese context and with a kind of long-term practice, Chinese medicine comes into being and develops in the traditional culture. Studying Chinese medicine should not be overly concerned with the universal norms, but focus on the specific situation in the formation of this medicine. The contexts of Chinese medicine, formed under the context of the specific historical conditions and a particular cultural background, should be studied by way of a set of particular medicine values, positions and sights.

Chinese and western medicines both belong to the field of medicine. They both face the problem of health and disease. However, their conclusions are different owing to different understandings of human body, research methods and tools, especially different medical practice and environment. Western medicine is good at using anatomical methods to study structure and function of the human body, while Chinese medicine pays more attention to use observation to study extensive relationships within the human body as well as the ones between body and the external. Therefore, western medicine includes anatomy, physiology, biophysics, biochemistry, molecular biology, which are not the contents of Chinese medicine. Chinese medicine studies the relationship between the body, the natural world, and its surrounding environment (such as climate, the sky, etc.). It also studies the relationship between Human organs and seasons climates, the relationship between Human disease, acupuncture and massage, the important role of the four diagnoses, the compatibility of medicine in order to reduce toxicity and increase effectiveness, and prescription and medication. All of those are not the contents of western medicine.

As is known, the way of holistic thinking has been emphasized in Chinese medicine, but this is not to say that the western medicine does not pay attention to holistic view and intention. Instead, western medicine also attaches importance to holistic view, though its essence and meaning of holistic view is different the one from Chinese medicine.

Due to different academic sources, Chinese and western medicine inherit different comprehensions of human holistic existence. Chinese medicine absorbed the thought of the Book of Changes, *Taoism* and *Confucianism*. It understands holistic from the view of Vitality theory. On one hand, it stresses the unified integrity between bodies and environments, and treats human being as the product of evolution in nature. Chinese medicine thinks of man as evolved from the natural world. People communicate not merely with each other but also with the world. Thereby, it elucidates main and collateral channels and the viscera doctrine of Chinese medicine.

In contrast, since ancient Greece Western medicine has interpreted human integration from the viewpoint of element and atomistic theory. The 'atomism' of Democritus takes everything in the world as a combination of the smallest and fundamental sub-atomics. That academic thought originated in ancient Greece, and had revived in modern Europe after medieval transformation. It becomes more popular and in-depth with the establishment of Newton's classical mechanics and the development of the machine-age of industrialization. Western modern science follows this way of thinking. Following the reductionism and expanding at all levels within human body, we construct modern physiology and pathology at the level of organs, tissues, cells, and even molecules. Logically, as long as the whole is decomposed and restored to its primitive atoms or elements, then the overall can be explained in the end. This thought is just completely different from the essential connotation of holism in Chinese medicine. One places extreme emphasis on the relationship between environment and the system; the other takes an overall view of the human body as a decomposable part or subsystem, and studies it in sections.

Chinese and western medicines involve different local characters or features, embodied in the perspectives of understanding, research methods, and ways of thinking. They emphasized the ways determined by their specific cultural contexts and actual practices separately. Chinese traditional culture expresses rich and diverse cultural forms, and cannot be mastered by western systems of knowledge, concepts, and terms. From the living experiences and characteristics of Chinese medicine, it constructs a way of life for themselves including practical skills and knowledge system.

15.6 The rationality of Chinese medicine

Chinese medicine is a kind of local knowledge. It's not only a kind of unique charming Chinese culture but also regarded as a unique science parallel to Western medicine. The Main and Collateral Channels theory and acupuncture approach of Chinese medicine not only reflects the characteristics of the traditional Chinese thought, but also has a strong national identity as characteristics of scientific practice.

The academic system of Chinese medicine has a strong empiricism in a long period of practical exploration. It has developed a unique system of diagnosis and treatment after a lot of careful practice. Such a clinical diagnosis and medical classification system is completely different from Western medicine. During the development of Chinese medicine, it is influenced by traditional Chinese culture and forms a specific standardized mode under the specific cultural context in China. Usually people only focus on the consistency between theoretical generalization model of Chinese medicine and the theory of Yin-Yang and Wu-Xing, and think that the medicine is a mere metaphysics and witchcraft. In fact, in this standardized model, it shows that the practical mode is still dominant in Chinese medicine clinic. In addition, traditional Chinese knowledge has concerned a lot for complex systems without giving up such an understanding of the natural and social complex systems. It forms its own unique understanding of complex systems. Accordingly, Europe is the tomb of complex systems research while China is a good example of complex systems research. China never stops exploring complex systems. There are two ways of thinking about human beings which are different and complementary – reduction method and holistic way. As previously proposed, Chinese medicine focuses on the holistic way of thinking. It considers various functions of the human body that must maintain coordination and balance. The treatment of disease is put before falling ill. It is necessary to resist the disease outside the human body. The treatment lies in adjusting living organisms, maintaining and restoring dynamic equilibrium. Western medicine uses reductionist approach to treat human body. It is dedicated to find out disease suffered. It seeks to identify the causes of the disease and eliminate it in order to make the body back to normal. The difference between Chinese and western medicine is that Chinese medicine focuses upon mutual relations while Western medicine focuses on the structure of entity. From the perspective of integration and mutual relations, Yin-Yang and Wu-Xing theory in Chinese medicine is actually used to describe basic forms of human complex systems, as well as the normative model of mutual transformation. From such a perspective on Chinese medicine, the research can be incorporated into the column of Complexity Science. However, it is regrettable that many people still regard Yin-Yang and Wu-Xing theory used in Chinese medicine as fundamental theories of the nature of witchcraft in Chinese medicine, treating it as Alchemy. It is not difficult to find that Yin-Yang and Wu-Xing theory is the philosophical basis rather than fundamental theories of Chinese medicine. According to the standards of western knowledge, it cannot confuse philosophy with science as a single category. While in terms of the tradition of Chinese knowledge, philosophical thought and the way of thinking of Yin-Yang and Wu-Xing are integral parts of Chinese medicine. The fundamental theories of Chinese medicine are Viscera Doctrine and Main and Collateral Channels theory. They have been implied in our earliest extant medical classic *Huang Di Nei Jing* since thousands of years ago. Yin-Yang and Wu-Xing theory provide essential and effective philosophical thoughts and way of thinking for the foundation of Viscera Doctrine and Main and Collateral Channels theory.

Huang Di Nei Jing, the classic literature in Chinese medicine, not only explains Main and Collateral Channels and Viscera theory in detail but also permeates with philosophical thought of Yin-Yang and Wu-Xing. Yin-Yang and Wu-Xing theory are the fundamental theories of Chinese medicine; they're also the key of understanding Chinese traditional knowledge in depth. They represent a different understanding line from Western science. After Yin-Yang and Wu-Xing theory have been introduced into the medical field, it succeeds in creating a unique way of thinking of Chinese medicine, and has profoundly impact on the formation, development, and application of Chinese medicine theory. It only has Yin-Yang theory without Wu-Xing theory in the early days of Chinese medicine. The reason of adding 'Wu-Xing' into 'Yin-Yang' theory is that Yin-Yang only explains both sides of things without explaining the nature of things. Only Wu-Xing have this condition. The essence of 'Wu-Xing' has two meanings: one is 'character' – the abstraction of the properties of things; the other is 'category' – the classification of things. It is a kind of scientific thinking to classify and make the awareness framework for stipulating the relationship, and then it constitutes systematic understanding. Yin-Yang and Wu-Xing theory are both the products of scientific thinking. Incorporating Wu-Xing theory into the Chinese medical system, it interprets of the organizational structure, physiological functions, and pathological changes of the human body by using allelopathy relationship. It can make a better network among organs, tissues, and meridian system, in order to keep the body healthy well. Yin-Yang and Wu-Xing theory work together to guide the medical direction of basic theory.

Chinese medicine also takes surrounding environment into the range of practical considerations. It embodies a harmonious whole complex view of life. It takes human being as the result of changes in the natural movement of the world. Thus, it can maintain the normal function of the body's state of life. Meanwhile, Chinese medicine also conveys a traditional humanistic spirit. It takes human being as natural person, even it is a whole with natural, social, and others. Since man has soul and spirit, which resides in the inner of body and not a puppet but the master of the body. It can be seen that the treatment measures are 'people-oriented' exactly in Chinese medicine. Chinese medicine judges the condition by means of four diagnostic methods, which is based on the natural evolution of organisms, embodied great exemplification of Chinese traditional science and the humanities spirit. Chinese medicine is out-ground harmony medicine between man and nature. It is a kind of humanistic medicine.

It is a convincing evidence of controlling and treating SARS epidemic sweeping the world in 2003. At that time, Severe Acute Respiratory Syndrome (SARS) outbroke in many parts of the world. In this global disease, western medicine captured 'corona-virus' by microscope as much as possible and then looked for methods to kill this virus. The process of looking for viruses and manufacturing viral vaccines were time-consuming. The way of western medicine was only for individual patient without conducive to large-scale outbreak control and treatment. From the holistic perspective and relationships among the prevailing climatic and environmental geography and the patient's syndrome performance, Chinese medicine did not look for 'corona-virus', while carried out treatment based on syndrome differentiation. It had achieved significant success. Chinese mortality of SARS patients is the lowest in the world. Since Guangzhou was the first to take Chinese medicine treatment, its mortality of SARS patients was the lowest in China. It is just a satisfactory respondent produced by Chinese medicine, which emphasizes an organic unity of the whole between the human body and the external environment. From a unique perspective on the nature and the law of life, it can cure disease by use of the unique theoretical system and abundant prevention experience. Although it does not find out external virus which caused disease, Chinese medicine tells us that a reasonable lifestyle is the best treatment for SARS. If there is no suitable host, the virus lost their homes and would not survive. Then there is no necessity to diagnose and treat individuals from SARS.

15.7 The dimension of local knowledge

Chinese medicine is a kind of charming traditional Chinese knowledge compared to western medicine. Chinese medicine is both a kind of medicine and a kind of culture. It's an important part of traditional Chinese cultural. Western medicine has come out from the western world, and expanded through the powers of modern time. In the process of standardization of Western medicine, Chinese medicine has been irrationally excluded in the metropolitan territory. However, understanding Chinese medicine is a way of learning Chinese traditional culture from the theory and practice of a specific area. It's a better way to understand Chinese medicine from the view of local knowledge.

Local knowledge is a kind of discourse weapon with a critique force. Its practical and constructive nature is the most valuable feature. At present, knowledge must be rooted in scientific practice instead of being completely abstract in representationalist theory. Chinese medicine looks for material characteristics and the structure of the Main and Collateral Channels in accordance with standard Western science. However, when people could not discover all the activities of the Main and Collateral Channels, or could not prove it to circulate and move in some material way, they arbitrarily denied its features. In fact, we should really change our perspectives to local knowledge. According to the view of Chinese medicine, Main and Collateral Channels is understood as a functional characterization of body-liquid and physical relationship. Then we do not need to confirm its physical structure but to deduce according to the research methods and results of the Main and Collateral Channels. Chinese medicine has some material basis and information transmission channels. It should be a holistic response to the function of physical interaction with the environment. It only needs to grasp the overall dynamic picture. At the same time, we should consider physics, chemistry, physiology, anatomy, and so on in western medicine as good at dealing with material structure and morphology. However, they have obvious defects in grasping disease from the whole of functions. Chinese medicine emphasizes a holistic view, by which it has the clear advantage of grasping Main and Collateral Channels and organs information. It's a solid foundation to grasp structures and functions of the Main and Collateral Channels. The inspections of Main and Collateral Channels should be based on the sight of local knowledge. We analyze and interpret by Main and Collateral Channels theory itself, by the way of philosophical thinking in Chinese traditional science, and by the thought of Yin-Yang and Wu-Xing. Of course, we do not exclude comparative study or the possibility of a complement to Chinese medicine by the local knowledge of western medicine.

Through conversion of the idea of local knowledge, there is no any purely universal truth or knowledge. Seemingly universal knowledge is merely a result of local knowledge whereas being standardized in essence. The notion of local knowledge provides wide range of space for the dissemination and exchange of knowledge. The confirmation of local knowledge plays a deconstructive and subversive role potentially. In the past, we do not have to think about the proof of 'axiom', nowadays it's inevitable that there are 'false' suspects if we impose a variety of local realities from top to bottom. This change of viewpoint on knowledge requires every researcher and his students learning to tolerate others and differences naturally. Theory of Chinese medicine is imbued with Chinese traditional culture. When it contacts with Western civilization, it is inevitably hard to eliminate barriers with modern science. However, it cannot be said to be unscientific when we use Chinese medicine to explain and diagnose disease which cannot be explained by western medicine. As a matter of fact, science is an only relatively true belief in a period of time, and any kind of the most advanced science has definite or potential 'error'. Nowadays it's reported that the US Food and Drug Administration issued a guidance document *complementary and alternative medicine products and FDA Administrator's Guide (draft)*. It separates traditional medicine (include Chinese medicine) from the 'Complementary and Alternative Medicine (CAM)'.Recognizing for the first time that the Chinese medicine is an independent scientific system with a complete theoretical and practical system, just as western mainstream medicine does, not merely a complementary to western medicine.

Through the above analysis, including the preliminary interpretations of practical activity, and basic theoretical knowledge about Main and Collateral Channels and viscera doctrine, we can clearly see that Chinese medicine pays attention to the holistic way of philosophical thinking, while western modern science emphasizes research approaches of empirical and analysis. They are two representative models of local knowledge in understanding the nature. They are not mutually separated and incompatible. In contrast, they are complementary and mutually reinforced. Up to now, the Main and Collateral Channels of *Huang Di Nei Jing* still effectively instructs clinical practices and theoretical studies of Chinese medicine. It plays a huge functional utility and serves for the purpose of the treatment and prevention of disease effectively, especially for the treatment of incurable diseases. Modern life sciences are also gradually getting rid of the limitations of Western reductionist thinking, just like physics getting rid of the particle theory, causal theory, and mechanist theory and prone to unified, non-mechanical, non-reducing organic natural idea. It begins to move toward holistic medication. Furthermore, the outstanding achievements of modern life sciences in regulating body functions also demonstrate the advantages and characteristics of Main and Collateral Channels in explaining human performance from holistic functions and dynamic transformations. No matter in academic thinking, theoretical study, and even practical experiences, Chinese and Western medicine should exchange and integrate with each other in the future.

16 Local knowledge (II)
Chinese theory of *Fengshui*

This chapter attempts to reconsider the practice and theory of Fengshui (geomancy) – a kind of 'local knowledge' – from philosophy of science. Meanwhile, it proves an analyzable case for the study of local knowledge issues. This chapter focuses on the exploration of practice characteristics of Fengshui by PSP as the tool of analysis, rather than appealing to whether ancient Chinese local knowledge is a science, or whether it contains a scientific component. We are not going to affirm scientific status of Fengshui, nor do we accuse practice activity of Fengshui without the understanding of its situation. We aim to test this Chinese traditional culture and to find out whether our long-term practice of dwelling activities lie in a reasonable practical knowledge and the reason for their mysterious trend, from the view of local knowledge in PSP. We just want to strip of their mysterious cloak, and to untie the practical characteristics.

16.1 Theories and practices of *Fengshui*

From a certain perspective of modern science, Fengshui in China have been regarded as superstition and witchcraft. The commonsensical understanding on them stayed on an early stage without further in-depth view. With huge impact, *Cihai*, the most convincing dictionary in China, defines Fengshui as follows:

> Fengshui, also known as "geomancy", is a superstitious thing in traditional China. It claims that currents and winds surrounding residential base or cemetery can lead to weal and woe of residents or dead people. It also refers to the means of testing a house or a cemetery'. Another important definition points out that 'Fengshui means the terrain and direction of homestead or cemetery. The brief judges good and bad fortune according to superstition.

Since 1980s, the understanding of 'Fengshui' undergone a subtle change along with the expansion of academic democracy, as well as with the impacts of changes and revolution in a variety of other thoughts, especially with the market-oriented economic development in social modernization process. In fact, for the Chinese people, 'Fengshui' (geomancy) is not unfamiliar, since a variety of knowledge about *Fengshui* began to emerge as along with ideological emancipation. Fengshui has a long history, which is very popular in the folk. Despite of ups and downs, it still spread so widely. Today it is applied not only to the understanding of ancient Chinese architecture, but also to modern architecture and home life, experiencing rapid development. In theory, academic works in which the theory and practice of Fengshui has been systematically studied in recent years involve researches on architecture, architectural historians, geographers (including landscape design), folklore, philosophy, and sociology gradually.

Westerners are showing an increasing interest in oriental culture, especially Chinese culture.

> Fengshui is an authentic local culture, however, it affects surrounding areas and even the United States and Europe like other branches of Chinese culture. From 7th to 16th century, Chinese Fengshui mainly spread in "Chinese cultural circle" consisting of neighbouring countries and regions; later in the 16th century, it gradually spread to Europe and other places.
>
> (Liu Peilin, 2001, p. 289)

From the very start, Eenest J. Eitel, a British Christian missionary, published *Feng-shui: The Rudiments of Natural Science in China* in 1873, which was reprinted continuously. So far, western scholars have not only studied the practice and theory of Fengshui, but also made various attempts and explorations of theoretical survival and application in the west for those practices and theories developed in China.

In fact, we can find by careful analysis that we have different understanding on the nature of Fengshui with multiple levels. When we judge whether it is based on science or superstition, we should follow these elements. First, we need modern science to serve as a background framework to understand Fengshui. Second, we make our determination on the basis of philosophical thinking, especially the criteria of philosophy of science. Further analysis shows that the underlying basis involves considerable problems. For instance, is contemporary natural science the only legitimate reference to judge as knowledge when dealing with nature? Can contemporary philosophy of science give consistent criteria of science accepted by all philosophers and scientists? The answer to both questions might be negative, when considering the continuous development of science and philosophy of science.

In recent years, the studies on contemporary philosophy of science, sociology, anthropology, and especially hermeneutics which is a branch of continental European philosophy, have revealed the following characteristics of science. First, contemporary natural science is not the only type of science. Second, scientific knowledge is not the only way to understand truth, since when people seek out truth, they need to consider the whole nature, with which knowers and actors interact. All of these factors reflect the truth. Third, contemporary philosophy of science has been a process of evolution. Different approaches of philosophy develop their own theories along with their animadversions, who come up the contradictory point for demarcation of science. The subsequent perspective tends to consider 'science' as a process, a kind of social practice for the communication between human and the nature. The newly developed PSP clearly define science as a kind of practice skills rather than rational beliefs. For example, Rouse (1987, pp. xii–xiii) points out that,

> I advocate that we should try to understand the concept of science in the area of scientific practices.... We find out that some local and existing knowledge is based on the intelligible and possible practices of using equipment, technologies, social roles, etc.

These practices are 'not for the purpose of applications, but they reflect that practical skills and manipulations would achieve a decisive and expected goal because of their won characteristics' (Rouse, 1987, p. x).

More importantly, according to PSP, some new perspectives which are totally different from traditional philosophy of science are brought forward. First, science is classified by practical activities; second, scientific practices refer to definitely local and social practices. Therefore, scientific knowledge based on local practices is not universal at the very beginning, which reserve their own important local characteristics. The universal characteristics of scientific knowledge serve only as appearances, which result from decontextualization and standardization.

By PSP, we are somehow excited and doubtful at the same time. What makes us excited is that some of the so-call 'non-scientific knowledge' which is excluded by traditional philosophy of science seems to have practical characteristics of communicating with nature, which can be treated and studied as 'scientific' knowledge. What puzzles us is whether the traditional Chinese knowledge such as traditional Chinese medicine and Fengshui theory can be regarded as objects in philosophy of science. Fortunately, science is classified in practice activities as a kind of local knowledge, hence we can question further: Is Fengshui a way of communication between Chinese and the nature? The answer is 'yes'. Of course, the statement that 'science is classified in practice activities as a kind of local knowledge' does not logically indicate that all practical activities and local knowledge are science. However, since Fengshui is a special and local way for Chinese to communicate with nature, it at least indicates that Fengshui has some sort of parallel and similar qualities with contemporary science, which enable us to compare science with Fengshui in our research. For instance, why are the modern architectural design, landscape design, and geography still making use of Fengshui to a certain extent? Does Fengshui still play a non-fungible role? How many parts of Fengshui practice derive from practical

Local knowledge (II): Fengshui 267

activities, and how many of them derive from the concepts of Yin-Yang and five elements? We would gain a lot in our analysis and research by considering and solving those questions.

All those above offer us a rational base to further analyze and study the Fengshui in the view of PSP. Meanwhile, the study of Fengshui at least offers us a beneficial case for the relevant studies of local knowledge.

To clearly illustrate the analyzability of Fengshui theory in philosophy of science, we first review demarcation theory in philosophy of science. And we summarize main points of the PSP – scientific practices and local knowledge. At last, necessary analysis is made on practice and theory of Fengshui.

16.2 The demarcation of science: a variety of demarcation standard and its changes

We can only describe demarcation criteria and its problems summarily in a historical order. In the traditional philosophy of science, the demarcation of science has generally gone through four stages: an absolute standard of logic doctrine, a relative standard of Historicism, the elimination of science demarcation, and the rebuilding of the demarcation of diverse criteria. Besides, according to PSP, the effectiveness and success of practical activities is a new criterion.

'Experience' is the standard cornerstone of logical positivism. The ideas that belong to science must be confirmable or falsifiable by experience. And sciences are universal propositions set forth in the system, among which each proposition is required to answer a question with clear answers. By this standard, although practice and theoretical activity of Fengshui may be able to explain the situation of some experience, it is difficult for them to agree with experience everywhere, and provide a coherent explanation. Practical statements of Fengshui obtain the testability through a lot of intuitive experience, hence it is consequently vague and indirect. Therefore, in logical positivism, we can ascertain that Fengshui is not science.

The historicism has softened and even undermined the foundation of absolute standard. Sciences before and after revolutions could not communicate with each other because of their incommensurability. Therefore, Thomas Kuhn can only discuss the demarcation issue depending on substitution for previous normal forms and the practice of scientific communities. For Kuhn's view, science can be only the theory and practice which are chosen successfully. In this historicist notion of paradigm, Fengshui has been retained even after several modifications and evolutions with rationality. However, compared with Western science, the paradigm just serves as a degenerate research program, and therefore Fengshui 'once' was science too.

According to the demarcation of science, since there are many different scenarios, we also need to take some more time. Feyerabend advocated to give science, witchcraft, and astrology the same opportunities and power, and to let the community decide what to do. Laudan believed that sciences in different periods had 'epistemic heterogeneity', therefore demarcation problem may be a pseudo-problem. However, since Laudan demonstrates this in a naturalistic way, this heterogeneity existed just in time rather than spatial coexistence. Therefore, it is still a historical evolution which naturalists has been replaced or selected by the community as science. Robert Rorty believes science didn't enjoy the truth and power of discourse exclusively. Sciences are just like other kinds of cultures. And Kit Fine also claims that, since science is rooted in the commonsensical way of thinking, in which no homogeneity is necessary in organizing the past, present, and future, the demarcation is meaningless for science. If demarcation is meaningless for science, it is no longer meaningful to ask whether practice and theory of Fengshui is scientific.

However, no matter whether in historicist or in Laudan's view, there are also differences between scientific and other types of practice in fact. With respect to demarcation problem, it is not meaningful for us to conduct delimitation. In Western culture, the kind of deep-rooted Western centrism makes it difficult for knowledge from other peoples and nations to be regarded as science. And in the European philosophical tradition, especially in France, these situations have undergone some changes. However, all other types of natural knowledge differed from Western science are still known as the 'ethno science', or 'non-normal science'.

After Laudan's dismissing view, Paul Thagard and Mario Bunge tried to rebuild the demarcation criteria on the basis of pluralism. They held that scientific demarcation could be possible. Demarcation can be operated metaphysically rather than be a metaphysical issue. It is notifiable that the manipulation needed here is bound to practice. Therefore, the rise of historicism and the elimination of demarcation serve as a good critique for scientific essentialism. Taking 'understanding heterogeneity' into science will inevitably lead to a pluralist view. As a result, it lays a foundation for the assumption of practice point for a scientific demarcation standard.

PSP did not discuss demarcation problem. However, it inevitably led people to handle the problem implicated by demarcation. How to judge a research program as science? The answer is appealing to a reasonable and effective practice that conforms to regulations. In that philosophy of science, as previously mentioned, science is the practical activities and way of dealing with nature. Scientific practice is derived from daily practices, including discursive practice. In that theoretical framework, is Fengshui in China a reasonable and effective practice dealing with nature? Or, is it a normative practice?

From the development of the demarcation theory, Fengshui has at least the following characteristics: first, according to Feyerabend, we do not need to seek for scientific components in practice and theory of Fengshui, since both science and witchcraft have rights to exist. According to Laudan, sciences in different historical periods have to 'understand epistemological heterogeneity'. And before contacting with Western science, Chinese people deal with nature on the basis of Fengshui. So at least until the rise of Western science, it has a heterogeneous scientific character and function in the land of China. In other words, more accurately speaking, Fengshui has its existence in terms of legitimacy and rationality in the practice of natural inquiry in China. Second, and more importantly, at least until the 18th century, the reason why industrious and brave Chinese people can survive in this ancient land of China, dealing with the nature also difficult for the Western people, is to realize that the reconciliation of a quarter to half people all around the world at the time lies in (of course, including other knowledge) practical knowledge and skills in thousands of years.

16.3 The Practice and local knowledge of *Fengshui*

16.3.1 *Fengshui and science as practical knowledge*

Scientific knowledge comes from the process of interactive practice between human and nature. However, traditional philosophy of science considers the importance of scientific knowledge as not based on its origin but on its abstract essential as universal above practices. Practices are the origin of scientific knowledge which only work at the origin terminal. Consequently, scientific knowledge removes their appearances of practice and become a series of abstract and universal statements and concepts. All of our perspectives and opinions about scientific knowledge are based on this traditional philosophy of science.

However, PSP has a different point of view. Joseph Rouse, for example, points out his opinion in the prologue of *Knowledge and Power* that knowledge is not only an superficial idea (such as a text, a way of thinking or a chart), but an existing interactive mode. This kind of mode contains phenomena, presentational objects, and corresponding scenarios. Only in these scenarios can appearances be understood when they are connected with practice. Therefore, in the view of PSP, scientific knowledge has practice characteristics. It does not mean that scientific knowledge has not been abstracted from practical operations, but it indicates that all the abstracted practical operations and scenarios do not lack scientific elements, which are always associated with science. According to the PSP, knowledge based on modern science still has fundamental connection with practice. Fengshui as one of the traditional survival practices for Chinese are also a kind of communication with nature. Compared with western science, Fengshui has a more concrete pattern, a closer relationship with humanities, and a more extensive connection with civilians.

From the viewpoint of practical knowledge to study Fengshui, our attention should focus on how Fengshui form the initial criteria via many constructive siting practices, and how these criteria are constructed as a series of major modes which guide people to carry on Fengshui practice in return.

The long-standing history allowed us to study through just few information behind historical documents. Some scholars would have discovered legends and historical evidences from Chinese earliest historical classics on ancestors' position in planning and building human settlements, palaces, and towns. The *Book of Songs* is identified as a collection of earliest poetry dating from the West Zhou Dynasty to the Spring and Autumn period in ancient China. Its geomantic omen is also entitled 'Xu-yu (胥宇)'. Documents such as *Book of Songs* seemed to be able to show more or less some practical signs of Fengshui or the geomantic omen. For example,

> the gourds grow in long vines, our people was the first to rise. By Qi and Ju they lived age after age, the old duke Tan-fu came on stage. They made huts and caves, but no house nor naves. Tan-fu came at dawn, horseback he was on, galloping along the strand, at the foot of Qi he did land. There he met lady Jiang, to find a place to settle down. Fertile was Zhou's plain, with violets sweet as sugarcane. He began to plan and design, by tortoise-shell he did divine. It hinted: to stay here and now, so he house there with his frau.

The context recorded the course of activities to select a residence, which began with Xu-Yu, then the people divined by tortoise-shell and housed there in the end. Those early evidence showed geomancy was of particular practices: including inspect, comfort, mount, go, gaze, view, take the place, survey, and a range of other movements.

Thus we can infer from the above account that, although Fengshui or geomancy had been defined by subsequent classics and geomancers, it still kept the practicality by closely associating with natural environment. Whether a place is thought to be good or bad depends upon the place features itself instead of geomantic theories. Of course, the leading characters, such as dragon vein as well as fronting water and with hills on the back, has been gradually standardized by geomancers, forming a batch of fixed explanation modes. The afterward instructions have also begun to break away from the practice, found to be more mystical and powerful. Therefore, when introducing and explaining geomantic activities as well as corresponding theories, our incidental task is tend to draw the real practicality out of its intentional mystification.

16.3.2 Local knowledge and Fengshui

When talking about local knowledge, it does not mean any specific knowledge with local qualities, but a new notion of knowledge, in which 'local' or 'native' is not a concept based only on specific areas, but comes down to specific contexts where knowledge is created and advocated. This context contains values and ideals formed in specific historical conditions of culture groups and subculture groups, positions and perspectives determined by specific interest groups. The meaning of local knowledge is that: since knowledge is always created and advocated in specific contexts, we had better emphasize the detailed context when reviewing knowledge and universal principles

(Xiaoming Sheng, 2000, p. 36). Accordingly, Joseph Rouse thinks that, basically speaking, scientific knowledge is local, since they are reflected in practices, which cannot be abstracted completely as theory or formula that is separated from specific contexts for the purpose of application. Based on this concept, Fengshui theory and other traditional knowledge, as a kind of local knowledge, are gradually accepted by people.

It is sure that Fengshui is a kind of local knowledge. However, this affirmation is misapprehended in many ways. One misconstruction is initially from the attitude that Fengshui practice is merely a Chinese local knowledge. That is not right. The locality of Fengshui means that we need to track back to the root of knowledge and the specific context in which it is produced, applied, and advocated. The values and ideals formed in specific historical conditions of culture groups and subculture groups are also necessarily considered and explained. Therefore, the first question we need to ask is: what is the relationship between Fengshui practice and the daily life of Chinese people? Secondly, what is the relationship between Fengshui practice and the world view of nature in China? Thirdly, why are there so many mysteries in Fengshui? Is Fengshui a kind of scientific knowledge or a sort of witchery? All those questions should be studied and explained under the notion of local knowledge.

As for the first question, Fengshui is actually a way for Chinese to prudently inspect surroundings, to acclimatize nature, to make use of and change the world properly, and to create a harmonious living environment in which the climate, geography, and human are perfectly synchronized (Qiheng Wang, 1992, pp. 3–4). Chinese always put their greatest emphasis on surroundings. This kind of practices actually reflects a great concern about 'micro-environment and micro-topography'. What's more, Fengshui practice is not merely about nature, but the relationship between human and nature. In Fengshui practices people are able to make full use of their subjective initiative to create a comfortable and harmonious living environment. Heidegger's 'poetic habitation' image displays its practicality fully in Chinese Fengshui practices. For example, the influence of micro-topography on microclimate and its environment can be explained and recapitulated in Fengshui, in which the micro-topography is compared as 'aperture' that is often surrounded by hills for three or four sides, with topography of north high south low. It can be a restrained-type basin or mesa facing the sun and seating in the shade, or even a manual topography. That is the typical pattern of 'aperture', which is considered to be a place 'gathering winds' as Fengshui structure (see Figure 16.1). Some scientists make a comparative analysis of this topography in contemporary sciences, in which those explanation and cognition of relations among micro-topography, microclimate, ecology, and natural landscape are mostly involving scientific significances (see Figure 16.2). (For the purpose of antecedent comparison, we have to make use the viewpoint that modern science is a standardized knowledge.) (Qiheng Wang, 1992, pp. 26–32).

As for the second question, the answer could be displayed as follows. Firstly, it involves the Chinese culture, in which human being in a highly harmony with

272 *Local knowledge (II)*: Fengshui

nature is the main feature of the world view of nature in ancient China. There is always a fundamental distinction in the world view of nature between ancient China and western countries, which in ancient China emphasizes the relationship between human and nature: human is included in any physical nature, while in western world one emphasizes the nature itself and excludes human from physical nature. Since Han Dynasty, the nature described in landscape and pastoral poetries is always with human in it, rather than a pure physical non-human nature. Secondly, this question also involves the normativity of Fengshui practice in Chinese culture. In traditional way of thinking, the former practices carried out by typical characters are handed down and theorized by later generations. The later generations do not practice longer, or their practices subjected themselves to the theorized forms. It may be the same for Fengshui practice. Some ancient

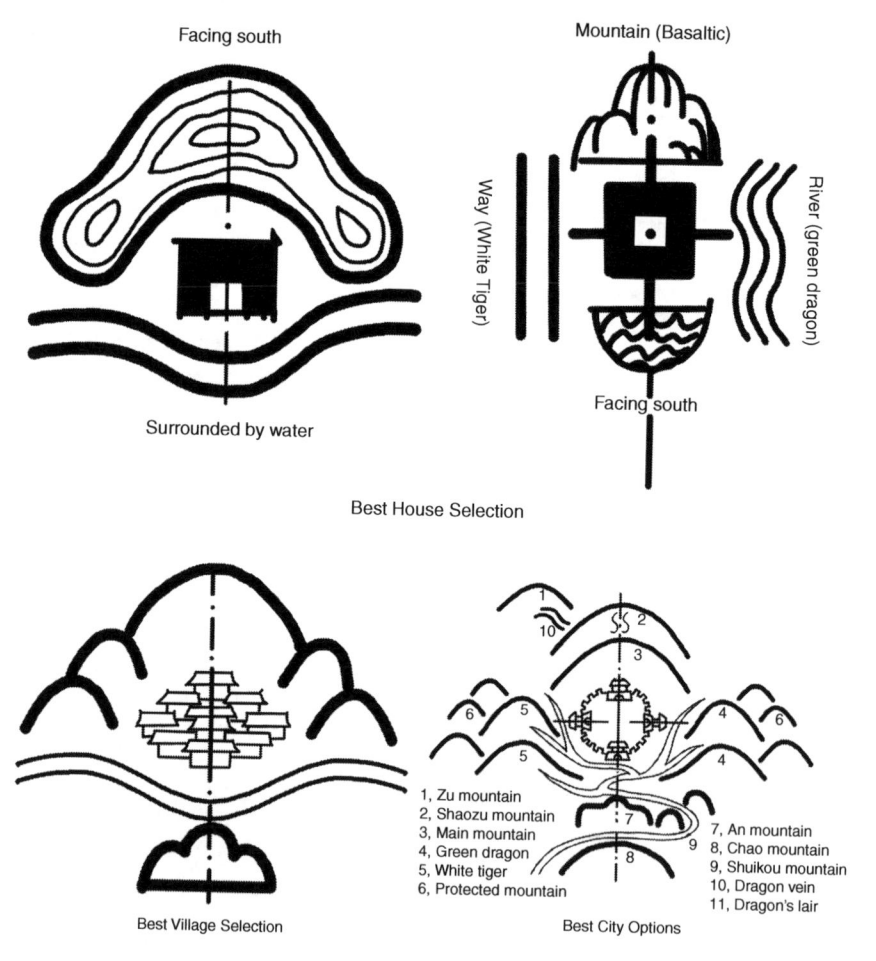

Figure 16.1 Optimal layout and sites of house, village, and city.

Local knowledge (II): Fengshui 273

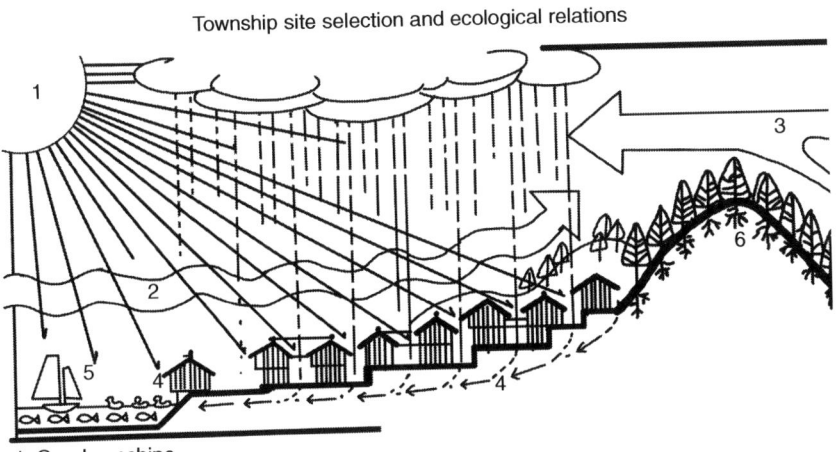

Township site selection and ecological relations

1, Good sunshine
2, Accept the summer southerly wind
3, Block the cold wind in winter
4, Good drainage
5, In order to facilitate contact on the water
6, Soil and water conservation in regulating climate

Figure 16.2 The scientific explanation of a Fengshui layout.

Chinese tribal chiefs carried out Fengshui practices and handed them down. Gradually, these practices are standardized for the latter generation to follow. The succeeding Fengshui theories can be regarded as the complementary or modification based on the former practices. The five elements theory, *Qi* theory and *Yin-Yang* theory are also important scenario factors in ancient Chinese culture, and Fengshui practice is more or less influenced by them. Moreover, Fengshui has a close relationship with *Qi*: the claim that 'gathering wind and *Qi* to gain water' is the standard to choose an optimal living place. These theories also follow the requirement that 'human being lives in a highly harmony with nature'. However, when it comes to the origin, we have to consider Fengshui practice as a kind of practice from people's daily lives. The understanding of our surroundings is neither a series of beliefs, object-oriented regulations, nor the traditional 'cognition'; it includes all the skills and abilities inside us (that is, the practical knowledge). Only by assembling ourselves, practical objects, and specific surroundings as a whole, can we fully understand our world and the real meaning of science. And then our interpretative behaviours can also be understood (Figures 16.1 and 16.2).

Why there are so many mysteries in Fengshui? Is Fengshui a kind of scientific knowledge or a sort of witchery? In the traditional classification of Chinese knowledge, Fengshui is named by 'Kan-Yu (堪輿)', which is a kind of so-called *Fang* technology, and which does not distinguish rational from irrational methods, and even confused Confucian and wizard in some ways. Here the 'wizard' does not mean something crazy or irrational, which is very different from its contemporary use. In addition, *Fang* technology tends to purposively make up things mysterious, thus it is not plausible to take *Fengshui* as irrational or mysterious merely according to its apparent discourse. The complexity of *Fengshui* makes us unable to tell which parts of it is science and which superstition. In a word, *Fengshui* is a kind of mixture which contains both scientific knowledge and wizardry.

16.4 The normative framework of *Fengshui* practice

Fengshui activity are already considered as complex and esoteric. In recent years, a systematic study of *Fengshui* has been carried out from architecture, architectural, history, geography, philosophy, folklore, and sociology, which is helpful for understanding its theoretical content and epistemic value, in order to figure out Fengshui essence completely.

According to the method of *Fengshui*, choosing inhabiting environment usually needs to consider four aspects: surveying dragon, observing sand, viewing the water and selecting the cave.

1 Survey 'Dragon'. Dragon vein is a mountain chain. Mountain was regarded as dragon, which looks like proton volting. The stretches of mountains mean the veins. *Fengshui* master's duty is 'seeking the dragon and finding the veins', also 'finding the dragon and watching the stretches' to refer to geographical situation. The dragons have different sizes (a simile of mountain sizes). First, finding the dragon depends on water resources. A big dragon is accompanied with a large pinch of rivers; while little dragon goes with streams. Large dragon couples with springs while small dragon is followed by branches. *Fengshui* survey on dragon vein actually refers to the setting of geographic relationship between mountains and water flow based on dragon imagery.

2 Observing 'sand pattern'. The 'sand' in Fengshui reflects conceptual relationship between mountains and residential building location. In *Fengshui* pattern, sand refers to mountains and hills surrounding cities, villages, and houses. It was said that

Dragon served as the doctrine for the king, while sand served as the doctrine for officials. It became a city if it's far, while a case if it's near. If eight winds lean close to each other, it's protected, while water gaps gather, it became a pass.

Surround city walls with sands in expressing clearly the meaning of protecting winds and guiding water. The geographical orientations expressed by west, south, east, and north refer respectively to Suzaku, basalt, Dragon, and White Tiger. However, geography regards front range as Suzaku, the ushiyama as basalt, the left mountain as Green Dragon, and the right mountain as White Tiger. It also made use of the four names to distinguish four mountains, in order to take mountain shapes as these things. It was said that Sand Mountain should 'protect its point guard zones from wind, force, discount and channel surrounded by love, so the dragon winds, white tiger was tamed', 'basaltic became depressed', and 'Suzaku become docile'. This concise imagery vividly describes the great significance of Sand Hill for exiting ecology and landscape, as well as the psychological feeling, and the perfect quality of the environment (Wang Qiheng, 1992, pp. 50–1). Therefore, the superficial mysterious language actually implies some long-term experience of ancient Chinese *Fengshui* masters. That experience is not only necessary for our interpretation of the relationship between architecture and environment of ancient China today, but also shows effective practice and specifications of the ways of Chinese people dealing with nature both in ancient time and today.

3 Surveying and estimate 'the flow of water'. It was believed in practices and theories of Fengshui that, 'the lucky land cannot do without water', 'The Tao of geographical features of a place lies in the locations of mountains and waters' and 'water first for the method of Fengshui'. Grounding on some scholars' research, we believe that water is considered to be vital in Fengshui not only because it's one of the five elements and source of lives, but also due to the following reasons, especially in the practice: first, water is closely linked to the environment, namely so-called Earth Qi or vitality; It goes as follows: 'Mountain features water as its blood vessel'; 'The biotic Qi is let out in case of disappearance of water and gathered in case of infusion'; and 'the people are wealthy at deep waters and poor at shallow; dense at plentiful waters and scarce at insufficient'. Water is essential to the existence and growth of lives (Wang Qiheng, 1992, p. 55). Second, water is also a better position used to transport and set risks. Third, water is also highlighted by Fengshui because of water hazards. And finally, water plays an important role in the view of building landscape.

4 Select site and 'pick dens'. The way of picking dens' rests on that the places, where mountains and water meet with each other and where Yin-Yang mix together, are thought to be beloved positions. In other words, the emphasis on the site selection for towns or other buildings should be put on the endocentric enclosure of multiple hurdles such as Dragon, Sand and Water. On the one hand, this situation can be exempt from the invasion of wind and sand, which embrace the sunlight with a back against mountains. The feminine-masculine harmony creates a well-behaved ecological microclimate. On the other hand, various landscapes like Dragon, Sand, and water, which are included in the dens, give humans the richest feelings accompanied by spiritual consignment to let the eye travel over the great scenes and let fancy free. 'Be sure to maintain its natural situation without going against its advantages of surrounding and safeguard' (Wang Qiheng, 1992, p. 58).

In a word, *Fengshui* practice is identified as a kind of emotional and visual process of observation and experience, as an on-site surveying process from outside to inside, from external to internal, and from the whole to parts. The process focuses mostly on water and mountains in the region itself followed by a variety of human feelings (visual feelings on the relation between human and nature, visual feelings on aesthetics, etc.).

The framework for practical normativity is likely found in its utilization by both Xingshi School and Liqi School, which are different approaches in Fengshui activities. After the theorization, Fengshui was valued as a significant part of Chinese culture, but at the same time marked a great growth in divided nature of mystery.

1. The theory of Yin-Yang and Wu-Xing claims that the world bases on substance which happens, develops, and changes under the efforts of the Yin-Yang. The theory for Yin-Yang basically includes its opposition, mutual rooting, growth, and decline as well as its transformation. The theory which was created by the application of Yin-Yang into Fengshui practice suggests that the north of mountains and the south of water is identified as Yin, while the south of mountains and the north of water is identified as Yang. The theory of Wu Xing argues that the necessary elements in the world can be classified as five most basic substances including 'wood, fire, earth, metal and water'. Wu Xing, which exist in the relationship of mutual transformation, interaction, generation and restriction, are moving and changing constantly, thus becoming the source of all things and giving an explanation to connection of them.

2. The theory of 'four seasons and five directions'. 'Four seasons' refer to spring, summer, autumn and winter. The theory of genethlialogy represents flourish, beauty, rest, exhaust, and death for Wu-Xing in manifestation of rise and fall in four seasons; 'five directions' refers to the east, south, west, north, and middle. The five elements (Wu-Xing) and the Chinese era they represent stand for their own directions.

3. Hetu and Luoshu are deemed as the source of Chinese culture, which is consequently entitled 'Heluo culture'. Although the origin of Hetu and Luoshu is mysterious, Fengshui has been valuing them as their own theoretical foundation. The Lord Fu-Xi derived from Hetu the Eight Trigrams, which were called 'connate Fu Xi trigrams'; At the ending of Shang dynasty and the beginning of Zhou dynasty, The King Wen of Zhou derived from Luoshu another kind of sequence, which were called 'postnatal King Wen sequence' or the Sixty-Four hexagrams. 'Connate Fu Xi trigrams' take charge of people's lives, including ordinary people life and production such as diet, dressing, birth, death, illness, old age, cultivation and plantation as well as metallurgy. 'Postnatal hexagrams' take charge of 'restraint', including ordinary people's sacrifice, weddings, funerals and other relevant activities.

4. The theory on 'seven luminaries and nine stars'. In *Fengshui*, 'seven luminaries' include the sun, the moon and five planets such as Mars, Mercury, Jupiter, Venus and Saturn. According to celestial equator and ecliptic,

the sky can be divided into twenty-eight constellations, which are further separated into four groups called 'four spirit beasts', namely Azure Dragon, White Tiger, rose-finch and tortoise on behalf of four directions. 'Nine stars' represents Big Dipper plus Uranus and Zi-wei Star. We have learned from the previous context that, in the practice norm of Xingshi method of Fengshui, Azure Dragon, White Tiger, rose-finch, and Tortoise are only representatives of hills in different directions surrounding potential sites. Therefore, the terminology in Fengshui was nothing but mystified by geomancers intentionally or a part of certain cultural system under the banner of Mysticism. Once demystification continues, the internal practicality is bound to be uncovered.

The development of the main theoretical tool includes ancient Chinese practice of elective residence in cultural system, gradually forming a mystical interpretation of the theory of Fengshui, and a mature culture system frame with mystical symbols. By theorizing, Fengshui practice is limited by false rigid specification to become an astrology tool servicing Chinese politics and culture in which the characters of natural knowledge and practice are greatly weakened. In fact, it is the process of theorization that imposes the practice of Fengshui more human characteristics, mystical and philosophical nature, among which natural and practical parts are further obscured. This combination of two characters in knowledge including nature and humanism may be known as the knowledge integrating theory and practice.

Fengshui has a purpose for the ideal achievement of human longevity through practices. And Fengshui theory has become a system by itself as one of two knowledge systems differed from modern science because of its unique assumptions and a series of relevant inferences.

16.4.1 The world view of 'the unity of heaven and human', and 'Harmony of Yin and Yang'

Fengshui is considered to be the Natural Unity, namely the harmony tween seeking for the best resident with the environment. Both Fengshui and Chinese medicine claims that the body has its own small universe of balance between yin and yang, and so does the geographical environment. Fengshui just combines the balance of yin and yang of the human body and geographical environment to achieve a Natural Unity.

16.4.2 The fortune of the Qi in Fengshui

'Qi Theory' is the core of Fengshui, which holds that 'Qi' determines the location as good or bad. All environments 'gathering Qi' are propitious, hence we should pay attention to multiply vigour and depression. 'Vigor' also became an important principle of Fengshui theory. It was said that the 'yin and yang of Qi ... goes on the earth when it's vigorous'. All specific measures of

Fengshui theory focus on how to find a rallying point of vigour, how to greet, sodium, gather Qi, and obtain the harmony with Qi in residence for them, thereby helping to improve the living environment and ensure human health.

Fengshui theory has divided into Xingshi School (Jiangxi school) and Liqi School (Fujian School) since the Tang and Song dynasties. The common place between these two schools lies in their attention on Qi. Xingshi School appeals to mountain situation and the ins-outs, rather than the auspicious of the position itself, which focuses on surrounding landscape, topography, and environment imagery in order to find dragon cave, watch dragon and check water. However, Liqi School focuses on the orientation of Qi, in emphasizing the use of the compass, and the importance of yin and yang, gossip Lunar, and 'seven luminaries and nine stars'. Liqi School does not pay attention to shape and size. Xingshi School is simple and easy to operate due to its attention to shape. Liqi School has relatively cumbersome and deep philosophy of *I Ching* since it focuses on inherent theory. Through the exploration of the two factions of Fengshui, we can find that there are two methods in Fengshui to judge good from bad fortune: astro-diagnosis and shape. They are used in Fengshui, making it more complex and obscure. In fact, the two schools of Fengshui have no complete difference, but merely different emphases. Excluding its complicated and mysterious matters, you can still find its practical and theoretical features.

Xingshi School is derived from the basis of experience with the humanities and interactions with social factors in practice. Taking shape as the main perspective, it determines the merits of residence and surrounding environment by the image, since this method is relatively straightforward. It makes Fengshui judgment by adopting analogy-oriented approach based on the consensus formed by the long-term production and life. This approach clearly has its empirical basis, and Fengshui masters continue to accumulate experience from their predecessors. But the inductive results are not once valid for all the time. Though most of the time they are effective, there will certainly be wrong some times. In the image astro-diagnosis, the emphasis on the image itself (determined from the empirical concept) constitutes a standard, so that the experience will be narrowed, or even get rid of the empirical constraints. The factors of Fengshui is mainly held by those masters, whose understanding in Fengshui practice often became the most important factor in affecting attitudes of others on Fengshui. The behaviour of Fengshui masters directly determine the reasonableness of their use, not only making it difficult to judge whether the practice and theory of Fengshui retain its empirical basis, but also directly affecting the way in which people think about the natural knowledge of Fengshui.

Liqi School determines the quality of living places mainly by making use of yin-yang, Lunar, gossip, astrology, and other symbolic measure orientation systems. That imposes a unique interpretation model of Chinese culture on the survey for nature in advance. Fengshui masters will make a deduction based on the Wu-Xing in order to make it more compelling. Their analysis and interpretation seem to be logically precise. However, since the interpretation system is not sufficiently sophisticated, Fengshui masters cleverly combine experience with theory to convey mysterious interpretations. In fact, for the purpose of making more money, some Fengshui masters in history often imposed Fengshui with some Taboo system formed by the combination of Wu Xing and auspicious locations. By doing this, Fengshui masters might conduct misinterpretation and destruction of these norms. Meanwhile, in order to cover up this practice, those masters will strengthen the mystery of Fengshui practice, by making it nonsense and sometimes harming practice of living, which mainly leads to the acknowledgement that Fengshui is a kind of superstition.

16.5 *Fengshui* in PSP

This chapter will describe Fengshui from the perspective of philosophy of science. An unavoidable problem encountered is the analyzability of Fengshui, which should be asked with two meanings: first, can the practice and theory of Fengshui be studied in philosophy of science? Second, what can be standardized in Fengshui? What affects its standardization? In fact, this chapter has made a practical answer to both questions. Fengshui practice and its theoretical activity could be interpreted in our PSP paradigm.

The characteristic of Chinese knowledge in integrating nature with human implies humanistic meaning, which were once considered non-scientific in the perspective of Western science. However, PSP claims that western science itself has cultural characteristics. Science belongs to the category of history, standards, and connotations which change along with different historical periods. If modern natural science is regarded as the only standard for science, it is a bit unfair. Knowledge is always local, in the sense of which there is no genuinely 'universal knowledge' at all. The determination of local knowledge can deconstruct and reverse traditional concept of knowledge and science. Therefore, as local knowledge, Fengshui can escape from marginalization and appear as an object of philosophy of science.

Our world is divided into different places. Each knowledge system in different places has their own value. The natural philosophy based on the Wu-Xing, Qi, and Yin-Yang theory constructs the scientific background of ancient China. We cannot ignore the natural parts of Fengshui just because of the blend of humanization, nor can we deny its strong characteristics of practice merely because of an inappropriate philosophization.

Researches on *Fengshui* provide cases for PSP. As local knowledge like Chinese medicine, Fengshui clearly has a different paradigm from modern science which is also local knowledge in the West. After standardization of western knowledge, how much value and meaning does Fengshui retain today? Compared with western knowledge, which has become the dominant scientific knowledge, what are the shortcomings and weaknesses of non-standardized Fengshui theory? If comparative studies of PSP and related theories (such as cultural research in laboratory by SSK, feminist philosophy of science, and postmodern philosophy of science) can be carried out, it will be a very meaningful project. I believe this study can promote researches on changes in demarcation criteria and PSP in philosophy of science. Researches on those issues could become an important part of researches in Chinese philosophy of science, which is also a particular aspect of Chinese philosophy of science contributing to philosophy in general.

Some believe that 'scientific evolution theory of post-Kuhn era moves from social studies of science to cultural studies of science (CSS)' (Sheng Xiaoming, 2003, p. 14). J. Rouse (1996, p. 238) tried to answer the question 'what are the cultural studies of science? I make use of the term broadly to include various investigations on practices, by which scientific understanding is articulated and maintained in specific contexts and extended into others.' CSS can be said as new syntheses on the basis of SSK, a study on the science in various cultures integrating history, philosophy, political science, sociology, anthropology, and even a full range of feminist perspectives or literary criticism. Therefore, CSS is more suitable for the study of local knowledge such as *Fengshui* theory.

In the view of CSS, all scientific knowledge involves a cultural dimension, no matter whether it is natural sciences or local knowledge. Natural science is less obvious on this since its cultural dimension has already hidden through standardization. As Nancy Cartwright claims, the nomological machine served as a local condition for contemporary science, as well as the condition to extend to other contexts. That is to say, empirical knowledge must act through local cultural factors. Therefore, science is not as sacred as before, while local knowledge is not actually humble. In fact, when Westerners choose their residence, they consider also cultural factors. We can find distinctive features and human factors behind of choosing residence from the differences between those who like to live on the top of the mountain and others who seek for enclosing environment.

Today, when understanding the natural world in its diversity, modern science should be on an equal status to some extent with other cognitive ways, which is a necessary implication from the study of local knowledge. In the process of choosing beliefs, modern science has occupied a privileged position in epistemology. However, the current situation is that people need to transform the concept based on traditional culture to a way of understanding modern science, with a comparison on which one is more important. That is even not enough. In every nation or continents, we should not dismiss knowledge, beliefs and ideas of others, but respect the effectiveness of local knowledge. If the scientific community can participate in communication about local knowledge in different regions and ethnic groups, the study of local knowledge will be greatly promoted. This is also one of the reasons that we study Chinese *Fengshui* from the point of view of PSP.

17 Local knowledge (III)
Ethnobotany

With the perspective of PSP and examples from ethnobotany and Mongolian natural knowledge, this chapter especially studies cases from particular types of Mongolian natural knowledge, including folk culture, ethnobotany, and medicine. It indicates that local knowledge has not only significant importance but also strong scientific and cultural significances.

17.1 Local knowledge from the perspective of PSP

In the past, the knowledge generated in Chinese minority cultures interacting with nature could not be taken as an object of philosophy of science, since non-Western knowledge of nature might not be reflected in the academic mainstream. Western philosophy of science thus puts forward important criteria to define 'knowledge'. However, with the rise of historicist philosophy of science, followed by a variety of anthropological researches, SSK and other post-modern thoughts gradually develop, and the mainstream philosophy of science has disintegrated. An important consequence of that collapse, namely, is the recognition of cultural relativism.

In addition to that background, through absorbing SSK and results of feminist studies, PSP further evolves from a theory-dominant to practice-dominant position. In this change, science in the first place is the practice interacting with nature, rather than a net of statements and proofs. Secondly, scientific knowledge is changing from the recognition of universal knowledge to local knowledge. Scientific knowledge is local and there is no purely universal knowledge. Universal knowledge is constructed as a dominant ideology in philosophy of science. Illusion of universal knowledge rises just because some local knowledge has achieved the status of superiority in an ideological sense, and then turns itself into a standard in applying to other non-Western cultures. However, for the PSP, there is no certain knowledge superior or inferior to others. All knowledge is effective based on the process of interacting with nature.

Since science in the first place is a kind of practice, local knowledge has become more and more of a priority. We are not going to require natural knowledge from ethnic groups to satisfy a unified standard, but to explain the way they are dealing with nature in terms of their own local knowledge, and to explain how to generate their own kind of 'science'.

Therefore, non-Western ethnic knowledge of nature has undoubtedly become a possible subject to study. In the mainstream of philosophy of science, it is impossible to take such an object for research. In other words, local knowledge and PSP support the study of natural knowledge in minority cultures, in order to promote the variety of ways of dealing with nature for the reasons of cultural diversity.

17.2 Mongolian natural knowledge

Early Mongolian local knowledge in dealing with nature may not have a clear independent form, since it may be reflected in its folksongs, geographic names, botanical name, unique dwellings, and many other aspects. What is important here is not only exploring but also displaying the importance of local knowledge in case studies. The following are some examples.

17.2.1 Natural knowledge embodied in Mongolian folksongs

Mongolian folksongs embody the Mongolian lifestyle. It is said that the Mongolian plateau is the home of song. For example, a folk song named Hang'gai Mount goes: '... The Mount Hang'gai on which ridges we can graze in summer/ The Mount Hang'gai that benefits us all/ The Mount Hang'gai on which hills we can graze in spring/ The Mount Hang'gai that benefits us all/ The Mount Hang'gai on which sunny sides we can graze in winter/ The Mount Hang'gai that benefits us all/ The Mount Hang'gai on which back side flows clean streams/ The Mount Hang'gai that benefits us all ...' That song, for instance, includes some basic knowledge how to survive in the environment and knowledge of ecological conditions, such as in particular seasons where is suitable for grazing, and where people can find clean water, etc.

Grazing activity is one of the most common practices of Mongolians. The grazing practice, which is highly relevant with the tribal livelihood, requires

Table 17.1 'Large and small Hang'gai mountain': local knowledge in Mongolian folk songs

Time	Spring	Summer	Autumn	Winter
Place of practical activity: large and small HangGai mountain	Suitable plants grow on hillside	Suitable plants grow on ridge	Suitable plants grow on the lowland of the mountain	Suitable plants grow on the sunny side of the mountain
Important resources: water source location	The source of clean water (clear springs) on the back of the mountain			

solidarity spatial knowledge, calendar knowledge (seasonal knowledge), botanical knowledge, and other kinds of knowledge. Through the form of Mongolian folksongs, the most effective form of common practiced, the relations between Mount Hang'gai and grazing practices become very clear and impressive in herdsmen's mind (Table 17.1).

In fact, it is also an imperative. By regularly chanting, people gain the knowledge of the suitable place for grazing in each season. Therefore, the herdsmen realize that grazing in other time or place are wrong, thus forming a taboo in grazing practices. In fact, there is much other ecological and livestock knowledge in Mongolian folksongs, awaiting further exploration.

17.2.2 Mongolian botanical and ecological knowledge

As for ethnobotanical study, some important results have been produced both domestically and abroad. At the same time, Mongolian ethnobotany has been either influenced by other theoretical and practical studies, or completely independent research, which also brings out some preliminary but very important achievements.

Mongolian people in their living and production practice has accumulated a lot of knowledge and generated a specialized botanical culture.

Chen Shan, an ethnobotanist, has pointed out that 'the original Mongolian names for plants have profound scientific connotations. Some names are more accurate than those given by the modern plant taxonomists'. He raised an example of the original Mongolian name for the Splendens genus 'inebrians', which is 'deresunhoor'. 'Deres' is the original Mongolian name for the representative species – Achnatherum splendens [A. Splendens (Trin) Nevski], which can be also understood as a generic name. But for inebrians, Mongolians formerly put 'hoor' after its genus name 'deres'. The meaning of the Mongolian word 'hoor' is 'poison', so its Mongolian name pointed out the toxic nature of inebrians (Hasi and Pei, 1999, pp. 42–7). Using its name to allege its nature, especially to point out its harm for horses, which are the critical means of living and production for Mongolians, is of strong practical meaning for herdsmen to recognize directly from language the relations between plants and human activities.

Some scholars study the meanings contained in the Mongolian names of plants. For example, for the Allium plants growing on the Mongolian steppe, some scholars have found that the number of names collected form the folks is 19. According to the evidence of identification and classification of specimens, it is found that the 19 names have a corresponding connection with the plant taxonomic 10 names (Table 17.2). Therefore, Mongolian names for the Allium genus involve different patterns, including one-to-one mapping, one-to-double mapping and one-to- many mapping.

The original name is a semantic unit, which has only one meaning. In Mongolian vocabulary, the original name for Allium plants means a botanical name with no other meaning. Mongolians use different words for different Allium species, which suggests a certain level of classification. In other words, the diversity of the Allium genus also enriches the Mongolian language. The name is derived from the original name plus some descriptive qualifier. For derivative names, 'songgin' and 'gogd' are simple original name. When they are combined with qualifier like 'wumhei' (smelly), 'herin' (wild), 'zherlig' (wild), 'har' (black), and 'henzhe' (nocturnal), they become plant names. A derivative name contains information of a plant's odour, colour, habitat, and growing season. For Mongolian names of Allium plants, 'senghus' is neither an original name nor a derivative name, while having the semantic meaning of 'shaking' or 'shake', which is relevant to its posture in wind. That kind of information is the most direct and recognizable feature of the Mongolian nomadic people (Figure 17.1).

In the 1950s Wang Dong, a pasture expert, noted the traditional knowledge and experience of Mongolian herders in using Allium forage. He recorded the Mongolian names of Allium plants, such as 'Min'ge' (Hill leek) and 'Tana' (alkali leek), and described that 'various livestock like eating hill leek, and less likely to lose weight. Hill leek is a fine grass. In summer, herders specially look for the place where that pasture is growing to graze' (Wang Dong, 1995, pp. 85–6). In the study of forage, the majority of Allium species are considered to be of fattening type or fatten pasture. According to the classification of forage plants in Mongolia chives, Allium base, mountain leek, wild

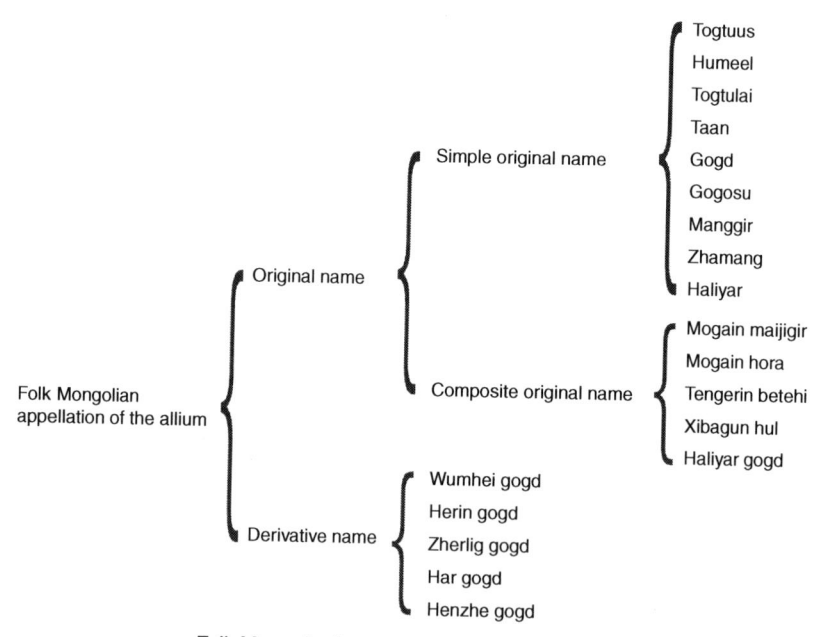

Folk Mongolian appellation of the allium:
- Original name
  - Simple original name: Togtuus, Humeel, Togtulai, Taan, Gogd, Gogosu, Manggir, Zhamang, Haliyar
  - Composite original name: Mogain maijigir, Mogain hora, Tengerin betehi, Xibagun hul, Haliyar gogd
- Derivative name: Wumhei gogd, Herin gogd, Zherlig gogd, Har gogd, Henzhe gogd

Folk Mongolian's named relationship of the allium plants

Figure 17.1 Folk Mongolian's named relationship with the Allium plant.

Table 17.2 The relationship between Mongolian and academic names for Allium

Mongolian name	Academic name	Chinese name	Nick name
mogain maijigir	Allium anisopodium	矮韭	矮葱
mogain hora			
wumhei songgin	A. condensatum	黄花葱	
togtuus	A. macrostemon	薤白	小根蒜
tengerin betehi			
humeel	A. mongolicum	蒙古韭	蒙古葱,沙葱
togtulai	A. nerniflorum	长梗韭	花美韭
senghus			
taan	A. polyrhizum	碱韭	多根葱,碱葱
gogd	A. ramosum	野韭	
gogosu			
herin gogd			
zherlig gogd			
har gogd			
benzhe gogd			
manggir	A. senescens	山韭	山葱,岩葱
zhamang	A. tenuissimum	细叶韭	细叶葱
xibagum hul			
haliyar	A. victorialis	茖韭	
haliyar gogd			

grasses, dwarf grasses, Zoysia grasses, bald leek (A llium leuco2 cephal um), sand leek (A. Bidentatum) and Allium przewalskianum (A przew alskianum) are excellent forage plants. Therefore, in this sense, it is proved from the perspective of modern science that Mongolian knowledge and experience – the use of Allium plants for grazing – have a consistent core in common with modern science. Their proto-typical knowledge corresponds to the evaluation of modern agrostology, which provides practical knowledge for pasture study with extensive folk experience.

In fact, the Allium plants, which are widely distributed in the prairie region and closely connected with Mongolian people's livelihood, also have some influence on culture and arts. In Mongolian folksongs of praise to grasslands and pastures, Mongolia leek, wild leek, mountain leek and alkali leek often appear in lyrics as symbols of beautiful prairies and fertile pastures. Allium plants are also related to civil taboos. For example, among Arhorchin Mongolian tribes, it is believed that God planted Allium. Human can only collect a suitable amount for one-time consumption, while being forbidden to pluck lavishly; Xiebai bulbs are called *tengerin betehi*, translated as 'the tumor of God'. Some say that people cannot eat Xiebai bulb before spring thunder. These old sayings are somehow related to religious belief, which is a performance of Mongolian primitive worship of nature on the local plants. Although they are considered religious, those taboos contain more or less some simple ideas of sustainability in using plant resources, and meanwhile play a role in the ecological maintenance.

17.2.3 Mongolian knowledge of medicine

According to some statistical data, for Mongolian plants which can be used as medicinal materials, there are 203 kinds of seeds, 231 kinds of roots and tubers, 256 kinds of grass, 54 kinds of leaves, 83 kinds of flowers, 35 kinds of leather, 36 kinds of vine woods and 14 kinds of resin. That constitutes the main content of Mongolian knowledge of local plants and livestock.

Some scholars have studied the Arhorchin Ploygonatum in Inner Mongolia region. There are three kinds: P. sibiricum, P. odoratum and P. macropodium, which is an objective fact. Local herders collect their root-like edible stems, which are believed as good for health and longevity in a long-term consumption. What is important here is how local people refer to them (ie, local cognitive situation). Herders call the P. sibiricum as 'imaan orhodai', while both the P. macropodium and P. odoratum as 'ebesun orhodai'. It is worthwhile to notice that *Panax ginseng*, a very rare kind of medicinal herb, is also called 'orhodai'. It is believed that this reflects the importance attached by Mongolians to this type of local plant. In Mongolian medicine, P. sibiricum is called 'Cha'gan-Huo'ri', and P. odoratum is called 'Mao'haori-Huri'. Mongolian understanding of P. sibiricum is similar to the description of *Yinshan Zhengyao* (Hasi and Pei, 2001, pp. 83–5). The document also records the wild plant which could defeat scurvy, the plants treating hypertension, and plants which can suppress the disease of Hoey.

Here is the ethnobotanical survey on categories of local medicinal plants in Arhorchin region (Hasi and Pei, 2003, pp. 44–8). From this table, we can clearly see the traditional doctor-patient dependency of Mongolians on grassland plants (Table 17.3).

17.2.4 Architecture and geographical knowledge in Mongolian houses

Interaction between people and natural environment is embodied also in Mongolian buildings. The reason why nomadic Mongolians created shelters in the form of the Mongolian yurt lies in their lifestyle of migrating to any place where there is water and grass. Therefore, a light-weighted, easily packed, and portable housing style has become the most important choice for nomadic people. A. Rapoport, an architecture scholar, points out that yurt is a very wise solution for buildings of the nomadic residents interacting with the natural environment (see Figures 17.2 and 17.3)

Aobao, which is piled up on the Mongolian grasslands, has been used to divine direction. Why do other peoples not have it? It is because the prairie is so vast and endless that people can rarely find obvious signs to know direction. Therefore, Mongolians have created a way of accumulate nearby stones on the relatively high spot of prairie, and then constitute some tradition and culture which tends to build the sign together. And later, the knowledge of Aobao can indicate significant characteristics of nearby environment. So that it evolved into the famous Mongolian mark: 'Meeting in Aobao'.

Table 17.3 The catalogue of national botany of Mongolian folk medicinal plants at Aru Corqin

Civil name	Scientific name	Chinese name	Part	Treated disease	Medicinal method	Evidence specimen number
sorgolo	Arenaria juncea	灯芯草蚤缀	root	cough	decoct with water to drink	089
xi bag	Artemisia brachyloba	山蒿	Root bark	arthritis muscle ache	Apply the ointment to the affected part	451
agi	Afrigida	冷蒿	overground part	skin disease	decoct with water to rinse	438
Jirbin mod	Berberis poiretii	细叶小檗	Fresh branches	Eye disease	Squeeze the juice drop in eyes	123
altagana	Caragana microphylla	小叶锦鸡儿	Flower (autumn)	poliomyelitis	decoct with water to drink	199
beri checheg	Delphinium grandiflorum	翠雀	the whole grass	Insecticide wipe off head lice Treatment of hair loss	decoct with water to wash hair	120
baxig	Dianthus chinensis	石竹	flower	liver-heat	decoct with water to drink	097
baxig	D. chinensis var. subulifolius	蒙古石竹	flower	liver-heat	decoct with water to drink	098
biriyanggu, bugesumem, ximdag	Dracocephalum maldavica	香青兰	seed overground part	pinkeye hepatopathy	put in the eyes. decoct with water to drink	342
zhegergen	Ephedra sinica	草麻黄	stem	skin disease cat fever	decoct with water to rinse and drink.	016
malagainzhala	Euphorbia humifusa	地锦	the whole grass	urodialysis	decoct with water to drink	253

(continued)

Table 17.3 (Cont.)

Civil name	Scientific name	Chinese name	Part	Treated disease	Medicinal method	Evidence specimen number
digda	Gentiana dahurica	达乌里龙胆	flower	abatement of fever	decoct with water to drink	309
digda	G.macrophylla	秦艽	flower	treat a cold abatement of fever	decoct with water to drink	308
xihir ebes	Glycyrrhiza uralensis	甘草	root	treat a cold cold, cough, tracheopathy	decoct with water to drink	202
langdangs	Hyoscyamus niger	天仙子	seed	toothache, decayed tooth	Plug in the wormhole	354
imaan eberebes	Incarvillea sinensis	角蒿	overground part	scabies	decoct with water to drink	371
chahildag	Iris Lacteal var. chinensis	马蔺	seed. root	enterobiasis	decoct with water to drink	632
durbeljinebes	Leonurus sibiricus	细叶益母草	overground part	canker sore eczema	stir the ash form burning with butter and smear on wound	346
manu	Paeonia lactiflora	芍药	root	blood-heat	decoct with water to drink	122
amu checheg	Papaver nudicaule	野罂粟	root	intestinal colic	decoct with water to drink	125
mungen digda, has hadhur checheg, banjingarbu	Parnassia palustris	梅花草	the whole grass	bring down a fever. treat a cold. swollen sore throat	decoct with water to drink	146
wuherin hele, nagaram chagaram	Plantage asiatica	车前	seed	diarrhoea	decoct with water to drink	375
wuherin hele, nagaram chagaram	P. depressa	平车前	seed	diarrhoea	decoct with water to drink	374

gejige ebes	Polygonum aviculare	萹蓄	the whole grass	oedema	decoct with water to drink	046
guilesu, haragan	Prunus sibirica	西伯利亚杏	flower, seed	poliomyelitis canker sore	decoct with water to drink. smear on wound	184
munghe chai, hadan chai, yajima harabor	Pyrrosia davidii	华北石韦	the whole grass	migraine	make it to be a medicine pillow	014
	Rhododendron micranthum	照山白	branch and leaf	skin disease	decoct with water to rinse	299
xira bargas, chagan bargas gandagai	Salix gordejevii	黄柳	Fresh branches	Tinea disease	use the juice to smear on wound	020
	Sambucus latipinna	宽叶接骨木	Stems and branches	cold, cough	decoct with water to drink	380
matarin homusu	Selaginella tamariscina var. ulanchotensis	舌尖卷柏	the whole grass	charley horse	decoct with water to drink	001
dogol ebes	Sphora flavescens	苦参	root	abatement of fever treat a cold	decoct with water to drink	187
guzhege bebs, hongorchog ebes	Sphaerophysa salsula	苦马豆	seed	oedema	decoct with water to drink	197
horlo ebes	Syneilesis aconitifolia	兔儿伞	overground part	swollen sore	smear on wound. bind	453
huji ebes, wunurt ebes	Thymus serpyllum var. asiaticus	亚洲百里香	overground part	sore, skin disease oedema	for external use. decoct with water to drink	349

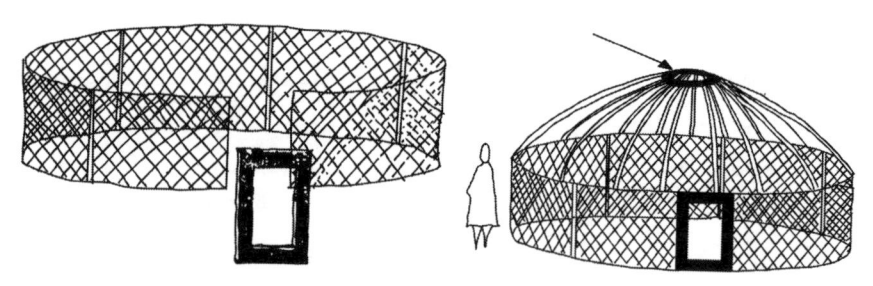

Figure 17.2 The structure of a yurt.

Figure 17.3 Yurts on Inner Mongolia's grass.

17.3 Problems and significance

Currently, the ethnobotany of the southeast, the southwest, and the Mongolian regions of China has become relatively abundant, and there are other studies on the relationship between humanity and nature according to other knowledge from the viewpoint of the Mongolian people. However, from the perspective of PSP, those studies are not complete at all.

There are important issues with anthropological meaning which have not yet been studied. Take Mongolian folk songs as an example: How do folk songs reflect the relationship between human and nature? Especially how do they display the plants and animals with which Mongolians live? Is there any uniqueness of Mongolian songs different from songs of nature, life and other activities relevant to lives in other ethnic groups?

In addition, we also have the following suggestions: first, we hope the relevant studies to extend original landscape, with strengthening theoretical constructions. But case studies are still necessary, especially in the study of local knowledge. Second, multidisciplinary and comprehensive study is most important for our purpose. In a sophisticated society, such issues and problems should be studied on the basis of multi-perspective and multi-discipline. Otherwise the study is likely not only to be biased, but also hardly continuable.

The significance of studying minority cultures on its local knowledge of nature is not only limited within the ethnic group. The meaning is very profound and significant.

First of all, the premise of this study is to fully recognize that all nations have their own way of treating nature, which is in their own viewpoint also science. This recognition is based on cultural relativism, which claims that there is no cultural hierarchy among different ethnics. Therefore, this research has played an important role in preserving and protecting national heritage of local knowledge. Since that local knowledge cannot be separated from its ethnic culture, we can not evaluate it merely from the Western perspective of science. (Of course, this is not to say we do not need Western knowledge of nature as a framework of reference.)

Second, this study opens a new possibility for the communication among different cultures and natural knowledge. It breaks down cultural chauvinism, and also an excessive scientism. Thus, the contemporary criteria of science – in fact it is also a local knowledge, according to J. Rouse, but a local knowledge standardize itself as a universal knowledge – can also lean down to absorb knowledge from the ethnic knowledge of nature, and to develop itself and exchange with others. On the other hand, the local knowledge of all nations will also learn a great deal from modern science. It is beneficial for both Western and non-Western knowledge of nature.

Finally, that study brings further empirical and conceptual challenges to the current theories of philosophy of science. Therefore, it will open a new chapter in the PSP. Furthermore, the study of local knowledge has blazed a new road for internationalizing Chinese philosophy of science, which is also a necessary step the Chinese philosophers of science must take.

The focus of philosophy of science on local knowledge has theoretically opened a new world, where more scholars begin to pay much attention to the evolution of local knowledge. Now might it be the best time for studying local knowledge. We should take the responsibility to make real theoretical progress. The next section will analyze a successful case in order to suggest how local knowledge of ethnobotany be internationalized.

17.4 Ethnobotany as a combination between natural and social sciences

Ethnobotany is a subject, though successfully integrated in modern science, which still retains a sense of locality. It is a system of local knowledge in the process of standardization. We can learn some important things from the case of Ethnobotany. According to the ideas of Chinese and foreign ethnobotanists, Ethnobotany focuses not only on the comprehensive relations between people and plant (including all plants which value in the economic, culture and other aspects) in certain areas, but also on the interaction between social structure, behaviour, and plants.

Ethnobotany plays a significant role in exploring sustainable employment of plant resources. Because of the latest development in China, the academic circle and the public might have several misunderstandings in Ethnobotany. Some have summarized that many misunderstandings for Ethnobotany can be mainly specified in six aspects:

1. Ethnobotany is to study how the minority people to use plants;
2. Ethnobotany is to do researches on the books recording the plant in the history;
3. Ethnobotany has no quantitative methods;
4. The development of ethnobotany is always limited, since every nation could be merely familiar with limited uses of plants;
5. Ethnobotany is purely a social science;
6. Ethnobotany selects remote areas as objects of research, so that the significance of research is limited within those areas.

(Huai, Khasbagan, and Wang, 2005, pp. 502–9)

It seems that Ethnobotany is a kind of universal knowledge rather than merely the knowledge of minority. Ethnobotany should not be misunderstood as a 'bat', which belongs to neither natural or nor social science, and therefore loses resources to develop itself. From the point of sociological view, this appeal is very important, since moving closer to natural science can convince academic authority to fund agency to obtain more standardized resources. And it is necessary to explain that appeal to natural science by revealing the nature of local knowledge. In a word, Ethnobotany has some particular advantages in the combination of natural and social sciences.

According to PSP, frequent changes have occurred on the concept of Ethnobotany, which reflects the process that Ethnobotany as local knowledge has stepped from scientific practice to a process of standardization.

The first definition of ethnobotany comes from Dr. Hanshbengen in the University of Pennsylvania in 1895, who published an article in the *Chicago Sun-Times* about aboriginal-botany, which is defined as a subject on the study of how the original indigenous peoples make use of plants. The official name of Ethnobotany appeared in 1896, when Dr. Hanshbengen published an article named 'Purposes of Ethnobotany' in *Chicago Plant*. He adopted Ethnobotany as a name of a discipline for the first time, which was defined as a subject on the study of how indigenous peoples use and trade plants. In this article, Hanshbengen points out that Ethnobotany is the study of how indigenous aborigines use plants, in order to elaborate the relevant Aboriginal culture, to reveal the history of distribution and propagation of plants, and to determine the ancient trade routes by providing certain reference basis for modern industry.

The concept of Ethnobotany has been defined once again in 1978. It is redefined as the study of the interactions between people and plants through a cultural phenomenon. In addition, Ethnobotany should also include theoretical research, as well as its relationship with studies in other fields such as the cognitive views of aboriginal people, utilization and management principles, ethnic communities, prehistoric economic and paleobotany. All those studies establish a complete status of discipline.

Contemporary ethnobotany is supposed to be traditional knowledge and experience, including historical and contextual features of economic utilization, medical utilization, ecological utilization and cultural utilization, as well as the process of dynamic changes of plant. Human beings are constituted in a large number of different ethnic groups with different cultural backgrounds of people. They are unevenly distributed on the Earth, living in very different natural environment with a variety of plants. And knowledge of plants in every ethnic cultures are impacted by lots of internal and external factors.

Thus, the experience and knowledge accumulated in the daily practice are of various kinds. The study of Ethnobotany is often based on minority peoples and their local relation with plants. But the utilization and protection of plants by 'minorities' are not the whole thing in the study of Ethnobotany. Many ethnobotanical researches focus on the 'minority' areas, mainly because they tend to preserve the traditional knowledge of botany in a relatively complete way.

Similarly, it is necessary to combine the research on plants with the one on humans. For one thing, it usually learns methodology from cultural anthropology, such as structured interview, key person interviews, and participatory rural appraisal. For another, it also adopts methods from natural science. Ethnobotany is supposed to explain the rational utilization and conservation of plant, so that it has to involve some modern scientific methods, such as specimen collection, identification of evidence, and determination of sample, which are the main parts of natural science. To be precise, Ethnobotany is an interdisciplinary field overlapping natural science, humanities, and social sciences, since it indicates characteristics of all those different methods of research. In particular, Ethnobotany refers also to hermeneutics, which turns out that it is wrong to claim Hermeneutics merely employable for Humanities. Hermeneutics can be also helpful in Ethnobotany which is already a kind of natural science as well.

The first heuristic idea from Ethnobotany is that local knowledge tends to expand from one aspect to another in both scope and depth of research. From the view of PSP, hermeneutics in Ethnobotany and local practices is very important for the development of human science. Only based on local practice Ethnobotany can continue to develop and gradually expand to a standardized academic area.

The second one is that, since there is such a discipline to closely combine natural sciences with humanities, hermeneutics can be applied to explain the phenomena of all over those subjects. On the view of PSP, local knowledge in China should be studied in a way that other interdisciplinary studies would like to take after trying various directions in a long run. That can guarantee acquisition of helpful ideas in forming a specialized philosophy of minority practices, which will be subject to academic norms like other disciplines, while holding itself at the same time.

The study of local knowledge promotes the maintenance of cultural diversity. A culture, no matter how strong and attractive it is, should neither charge easily other cultures as disadvantaged or underdeveloped, nor hope to transform other cultures. There is a serious difference between culture and civilization. A nation might be not completely civilized as contemporary western countries, but that does not mean its culture is also inferior to Western cultures.

There seems to be no need to compare different regions and cultures according to a single unified standard. We should respect the difference among the variety of cultures, which is not the difference between developed and underdeveloped civilizations. The culture difference is another kind: namely, an incommensurable difference. According to the view of local knowledge, many concepts used in comparing different cultures might be mainly rooted in Western cultures, which tends to be reducible to purely economic standards. That is an invasion by economic and cultural powers from the Western world without enough awareness. However, different living environment might have different cultural measurements. Though they might be comparable in some aspects, cultures are indeed incommensurable in their very essences. Adopting some unified criteria to measure different cultures might lead to improper results, which is actually forcing other cultures into an alien form, and which might be injustice those cultural communities.

In a word, anthropological knowledge gives us not only a new perspective of understanding, but also a view of local knowledge to promote new grasp of the relationship between nature and society. Harmonious development of every nation should begin with local knowledge in their own views of nature. A unified and non-local norm might be given to a non-western culture within a beneficial motivation. However, that might bring about a disaster, which could extinguish the whole possibility to hold the self-identity of the non-western culture. Therefore, for PSP to promote a study of local knowledge is not merely a task for theoretical construction, but, more importantly, a practical purpose in preserving cultural variety for humans.

18 Conclusion
Scientific practice in ongoing and unlimited process

As a matter of fact, there is a practical turn in Philosophy of Science and its studies. In the Philosophy of Scientific Practice (PSP), we do our research based on: various forms of scientific practice; its connotations; descriptions and normalization; power characteristics; the relationship between PSP and New-experimentalism, New-empiricism, etc. We have carried out our research within Chinese knowledge and practice. Though it seems to be very comprehensive, complete, and in-depth to our readers and researchers, we realize that it is still very basic. There are a lot of valuable issues worth studying. For example, we lack understanding of the real and basic direct activity of scientific practice. Our research remains Philosophy of Science. In addition, the situation in which the internal views of the new experimentalists are not uniform and the positions are not consistent need further study. In sociology, Bourdieu's concept of a practical field has not been fully utilized. But currently the weakest aspect of our research is the paucity of hermeneutic and phenomenological studies. Problems are all-important to the study of scientific practice: for example, why Rouse's PSP is characteristically hermeneutic; why hermeneutics is a necessary and sufficient condition for scientific practice; and so on. We all think that further studies in this field should focus on these following aspects.

18.1 Attention or contact with direct science practice

In terms of direct activity, scientific practice has extremely complex features. Rigid philosophy of science conclusions may not be accurate when considering complex scientific practices, and may not be able to explain them. It is vital to the philosophy of science, especially for the PSP studies, that we can directly or indirectly delve deep into direct scientific practice. Since, from the early stages of abundant scientific practice coming into the field of philosophy of science, scientific practice still plays a major role in critiquing traditional philosophy. Moreover, practice informs PSP, correcting possible errors and filling in the blank spots of some research. Therefore, the wealth of scientific practice is an important foundation and a basic pre-condition for further development of the philosophy of science. However, lack of understanding and research of direct scientific practice may be the bottleneck that restricts PSP's development.

Currently, in Tsinghua University, we are making contact with direct scientific practice in three ways:

First, by guiding the philosophical doctoral coursework of science and engineering, we can understand their frontier research and engage in scientific practice.

In Tsinghua University, we have a special institutional system for contact with research in science and engineering. All doctoral students in these fields need to make philosophical summaries for their own research, before their doctoral dissertations have been completed: they can sum up the methodology of their research, or its significance to national economic construction and the development of society. The guidance for these doctoral dissertations comes from our teachers in the Institute of Science, Technology and Society. The procedure requires us to contact doctoral students as early as possible, with a view to getting a good understanding of their work. Then we provide guidance on subject matter, content and how to write a thesis. After several rounds of consultations we are able to determine the content (with much electronic help, by email, phone, etc.). In this process, we require students to submit their doctoral dissertations, completed or in part, for us to read, as long as these students are not involved in confidential work. Thus, we analyse in advance the results of practical scientific research via a doctoral dissertation on the forefront of scientific research. If we are interested in their work, we will usually contact the author of the thesis to tutor them in scientific practice. We will require the author to elaborate the details of some practice activities in their philosophical treatises, so that some real scientific practice is revealed. After we expand the PSP, we will study the science of doctoral students' cutting-edge research for inspiration and acclimatization. Doctoral students are also required to pay more attention to the relationship between theory and experiment, the nature of scientific practice, refining their methods through practice, etc. By following this guidance, doctoral students' philosophical treatises are more concerned with scientific practice. For example, in this book there we studied a case about the research of bionic drone aircraft, which was refined by us. The research has been modified to the study of the measurement systems in the aircraft, as in the case of change and shrinking objectives (see Chapter 12). This is a typical PSP case. And it shows how science study begins with opportunity and is amended by opportunity. I have been responsible for directing doctoral dissertations from the department of Precision Instruments. Recently, I began to be responsible for philosophical theses written by doctoral students from the department of Hydraulic Engineering. There are a lot of differences in science studies and the ways of scientific practice between these two departments. For example, students from the Department of Precision Instruments attach more importance to their theoretical studies; their scientific practice usually begins in the laboratory. But the objects researched by students from the department of Hydraulic Engineering are often actual rivers, so their scientific practice features a field-base and a material laboratory model, and a more theoretical characteristics of scientific practice. Just like the triple helix, they connect closely with each other and promote the development of science. If we do not actively study such typical scientific practices, our study of PSP will be a kind of false research with no practical significance. We'd just be talking to ourselves.

Second, we also engaged in the opening and summarizing of scientific experimental programs. By the way of summary in methodology, we can explore the characteristics and different types of scientific practice.

For example, we have encountered the experiment course 'river model experiments and numerical simulation' set up by Danxun Li, Shanghong Zhang and Jiahua Wei, who are professors from the National Key Laboratory of Hydroscience and Engineering of Tsinghua University. They divided river studies into three forms: the actual river, a model river, and a digital river. The first is the true and actual river itself. Researchers are still required for the possession of actual investigation and practical experience, regardless of high technologies development to what extent. That is because some of the research results will eventually have to face the problem of the actual river. The river model is built up in the laboratory, based on the actual river itself. The model is organizing miniatures of the real river; it possesses the structure of the actual river and is built up in proportion to the miniature real river. This model is tested in the river model, and the model is generally a prototype which is reduced (or enlarged) in geometric size. River model tests are studying the evolution characteristics of rivers based on a certain geometric scaling of natural rivers (often a certain period of the river) down to the laboratory scale. Then we will restore the phenomena and laws which we have observed to the protocol-type river, so that you can get a good understanding of the movement of the prototypes. And the digital river is research on the digital simulation of rivers. According to their research, river digital simulation is a relatively general concept. This concept mainly refers to the comprehensive application of GIS, GPS, RS (remote sensing), Internet, multimedia technology and virtual simulation technology. Then, under the frame composed of data platform, network platform and applications platform, it accomplishes a digital reproduction of the real river system. Consequently, we can achieve these goals in scientific practice research techniques: for instance, hydrological automatic monitoring of the river, real-time scheduling of flood control programs, accurate prediction to the evolution of river basin, and modern management of watershed system, etc. It is a huge and difficult project to truly complete the digitization of a river. We cannot implement it overnight, but step by step. In addition, the concept of river digital stimulation is getting wider and wider with the development of technology advances. From current practice, the simulation covers the following aspects:

(a) watershed geomorphology

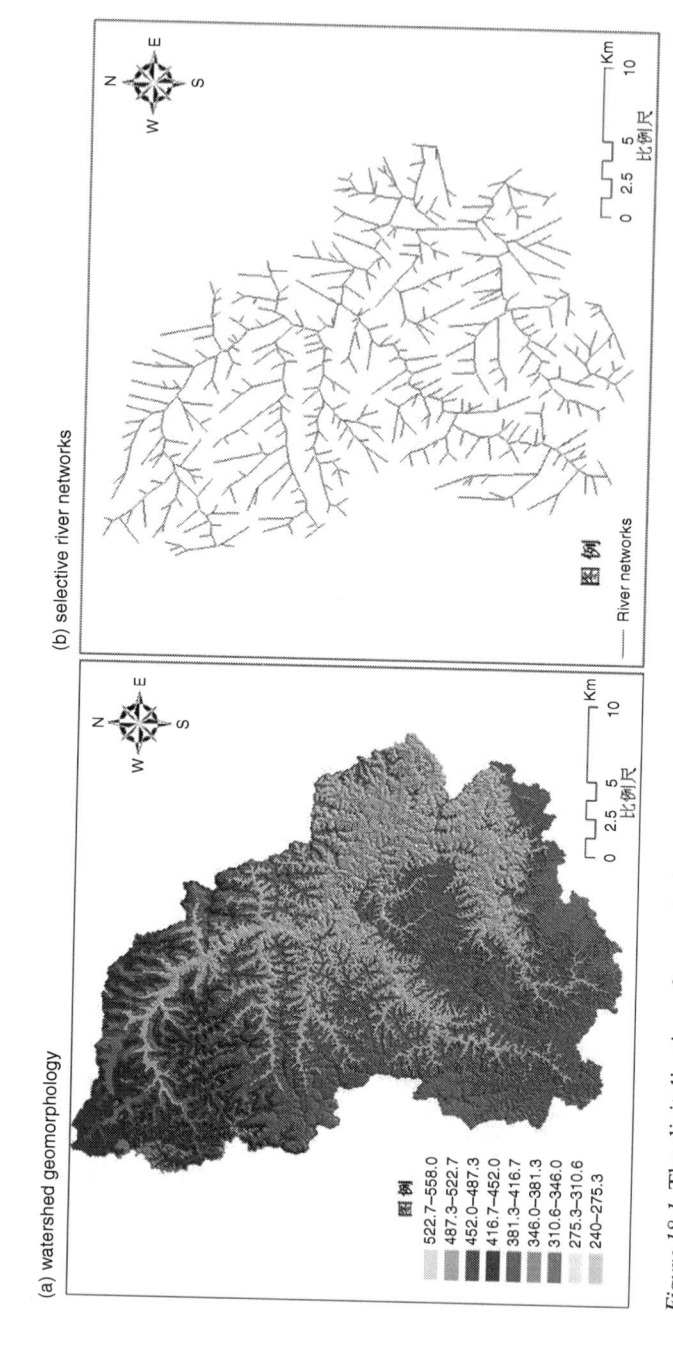

(b) selective river networks

Figure 18.1 The digitalization of watershed geomorphologic formation. (Taking Li Zixi, a tributary of the Jialing River, as an example. Quoted from the conclusion materials of Dan-xun Li's laboratory exploring lessons.)

Conclusion 299

1. Watershed integrated GIS. GIS is the basis to manage the distribution data of watershed geospatial date. The application of GIS technology in water conservancy industry has started from the 1970s or 1980s. At the beginning, GIS was mostly applied to flood control information system. Subsequently it was extended in many aspects, such as water resources, water environment, soil and water conservation, farmland water conservancy, and water conservancy planning and management. So far, this system has become the basic information platform to river digital simulation.

2. A distributed hydrological model which is based on DEM (Digital Elevation Model) divides the river basin into a network (see Figure 18.1). And this model is established according to rainfall process and ground topography, soil, vegetation, etc. This model can also predict runoff process and total pollutant amount of one river basin to achieve a point-line-surface integrated management of the river basin. It should be noted that distributed hydrological model's coverage generally is relatively large, and this model can describe the watershed hydrological processes accurately, but it is difficult to describe the evolution of the river's scouring and silting characteristics accurately.

3. Numerical simulation of basic equations of river water and sediment movement. This includes water dynamics equation, sediment transport equation, and river-bed evolution equation. And they mainly study discrete method, solving method, precision control, parameter selection of these equations, and these aspects are one of the numerical bases of river simulation. Currently, one-dimensional and two-dimensional flow and sediment mathematical models have been widely used, and the application of a three-dimensional mathematical model has gradually expanded. At the same time, there are many integrated models with functions of water movement, sediment transport, pollutant dispersion and many other contents.

4. 3D visualization. Both the watershed hydrological model and the model of river water and sediment previously mentioned output much numerical data. To show the results more intuitively and to bring convenience to the analysis and comparison of engineering program, not only is scientific simulation calculation required, but also the need to show results intuitively, graphically and in real-time. This is the model visualization. The simulation of the flow field is of great significance. Dynamic visualization of the flow field performance is not only beneficial to the overall expression of calculation results, but also makes the changing process of flow field easy and clear. And it helps to analyze in-depth and research the calculation results. Research on visualization of mathematical models has been ongoing for a long time, and this research is part of visualization in scientific computing research. As so far, numerical simulation of visualization in scientific computing has not been content with the performance of static results, but it is on the way for dynamic, three-dimensional, instantaneity, interactivity, etc.

300 Conclusion

5   Virtual simulation system. The 3D Virtual simulation system is an all-in-one system with GIS, database technology, calculation of professional model, and visualization. This system is equivalent to putting the river system protocol-type into a computer, thereby realizes a full digital expression and management to the river system. A 3D virtual simulation platform generally consists of four modules: 3D visualization module, 2D GIS module, computing module of mathematical model and database module. A database, as the warehouse of background data, is responsible for storing the data required to run the system, which accepts the data input by computing or monitoring equipment, at the same time providing display data for 2D GIS module and 3D visualization module, then providing calculation parameters for the mathematical model. Mathematical model as the core content of scientific simulation system is providing scientific calculation results for visualization platform for users to judge decisions, and receives feedbacks from visualization platform, always adjusts the calculation conditions, and calculating results of this model can also be stored in the database. As the display and interface, the 2D GIS and 3D visualization modules are the user-oriented windows of the whole system. The GIS module can complete the 2D display and statistical analysis, which connects with 3D visualization module to realize map navigation. The 3D visualization module can achieve the task of displaying scene in 3D. These modules finally are integrated to form a 3D virtual simulation platform (Figure 18.2).

Therefore, in accordance with the historical order, the three kinds of research method and technology are summarized as follows (Figure 18.3).

It is worth noting that the three scientific practices did not substitute for but complement each other. Practice is still the foundation and an important aspect which cannot be ignored.

Third, we can engage in direct scientific practice through participating in major projects in science and engineering research. Currently, there is not much experience to talk about because this area has just begun. It should be noted that this intervention may be the most direct one, and we hope to gain a better understanding in these interventional studies.

18.2 New-experimentalism and new-empiricism

The metaphysical foundations of PSP are local, pluralistic, and pragmatic so that PSP can assimilate much from other philosophical resources. There are many ideological resources available, such as advances in New-experimentalism in philosophy of science, other philosophers' research (as Hayes's research) that are in New-empiricist philosophy of science, views of New-pragmatism, and the progress in Checkland's methodology of soft system.

As for New-experimentalism, our research is still insufficient. Ian Hacking is the pioneer of New-experimentalism. Hacking believes that the dialectical

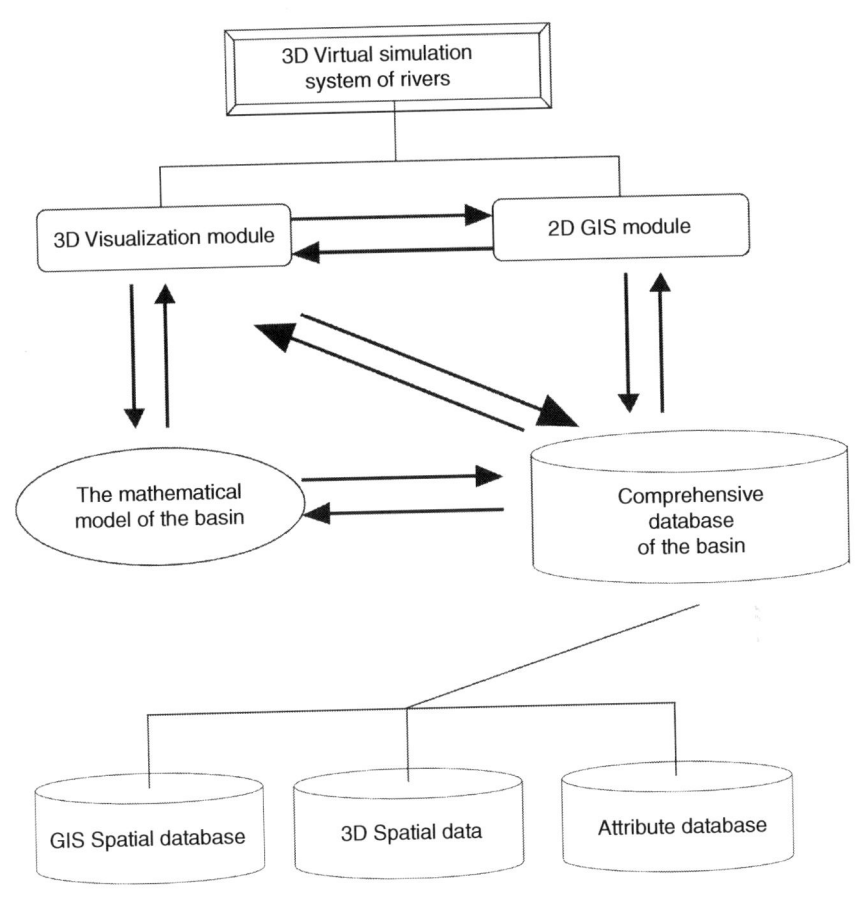

Figure 18.2 The structure of the 3D digitization simulation platform for a river (quoted from Dan-xun Li, etc. materials of laboratory exploring lessons).

interactions among scientific experiments or tools, experimental data, and theory have provided a solid basis for rational evaluation of science. He has pointed out that the mutual influence or 'plasticity' between tools, experimental model and theory shows the 'local stability' in experimental environment. This local stability provides a relatively solid foundation for reasonable evaluation or truth identification. Of course, this stability is not a summative one, it is just a result of the operation of specific tools and equipment. He insists on experimental realism, and our research on experimental realism is still absent.

The rise of New-experimentalism has ignited a flame of new philosophy of science on experiments. There are three so-called 'experimental' articles, written by Peter Galison, J.S. Rigden and Roger Stuwer, which are included

302 Conclusion

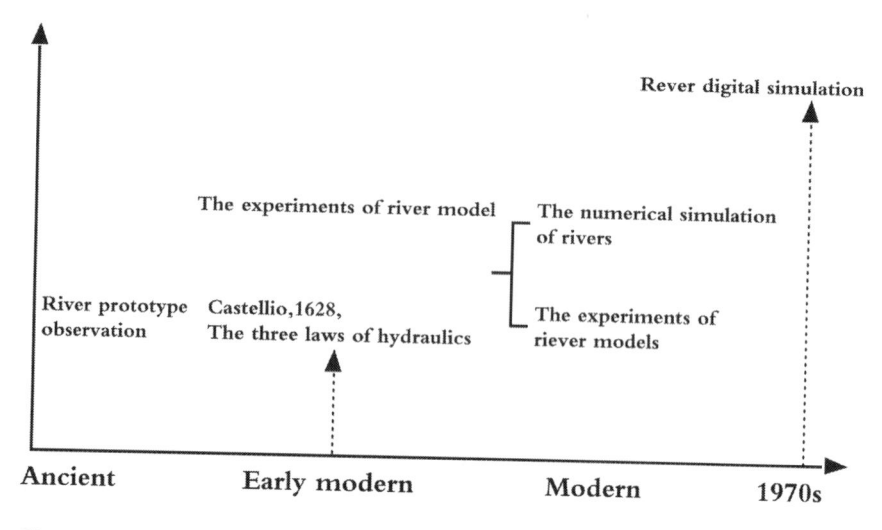

Figure 18.3 The evolution of scientific river research.

in 'Observation, Experiment and Hypothesis in Modern Physical Science' (P. Arcinstein and O. Hannaway, 1985). Related papers had increased sharply since then until 1989 around. For example, the meeting named 'the uses of experiment' which was held in Bath, England, in September 1985, was the first wave we had met on the philosophical study of the experiment. The conference papers which edited by Gooding, Trevor Pinch, and Simon Shaffer are included in *The Uses of Experiment: studies of Experimentation in the Natural Sciences*. The book collects articles of Hacking, Gooding, Pinch, Schaefer et al, and its main idea is that experiment is a scientific practice. And this practice is completely different from the theory, and they insist that the experimental activities need to be understood as a cultural practice rather than the logical structure. The book *Theory and Experiment* which was co-edited by Batens and van Bendegem in 1988 has explained the relationship between theory and experiment from a new perspective. Most of these papers are rich in cases of history of science. The same year, the famous history of science journal *ISIS* had published a group of articles on artificial phenomena and experiments, including articles written by Shapin, Hacking and Galison. They had also examined several experimental phenomena, experimental design and implement, the uncertainty in experimental design. We can find more basic researches on experiment and technology in Hans Radder's writing in 1988. However, New-experimentalism was somehow dreary after 1990s. What problems was New-experimentalism encountering? Is it the resistance of the priority of theory or the imperfection of New-experimentalism itself? All these issues require further studies. Only by studying these issues clearly can we use the research results and ideas of the New-experimentalist to advance the research of PSP.

There is little research in relationship between New-experimentalism and New-empiricism. For example, in Hacking's view, science is mottled, and it includes not only theory, but also experiments and diverse movement during the process from exploring the basic theoretical assumptions to using apparatus to transform nature (Hacking, 1988, p. 154). Cartwright also puts the world as a mottled one, but she regards science as a patchwork, not a unity. Hacking believes that contemporary culture is suitable, diverse and optional. This culture rejects fundamentalism but pursues stability. This stability ensures the coexistence of diverse interests. However, Cartwright demonstrates laws of science are true based on local realism through the patchwork of laws. Hacking considers that tolerance and respect rather than unified control are required by science. Science has been so prosperous, so that it can be stable without foundation, and can be shared without commensuration. Science is about the reality of physical world that has the most diversity and the least subjectivism. Science has become a field where consistent activities happen in a completely non-unified world. But Cartwright thinks that there is no need for such unification, who believes it is a problem that science chases this kind of unification. For another, Hacking (1983, p. xiii) focuses on how experimental knowledge is generated by the interference experiments in the laboratory. And he pays less attention on expanding knowledge outside the laboratory. He believes that the experiment could not be duplicated and the so-called duplication means trying to improve to produce the phenomenon regularly. Hans Radder is more concerned about the expansion of experimental knowledge and the normative questions during the expansion process. He thinks that 'local' word is often used to criticize the traditional universal proposition, but it does not mean we repudiate the existence of non-local mode or the continuity of technological development in development of science and technology. These modes include repeatability of experiments, standardization of theory, political dilation of techniques, etc. And these modes are not necessarily inevitable but accidental, and they have a historical trend. Moreover, there is a need of material interference and social intervention for generating and maintaining them. Although they are occurred in a context, they can be generated repeatedly, weakened, strengthened, or changed. During the process that the mode of experimental knowledge has developed from the local to the non-local, it involves normative questions. Differences not only exist within the New-experimentalism, but also among the surface of PSP and the New-experimentalism.

Also, on the normativity of philosophy of science, Rouse's notion of normativity is one of the attributes of practice. It is a natural phenomenon generated during the engaging process of practice, and antecedent to people's cognitive process. While Radder's notion of normativity is a procedural standardization, a mechanism in which experiments can be repeated. It occurs after we have known the objects. Rouse's notion of normativity has provided a basis of Radder's view. Of course, both of their understanding of normalizativity has something in common. They think that normalizativity is associated with human, action, and free will. On a moral level, the so-called norm is that people under certain constraints choose a certain behaviour with other options as a precondition. Normativity is a core concept, whether in Rouse's or Radder's philosophy of science. Therefore, further studies on the relationship between New-experimentalism and New-empiricism, and the one between PSP and its contents, are also an important part of improving the research on PSP.

18.3 PSP, phenomenology, and hermeneutics

To explain the hermeneutic nature of PSP is the weakest part of our research. In fact, this is the influence from continental philosophy to PSP. We have just dug into connection between Rouse's and Heidegger's thoughts. We also have studied the successive and critical relationship between Rouse's practical hermeneutics and Heidegger's hermeneutics which have some theoretical hermeneutic factors. There is insufficient research on the influence, role and significance of phenomenology of perception and other hermeneutics on the PSP. In the final part of this book, we cannot elaborate the current research in details but merely present the research clues.

18.3.1 Husserl in Rouse's works

About Husserl, Rouse in his book *Knowledge and Power* has quoted and discussed as follows:

First, in chapter four 'local knowledge', Rouse thinks that Heidegger's phenomenological description on theoretical attitude is a prejudice that carries the theory-dominance from tradition while he is criticizing Heidegger's practical hermeneutics. Rouse thinks the centre position adopted by Heidegger in the analysis of science have accepted Husserl's study on the theory-dominated (Rouse, 1987, p. 79). It should be noted that, Rouse accepts the main position of phenomenology, but he does not agree with Husserl's theory on apriority and theory-dominated.

Second, in chapter five 'beyond realism and anti-realism' (Rouse, 1987, p. 146). Rouse has spent almost an entire page to discuss the impact of phenomenology on the philosophy of science in realism.

Third, involved in the differences interpretation between Natural Science and Humanities (or Human Sciences), Rouse has pointed out that phenomenology to some extent dispelled the differences between the two, but the so-called methodological or ontological differences are related to Heidegger's phenomenology presupposition or the misreading to Heidegger's practical hermeneutics.

In the book *Engaging Science*, Rouse has referred Husserl four times, two of which are quoted notes, another two are annotated references. Among them, there is a discussion that Fine rejects Husserl's doctrine about the scientific essentialism (Rouse, 1996, p. 83), and about Frege and Husserl rejecting naturalism on the semantics meaning (Rouse, 1996, p. 154). Another two annotated quotations have discussed the source of classical traditions which social constructivism relied on and Rouse thinks Husserl is included. When discussing the view of how to apply Cartwright's mathematical theory to things in the world, Rouse also believes it is closely related with Husserl.

In the book *How Scientific Practices Matter*, Rouse has mentioned Husserl 14 times. If we say that Husserl is just mentioned accidentally in *Engaging Science*, but in *How Scientific Practices Matter*, Husserl made his debut as a villain who had influenced the development of philosophy in the 20th century, especially the view of naturalism, and he is the heavyweight that cannot be bypassed for Rouse to construct the philosophical naturalism. Husserl is mentioned firstly in the situation of tracing the influence of Husserl's rejection to the naturalism in the beginning of philosophy of science of the 20th century (Rouse, 2002, p. 5). And in the situation of Heidegger's reconstruction about Husserl's anti-naturalism criticism, and the view of contemporary naturalism resorting to contemporary naturalism are not better than Husserl's view of resorting to the form of life (Rouse, 2002, pp. 13–14). Particularly in chapter one 'the problem of manifest necessity', there is almost more than half of the section relates to Husserl's point of view. Especially, Rouse has made a comparison between Husserlian phenomenological traditions and the traditions of Fregean logical analysis. He also makes a comparative argumentation about the commonalities and differences between Husserl's and Carnap's standpoint of anti-naturalism.

In addition, Husserl's appearance also paves the way for Heidegger's access: Rouse thinks that where Husserl interprets consciousness as inner field of experience by means of phenomenology, Heidegger has replaced it with an excellent analysis of 'Dasein'. In comparison with Husserl (and Carnap) who tries to express and explain the world's objectivity by using formal structure terms, Heidegger insists that relevant connection of practical significance could not be fully understood as a formal structure which is abstracted from actual physical relationship. These are closer to the view of practical normative naturalism. They are also endorsed by Rouse and become the new sources of PSP to establish and develop.

18.3.2 *Follow-up study on the relationship between PSP and phenomenology and hermeneutics*

In the book *Knowledge and Power* and *How Scientific Practices Matter*, Rouse has been concerned with the after phenomenology and hermeneutics. He points out that some phenomenological scholars after Husserl had developed 'Horizonal Realism' or 'phenomenological realism' by using phenomenology and hermeneutics to discuss scientific phenomena and explanation. These scholars (mainly including Heelan, 1983, part II; Compton, 1983) believe that science makes new objects present on knowledge. Therefore, they emphasize the significance of perceptual hermeneutics to scientific understanding. Phenomenologist think that reality is just the thing which is presented by world to human perception, thus we must understand the reality in different ways. The primary goal of the science of realism is to reveal the structure of reality which cannot be reached by pre-scientific perception (Heelan, 1983, p. 174). Now, Heelan's book, especially his philosophy of science study about phenomenology and hermeneutics, is drawing more and more attention.

In the first part of Heelan's book *Space-Perception and Philosophy of Science*, he has paved a road for philosophy of science based on perceptual priority, by means of the construction and demonstration to the diverse space that is possessed by the original human visual space. Hyperbolic visual space is one of them, while Van Gogh's visual space has performed modality of the space, and carpenter's working right angle ruler turns the hyperbolic visual space into Euclidean space. While in the second part he has outlined the framework and main content of phenomenology and hermeneutics which are based on perceptual philosophy of science, then he proposes horizontal realism relating to the combined situation recognition (Heelan, 1983, pp. 173–91). By providing some examples including Van Gogh's painting, Heelan points out that visual space is a sort of hyperbolic space with curvature, people have corrected human visual space from hyperbolic space to Cartesian space through the carpenter's rectangular bar.

Rouse has also attached great importance to Heelan's work. Some ideas in *Space-Perception and Philosophy of Science* are cited in Rouse's book *Knowledge and Power* and *How Scientific Practices Matter*. For example, Rouse has indicated that Heelan thinks perception is hermeneutical, since an embodied subjective perceptual intention or perceptual system comprises a (system of behaviours or) praxis. Every perceptual practice, of course, exercised within a world by a community of human inquires and uses a descriptive language. Every perceptual praxies actualizes an intentional possibility of a body. Moreover, every perceptual praxies can make some horizon of a world present to perceivers. Reciprocally, a horizon of some world can impose its presence on perceivers only be causing some appropriate function of an intentional structure of a body to resonate in response to its presence (Heelan, 1983, p. 176). Especially, Heelan's relevant discussions on body posture, spatial perception and context. Rouse has discussed Heelan's 'carpentered environment' example twice. Heelan had argued that this orthogonal reconstruction of familiar surroundings effected a shift in the predominant spatiality of visual perception, from hyperbolic to Euclidean space (Rouse, 2002, p. 250n, 253n). The importance of carpentered cues is to establish a practical spatial orientation. Here, Rouse believes main parts of his PSP in hermeneutics are as same as this kind of phenomenological realism, while the difference is just different focus. Phenomenology, especially Phenomenology of Perception, emphasizes the perceptual factors in operating activities. However, PSP emphasizes practice which is not dominated by perception, but beyond perception. In the relationship between perception and practice, Rouse remains a consistent emphasis on the priority and leading position of practice, he has pointed out that 'things become "perceivable" in the course of our practical dealings with them and that the key effect of science is to enable us (but also compel us) to take account of them in our practices' (Rouse, 1987, p. 146).

By emphasizing practices, we can make a united understanding between PSP and phenomenology: our understanding of how things are presented to scientific practice; our understanding of how those things affect larger social context within which science is situated. This can dissolve the boundaries inside and outside of science. Moreover, the emphasis upon practice also avoids the confusions that result from traditional empiricist assumptions about perception. In a word, Rouse believes that if we place such a central role upon perception in epistemology, we cannot adequately understand that science is just a practical engaging with the world; therefore we will insufficiently understand the radical change that is brought by the practical engagement of science with the world. For phenomenology and hermeneutics, which places perception at the centre, they compare hermeneutics with PSP, which places the experience (perception, more specifically) as the central concept in epistemology, while another promoting practice in epistemology.

18.3.3 Relations between modern science, laboratory research, and Husserl phenomenology

Modern science is far away from natural nature world since Galileo. As Husserl said, Galileo is a genius not only for discovery, but also for cover. Galileo has discovered an abstract science world, but he also has covered an emotional life world. Modern science has built a world artificially rely on counterfactual conditionals.

In what circumstances can counterfactual conditions be valid? Maybe implemented in two cases. First, the conditions of similarity law may approximately exist in the sky. If there is no interaction between large pluralities of stars, we can calculate and infer both the distance and the interaction force between the stars. Second, in the artificial simulation laboratory, we can set up the relationship between every two factors through controlled conditions. As Cartwright said

> Sometimes the arrangement of the components and the setting are appropriate for a law to occur naturally, as in the planetary system; more often they are engineered by us, as in a laboratory experiment. But in any case, it takes what I call a nomological machine to get a law of nature'.
> (Cartwright, 1999, p. 49)

Cartwright substituted ordinary people then threw out a question, that is, what is the amazing success of science obtained through transformed the world around us? These successes have demonstrated that laws based on the business plan will work as true. With the help of social constructivists' research, Cartwright would like to argue as follows:

> Social constructivists are quick to point out that the successes are rather severely restricted to just the domain I mentioned – the world as we have made it, not the world as we have found it.... With a few notable

exceptions, such as the planetary systems, our most beautiful and exact applications of the laws of physics are all within the entirely artificial and precisely constrained environment of the modern laboratory.... Even when the physicists do come to grips with the larger world.... They do not take laws they have established in the laboratory outside, in miniature. They construct small constrained environments totally under their control. They then wrap them in very quick coats so that nothing can disturb the order within; and it is these closed capsules that science inserts, one inside another, into the world at large to bring about the remarkable effects we are all so impressed by

(Cartwright, 1999, pp. 46–7).

Therefore, the laboratory has become a prerequisite for the development of modern science. Without laboratories, modern science simply cannot be established. Sometimes, we also bring physics to the outdoors. And this time, the shield becomes even more important. According to Cartwright's statement, SQUID can accurately measure the magnetic pulsations to help find stoke. But to implement these tests, hospitals must get a Hertz box – a small space to isolate the magnetic field from environment made of completely metal.

Conversely, logically speaking, scientific research is not on the real nature, but an abstract, vacuum nature, and a nature transformed into artificial objects. This is the so-called nature which established under counterfactual conditions. If it is natural, there will not be counterfactual conditions. Therefore, the nature people studied is in fact an artefact. Just as Cetina said that scientists can bring this kind of nature home through the laboratory, so laboratory science can bring the object to the 'home' and operate them in their own way.

In this way, the world where we are now living in has been transformed into more and more artificial by science and laboratory. For instance, wild drosophilae are not suitable for laboratory studies. Wild drosophilae were transformed several generations into the laboratory drosophilae. Laboratory mice were transformed into pure and pure, even become a patent product of a company. (Laboratory mice are not natural creature, but scientific artificial creature.) Thus, Harré has called wildlife which are used in the laboratory as domesticated nature creatures or artifacts docking with natural creatures. He also said this version of natural things are semi artificial. It is not only biological transformation, but the inorganic nature also is transformed. Now, we can realize hydrogen nuclear explosion reaction in nature by tokamak machine. Tokamak device is the functional substitute of natural hydrogen nuclear reaction. Through the large-scale transformation of nature, Science has obtained the qualification of applied in the whole world, which makes its knowledge form universal in apparent. Therefore, we can give a descriptive definition of modern science which use modern physic as a representative from the significance of Cartwright: matured modern science, taking physics as the basis, is a kind of practical activity which is based on the conditions as the same as the anti-facts and laboratory studies. However, modern science has forgotten its meaning. This kind of activities which are transformed into artificial world has been highly concerned about as early as Husserl. But we did not have the insights to know the importance of the matter.

Husserl's conception of 'Lebenswelt' means:

1. The primitive character of the world. Primitive character means pre-scientific and non-theoretical. Clearly, modern science is a product of modern world, while daily living-world is prior to modern science, hence it is a world not be themed. Ever since Galileo, nature science, especially the continuous development of objectivism and scientism which are represented by mathematics and physics. At the same time, in their own development, they have constantly mathematicized and symbolized our living-world in which we are living and give the significance of our existence. That is the root of the European crisis of science. Such a 'scientific world' has imposed a conceptual appearance onto the 'life-world', which lead people to ignore the original 'living-world'.

2. The living-world is still a non-objective world. That is, in the living-world, the cognitive, affective and volitional, these three conscious functions of people are not divided. Practical purposes, desires, and attitudes can influence the tropism of thought. If the truth of the scientific world is inevitable (in fact, the different categories of science was also considered to have different boundaries without universal law covering. Therefore, the so-called universal law is only valid for specific situation. Just as Marx said, universal law is specific.), then, the truth of ours living-world is a local area which is for people in constantly generating and changing. Since the given meaning of existence to the living-world in advance is a subjective structure, it is a result of experiential and pre-scientific life. As for the real objective world, namely the world of science, it is a structure on a higher level, based on the pre-scientific experience and conceptual work, or based on the effectiveness of them.

For Husserl, the mathematized and scientific world can have a connection with the living world. It seems that if we could realize the existence and significance of living-world, we can return to the living-world. It seems that, this kind of origin meaning of thought, which is transformed by scientific experiments and laboratory studies, has grown into a genuine substantial force. And the living-world has been transformed almost along with it. After Rouse, Cartwright, and other scholars' deconstruction, we have realized the scientific world is not an organic whole, it has a localized nature. But almost all scientific fields have come to rely on the nomological machine, counterfactual condition, or other conditions that are the same as excluding differences, and towards a localized and mathematicized science world.

In short, phenomenology and hermeneutics have given us an understanding and description on the view of science which is like PSP. Science is a practical activity, a hermeneutical, narrative, historical, context-dependent activity of meaning construction with the living world. Philosophy of Scientific Practice (PSP) has put the interpretation advanced in the sense of practice, which deeply introduces the practical hermeneutics and phenomenology into philosophy of science.

We hope that in the future course of research, phenomenology and hermeneutics should give more attention to philosophy of science, also we hope that phenomenology and hermeneutics researchers pay more attention to philosophy of science to form a joint force to promote the study of philosophy of science and technology.

This book is coming to an end, but the research of PSP never ends. In the future, we will continue to focus on the study and issues of PSP. And we will actively explore the various questions raised in the last chapter. We will also continue to focus on the study of science knowledge in ancient and modern china with local characteristics, including the study of ethnologic natural knowledge.

Scientific practice is local, and the locality of science is also international. We hope that we will develop a new research path not only following international approaches, but also with Chinese characteristics in philosophy of science. As for whether our study could achieve the goal, it not only lies in our own ability and effort, but also in the evaluation of results. Since it is a time-open question as scientific practice, let the practice give us a result.

References

English references

Achinstein, P. and Hannaway, O. (eds.), 1985, *Observation, Experiment and Hypothesis in Modern Physical Science*. Cambridge, Mass.: MIT Press.

Ackermann, R., 1989, The New Experimentalism. *British Journal for the Philosophy of Science*, 40, pp. 185–90.

Apel, K-O., 1980, *Towards a Transformation of Philosophy*, trans. Adey, G. & Frisby, D. London: Routledge & Kegan Paul.

Apel, K-O., 1998, *From a Transcendental-Semiotic Point of View*, trans. Papastephanou, M. Manchester and New York: Manchester University Press.

Aristotle. 2009. The *Nicomachean Ethics*. L. Brown, D. Ross (trans.). Oxford: Oxford University Press.

Audi, R. (ed.), 1999, *The Cambridge Dictionary of Philosophy*. Cambridge: Cambridge University Press.

Babich, B.E., 2002, *Hermeneutic Philosophy of Science, Van Gogh's Eyes, and God, Essays in Honor of Patrick A. Heelan, S.J.* Dordrecht and Boston: Kluwer.

Barnes, B., 1982, *T. S. Kuhn and Social Science*. London: Macmillan.

Batens, D., and van Bendegem, J.P. (eds.), 1988, *Theory and Experiment: Recent Insights and New Perspectives on their Relation*. Dordrecht: Reidel Press.

Bloor, D., 1973, Wittgenstein and Mannheim on the Sociology of Mathematics. *Studies In History and Philosophy of Science, Part A*, 4, 2, pp. 173–91.

Bloor, D., 1991, *Knowledge and Social Imagery*. London and Chicago, Ill.: University of Chicago Press.

Bourdieu, P. 1990. *The Logic of Practice*, R. Nice (trans.). Stanford, CA: Stanford University Press.

Brandom, R., 1979, Freedom and Constraint by Norms. *American Philosophical Quarterly*, 16, pp. 187–96.

Brandom, R., 1994, *Making it Explicit*. Cambridge, Mass.: Harvard University Press.

Brandom, R., 2000, *Articulating Reasons*. Cambridge, Mass.: Harvard University Press.

Buchwald, J.Z. (ed.), 1995, *Scientific Practice – Theories and Stories of Doing Physics*. Chicago: University of Chicago Press.

Cartwright, N. (1983). *How the Laws of Physics Lie*. Oxford: Clarendon Press.

Cartwright, N., 1999, *The Dappled World: A Study of the Boundaries of Science*. Cambridge: Cambridge University Press.

Cetina, K.K., 1981, *The Manufacture of Knowledge: An Essay on the Constructivist and Contextual Nature of Science*. Oxford and New York: Pergamon.

Cetina, K.K., and Mulkay, M. (eds.), 1983, *Science Observed: Perspectives on the Social Study of Science*. Beverly Hills, Calif.: Sage.

Chalmers, A.F., 1990, *Science and Its Fabrication*. Minneapolis: University of Minnesota Press.

Chalmers, A.F., 1999, *What Is This Thing Called Science?* Buckingham: Open University Press.

Checkland, P., 2000, Soft Systems Methodology: A Thirty Year Retrospective. *Systems Research and Behavioral Science*, 17, S11–S58.

Compton, J. 1983. 'Natural Science and Being-in-the-world', Paper presented to the Pacific Division, American Philosophical Association, March 1983.

Davidson, D., 1980, *Essays on Actions and Events*. Oxford and New York: Oxford University Press.

Davidson, D., 1986, "A Coherence Theory of Truth and Knowledge", in LePore, E. (ed.), *Truth and Interpretation: Perspectives on the Philosophy of Donald*. New York: Basil Blackwell, pp. 307–19.

Davidson, D., 2001, *Inquiries into Truth and Interpretation*. Oxford and New York: Oxford University Press.

Devitt, D., 1996, *Coming to Our Senses: A Naturalistic Program for Semantic Localism*. Cambridge: Cambridge University Press.

Ford, R.I., 1978, *The Nature and Status of Ethnbotany*. Ann Arbor and NewYork: University of Michigan.

Franklin, A. 1990. *Experiment, right or wrong*. New York: Cambridge University Press.

Franklin, A., 1999, *Can That Be Right? Essays on Experiment, Evidence, and Science*. Dordrecht and Boston: Kluwer.

Fujimura, J. (1996). *Crafting Science: a sociohistory of the quest for the genetics of cancer*. Cambridge, Mass.: Harvard University Press.

Galison, P., 1997, *Image and Logic, A Material Culture of Microphysics*. Chicago and London: University of Chicago Press.

Giere, R.N., 1985, Philosophy of Science Naturalized. *Philosophy of Science*, 52, 3, pp. 339–40.

Gooding, D., 1990, *Experiment and the Making of Meaning: Human Agency in Scientific Observation and Experiment*. Dordrecht and Boston: Kluwer.

Gooding, D. 1992. 'Putting Agency Back into Experiment', A. Pickering, *Science as Practice and Culture*, Chicago: University of Chicago Press, 65–112.

Gooding, D., Scheffer, S., and Pinch, T. (eds.), 1989, *The Uses of Experiment: Studies of Experimentation in the Natural Sciences*. Cambridge: Cambridge University Press.

Habermas, J., 1971, *Knowledge and Human Interests*, trans. Shapiro, J. Boston: Beacon Press.

Hacking, I., 1983, *Representing and Intervening: Introductory Topics in the Philosophy of Natural Science*. Cambridge: Cambridge University Press.

Hacking, I., 1988, Philosophers of Experiment. *PSA: Proceedings of the Biennial Meeting of the Philosophy of Science Association*, 2, pp. 147–56.

Heelan, P.A., 1983, *Space-Perception and the Philosophy of Science*. Berkeley: University of California Press.

Hesse, M., 1980, *Revolutions and Reconstructions in the Philosophy of Science*. Bloomington: University of Indiana Press.

Huxtable, R.J., 1992, The Pharmacology of Extinction. *Journal of Ethnopharmacology*, 37, pp. 1–11.

Janvid, M., 2004, Epistemological Naturalism and the Normativity Objection or from Normativity to Constitutivity. *Erkenntnis*, 60, pp. 35–49.

References 313

Jinlong, Y. 2007. 'A Study on the Normativity of Scientific Practice'. Doctoral dissertation. Beijing: Tsinghua University.

Jones, V.H., 1941, The Nature and Scope of Ethnobotany. *Chronica Botanica*, 6, pp. 219–21.

Jasanoff, S. 1995. *Science at the Bar: law, science, and technology in America*. Cambridge, Mass.: Harvard University Press.

Kant, I., 1997, *Critique of Pure Reason*, trans. Guyer, P., & Wood, A. New York: Cambridge University Press.

Keller, E.F., 1985, *Reflections on Gender and Science*. New Haven, Conn.: Yale University Press.

Kertesz, A., 2002, On the De-naturalization of Epistemology. *Journal for General Philosophy of Science*, 33, pp. 269–88.

Knorr-Cetina, K. 1995. 'Laboratory Studies: the cultural approach to the study of science'. S. Jasanoff et al, *Handbook of Science and Technology Studies*. London: Sage Publications, 140–166.

Kuhn, T., 1970, *The Structure of Scientific Revolutions*, 2nd ed. Chicago: University of Chicago Press.

Kuhn, T. (1977). *The Essential Tension: selected studies in scientific tradition and change*. Chicago: University of Chicago Press.

Latour, B., 1987, *Science in Action: How To Follow Scientiists and Engineers through Society*. Cambridge, Mass.: Harvard University.

Latour, B., and Woolgar, S., 1986, *Laboratory Life: The Construction of Scientific Facts*. Princeton, N.J.: Princeton University Press.

Livingston, E. 1986. *The Ethnomethodological Foundations of Mathematics*. London: Routledge and Kegan Paul.

Lynch, M., 1985, *Art and Artifact in Laboratory Science: A Study of Shop Work and Shop Talk in a Research Laboratory*. London: Routledge & Kegan Paul.

Lynch, M., 1993, *Scientific Practice and Ordinary Action – Ethnomethodology and Social Studies of Science*. Cambridge: Cambridge University Press.

Lynch, M., Livingston, E., and Garfinkel, H., 1983, "Temporal Order in Laboratory Work", in Cetina & Mulkay, 1983, pp. 205–38.

MacIntyre, A. 1987, "Relativism, Power, and Philosophy", in Baynes, K., Bohman, J., & McCarthy, T. (eds.), *After Philosophy: End or Transformation*. Cambridge, Mass.: MIT Press, pp. 385–411.

Maffie, J., 1990, Recent Work on Naturalized Epistemology. *American Philosophical Quarterly*, 27, 4, pp. 281–93.

Mayo, Deborah G., 1996, *Error and the Growth of Experimental Knowledge*. Chicago: University of Chicago Press.

McDowell, J., 1994, *Mind and World*. Cambridge, Mass.: Harvard University Press.

Minnis, P.E. (ed.), 2000, *Ethnobotany: A Reader*. Norman: University of Oklahoma.

Peirce, C., 1998, *Collected Papers*, Vols. II and V, eds. Hartshorne, C., & Weiss, P. Bristol: Thoemmes Press.

Pickering, A., 1984, *Constructing Quarks: A Sociological History of Particle Physics*. Oxford: Edinburgh University Press.

Pickering, A. (ed.), 1992, *Science as Practice and Culture*. Chicago and London: University of Chicago Press.

Pickering, A. (1995). *The Mangle of Practice: time, agency, and science*. Chicago: University of Chicago Press.

Quine, W., 1960, *Word and Object*. Cambridge, Mass.: MIT Press.

Quine, W., 1969, *Ontological Relativity and Other Essays*. New York: Columbia University Press.
Radder, H., 1988, *The Material Realization of Science – A Philosophical View on the Experimental Natural Sciences*. Assen and Maastricht: van Gorcum Press.
Radder, H., 1992, Experimental Reproducibility and the Experimenters' Regress. *PSA: Proceedings of the Biennial Meeting of the Philosophy of Science Association*, 1, pp. 63–73.
Radder, H., 1995, Experimenting in the Natural Sciences: A Philosophical Approach, in Buchwald, Jed Z. (ed.), *Scientific Practice: Theories and Stories of Doing Physics*. Chicago: University of Chicago Press, pp. 56–86.
Radder, H. (ed.), 2003, *The Philosophy of Scientific Experimentation*. Pittsburgh, Pa.: University of Pittsburgh Press.
Rheinberger, Hans-Jörg, 1997, *Toward a History of Epistemic Things*. Stanford, Calif.: Stanford University Press.
Rorty, R., 1982, *Consequences of Pragmatism*. Mineapolis: University Minnesota Press.
Rorty, R., 1986, "Pragmatism, Davidson and Truth", in LePore, E. (ed.), *Truth and Interpretation: Perspectives on the Philosophy of Donald*. New York: Basil Blackwell, pp. 333–55.
Rotenstreich, N. 1977. *Theory and Practice: an essay in human intentionalities*. The Hague: Martinus Nijhoff.
Rorty, R., 1991, *Objectivity, Relativism, and Truth*. Cambridge: Cambridge University Press.
Rouse, J., 1987, *Knowledge and Power: Toward a Political Philosophy of Science*. Ithaca, N.Y.: Cornell University Press.
Rouse, J., 1996, *Engaging Science: How to Understand Its Practices Philosophically*. Ithaca and London: Cornell University Press.
Rouse, J., 2002, *How Scientific Practices Matter, Reclaiming Philosophical Naturalism*. Chicago and London: University of Chicago Press.
Rouse, J., 2003, Remedios and Fuller on Normativity and Science. *Philosophy of the Social Science*, 33, 4, pp. 464–71.
Schatzki, T.R., Cetina, K.K., and Savigny, E. von (eds.), 2001, *The Practice Turn in Contemporary Theory*. London and New York: Routledge.
Suzuki Jun, 2002, Two Time System, Three Patterns of Working Hours. *Japan Review*, 14, pp. 79–97.
Traweek, S. 1988. *Beamtimes and Lifetimes: the world of high energy physicists*. Cambridge, Mass.: Harvard University Press.
Wittgenstein, L. 1953. *Philosophical Investigations*, G. E. M. Anscombe (trans.). Oxford: Blackwell.

Chinese references

Aristotle, 2003, *Nicomachean Ethics* (亚里士多德：《尼各马科伦理学》，苗力田译，北京：中国人民大学出版社, 2003).
Pierre Bourdieu, 2003, *The Logic of Practice* (布迪厄, P.：《实践感》，蒋梓骅译，南京：译林出版社, 2003).
David Bloor, 2001, *Knowledge and Social Imagery* (布鲁尔, D.：《知识和社会意象》，艾彦译，北京：东方出版社, 2001).

References

Ulrich Beck, 2004, *The Risk Society* (贝克：《风险社会》, 何博闻译, 南京: 译林出版社, 2004).

Cai, D.F., 1994, *The History of the Feng Shui* (蔡达峰：《历史上的风水术》, 上海: 上海科技教育出版社, 1994).

Cai Zhong, 2002, Relativistic Features of 'Strong Program'SSK (蔡仲: 《"强纲领"SSK的相对主义特征》, 载《自然辩证法研究》, 2002(3)).

Cai, Z.J., 2005, Insect Flight Tracking Measurement System Based on Magnetic Field Sensing Coil (蔡志坚：《基于磁场传感线圈的昆虫飞行跟踪测量系统》, 清华大学博士论文, 2005–10).

Cartwright, N., 2006, *The Dappled World, A Study of the Boundaries of Science* (卡特赖特, N.：《斑杂的世界，科学边界的研究》, 王巍、王娜译, 上海: 上海世纪出版集团, 2006).

Chen Shan, 1994, *Resources of Chinese Grassland Feeding Plants* (陈山：《中国草地饲用植物资源》, 沈阳: 辽宁民族出版社, 1994).

Chen Shan, 2002, Knowledge and Research Value of Mongolian Traditional Botany (陈山：《蒙古族传统植物学知识及其研究价值》, 载《中国医学生物技术应用杂志》, 2002(3): 22–5).

Chang, C.L., 2007, Re-examine the Classification of Cognitive Relativism (常春兰: 《重新考察认知相对主义的分类》, 载《安徽大学学报》(哲学社会科学版), 2007(3): 33–7).

Cui, X.Z., and Pu, G.X., 1997, On the Symbiosis of Experiment and Theory – From the Discovery Process of "Neutral Flow" (崔绪治、浦根祥: 《试论实验与理论的共生关系——从"中性流"的发现过程谈起》, 载《学海》, 1997(4): 41–6).

Anthony B. Cunningham, 2004, *Application of National Botany: Utilization and Protection of Human and Wild Plants* (坎宁安：《应用民族植物学: 人与野生植物利用和保护》, 裴盛基、淮虎根编译, 昆明: 云南科技出版社, 2004).

Charles Robert Darwin, 1982, *Darwin's Memoirs* (达尔文：《达尔文回忆录》, 毕黎译注, 北京: 商务印书馆, 1982).

John Dewey, 2004, *Deterministic Seeking – A Study on the Relationship between Knowing and Doing* (杜威：《确定性的寻求——关于知行关系的研究》, 傅统先译, 上海: 上海人民出版社, 2004).

Friedrich Engels, 1984, *Dialectics of Nature* (恩格斯：《自然辩证法》, 于光远等译编, 北京: 人民出版社, 1984).

Fan, X.Z., 1986, *A Brief History of Chinese Medicine* (范行准：《中国医学史略》, 北京: 中国古籍出版社, 1986).

Feyerabend, P.K., 1990, *Science in a Free Society* (法伊阿本德：《自由社会中的科学》, 兰征译, 上海: 上海译文出版社, 1990).

Feyerabend, P.K., 1992, *Against Method: An Outline of Anarchist Knowledge* (法伊阿本德: 《反对方法: 无政府主义知识论纲要》, 周昌忠译, 上海: 上海译文出版社, 1992).

Michel Foucault, 1997, *The Eye of Power: An Interview with Foucault* (福柯：《权力的眼睛: 福柯访谈录》, 严锋译, 上海: 上海人民出版社, 1997).

Michel Foucault, 1998, *Foucault Collection* (福柯：《福柯集》, 杜小真编选, 上海: 上海远东出版社, 1998).

Michel Foucault, 1999, *Society Must Be Defended* (福柯：《必须保卫社会》, 钱翰译, 上海: 上海人民出版社, 1999).

Michel Foucault, 2003, *The Archeology of Knowledge* (福柯：《知识考古学》, 谢强、马月译, 北京: 三联书店, 2003).

Joan H. Fujimura, 2001, *Grafting Science: A Sociohistory of the Quest for the Cancer* (藤村:《创立科学: 癌症遗传学社会史》, 夏侯炳、刘喜申译, 南昌: 江西教育出版社, 2001).

Hans-Georg Gadamer, 2005, What is Practical Philosophy – An Interview with Gadamer" (伽达默尔:《什么是实践哲学——伽达默尔访谈录》, 载《西北师大学报(社会科学版)》, 2005(1): 7–10).

Gan, S.P., 2002, *A Study on the Frontier of Applied Ethics* (甘绍平:《应用伦理学前沿问题研究》, 南昌: 江西人民出版社, 2002).

Clifford Geertz, 2004, *Local Knowledge-Interpretation of Anthropological Essays* (吉尔兹:《地方性知识——阐释人类学论文集》, 王海龙、张家宣译, 北京: 中央编译出版社, 2004).

David Griffin, 1985, *The Reenchantment of Science* (格里芬:《后现代科学》, 王治河译, 北京: 中央编译出版社, 1985).

Hasi, B., 2000, Ethnic Botany Research on Inner Mongolia Aru Corqin (哈斯巴根,《内蒙古阿鲁科尔沁蒙古民族植物学的研究》, 载中国博士论文库, 2000).

Hasi, B., Pei, S., 1999, Ethnic Botany Research on Inner Mongolia Grassland Genus (哈斯巴根、裴盛基,《内蒙古草地葱属植物的民族植物学研究》, 载《中国草地》, 1999(5): 42–7).

Hasi, B., Pei, S., 2001, Mongolian Folk Wild Therapeutic Plants in Aruqalqin (哈斯巴根、裴盛基:《阿鲁科尔沁蒙古族民间野生食疗植物》, 载《中药材》, 2001(2)).

Hasi, B., Pei, S., 2003, Mongolian medicinal plants in Aruqalqin (哈斯巴根、裴盛基:《内蒙古阿鲁科尔沁蒙古族药用植物》, 载《中国医学生物技术应用》, 2003(2): 44–8).

Hanson, N.R., 1988, *Patterns of Discovery* (汉森:《发现的模式》, 邢新力、周沛译, 北京: 中国国际广播出版社, 1988).

Stephen Hawking, 1992, *A Brief History of Time* (霍金:《时间简史》, 许明贤、吴忠超译, 长沙: 湖南科学技术教育出版社, 1992).

Huai, H.Y., Ha, S.B.G., and Wang, Y.H., 2005, Some Misunderstandings in the Understanding of Ethnic Botany (淮虎银、哈斯巴根、王雨华等,《民族植物学认识的几个误区》, 载《植物学通报》, 22(4): 502–509).

Edmund Gustav Albrecht Husserl, 1997, *Husserl Anthology* (胡塞尔:《胡塞尔选集》, 倪梁康编, 上海: 上海三联书店, 1997).

Sheila Jasanoff, 2004, *Handbook of Science and Technology Studies* (贾撒诺夫等:《科学技术论手册》, 盛晓明等译, 北京: 北京理工大学出版社, 2004).

Jiang, J.S., 2003, A Rational Ddevil or Arrogant Angel: Does Science Need Supervision and Balance? (蒋劲松:《理智的魔鬼抑或狂妄的天使: 科学是否需要监督和制衡?》, 载《自然辩证法通讯》, 2003(1): 1–3).

Jiang, T.J., 2007, The Problem of Relativism (江天骥:《相对主义的问题》, 李涤非译, 朱志方校, 载《世界哲学》, 2007(2): 32–40).

Jin, W.L., 2000, Scientific Research and Science and Technology Ethics (金吾伦,《科学研究与科技伦理》, 载《哲学动态》, 2000(10): 4).

Immanuel Kant, 2000, *Critique of Judgment* (康德:《判断力批判》(上卷)., 宗白华译, 北京: 商务印书馆, 2000).

Immanuel Kant, 2004, *Critique of Pure Reason* (康德:《纯粹理性批判》, 邓晓芒译. 杨祖陶校, 北京: 人民出版社, 2004).

Sasaki Katsuhiro (佐々木勝浩), 2003, 'History of Mechanical Clock in Japan' (日本の機械時計の歴史), National Museum of Nature and Science News (国立科学博物館ニュース), 2003(414).

Karin Knorr-Cetina, 2001, *The Manufacture of Knowledge – An Essay on the Constructivist and Contextual Nature of Science* (塞蒂娜,卡林·诺尔:《制造知识:建构主义与科学的与境性》,王善博等译,北京:东方出版社,2001).

Thomas Sammual Kuhn, 1981, *The Essential Tension* (库恩:《必要的张力》,纪树立、范岱年、罗慧生等译,福州:福建人民出版社,1981).

Thomas Sammual Kuhn, 2003, *The Structure of Scientific Revolution* (库恩:《科学革命的结构》,金吾伦、胡新和译,北京:北京大学出版社,2003).

Hugh Levers, 2003, Intimate Knowledge (拉弗勒斯,H.,《亲密知识》,载《国际社会科学》(中文版),2003(3): 47–59).

Li, X.M., 2003, Between Science and Technology (李醒民:《在科学和技术之间》,载《光明日报》,2003年4月29日,B4版).

Li, X.M., 2004, Answering Questions about Scientism and Anti-Scientism (李醒民:《就科学主义及反科学主义答客问》,载《科学文化评论》,第1卷2004(4): 94–106).

Lin, D.H., 2002, Interpretation of 'Double–edged Sword' (林德宏:《"双刃剑"解读》,载《自然辩证法研究》,2002(10): 34–6).

Liu, H.J., 2004, Relativism is Superior to Absolutism (刘华杰:《相对主义优于绝对主义》,载《南京社会科学》,2004(12): 1–4).

Liu, L.H., 2002, *Thinking Chinese Medicine: An Introduction to Typhoid Fever* (刘力红:《思考中医:伤寒论导论》,桂林:广西师范大学出版社,2002).

Liu, P.L., 2001, *Feng Shui – The Chinese People's View of Environment* (刘沛林:《风水—中国人的环境观》,上海:上海三联书店,2001).

Liu, Y.X., 2003, Why is Science and Technology a Double-edged Sword? (刘怡:《科学和技术何以会是双刃剑》,载《科学时报》,2003-03-20).

Liu, Y.D., 2002a, On the Uncontrolled Growth of Scientific and Technological Knowledge (Ⅰ) (刘益东:《试论科学技术知识增长的失控(上).》,载《自然辩证法研究》,2002a(4): 39–42, 48).

Liu, Y.D, 2002b, On the Uncontrolled Growth of Scientific and Technological Knowledge (Ⅱ) (刘益东:《试论科学技术知识增长的失控(下).》,载《自然辩证法研究》,2002b(5): 32–36).

Jean-Francois Lyotard, 1996, *The Postmodern Condition: A Report on Knowledge* (利奥塔:《后现代状况》,岛子译,长沙:湖南美术出版社,1996).

Ma, Shi, 1998, *The Commentary on the Chapters of the Yellow Emperor* ([明]马:《黄帝内经素问注证发微》,北京:人民卫生出版社,1998).

Karl Heinrich Marx, 1972, *Selections (Anthologies). of Marx and Engels* (马克思:《马克思恩格斯选集》,中共中央编译局,北京:人民出版社,1972).

Karl Heinrich Marx, and Friedrich Engels, 1957, *Complete Works of Marx and Engels* (马克思、恩格斯:《马克思恩格斯全集》(第2卷).,北京:人民出版社,1957).

Meng, D.L., 2005, *Residential Fortune* (孟东篱:《宅运》,西安:陕西师范大学出版社,2005).

Jacques L. Monod, 1977, *Occasional and Necessity: On the Natural Philosophy of Modern Biology* (莫诺:《偶然性与必然性:略论现代生物学的自然哲学》,上海外国自然科学哲学著作编译组译,上海:上海人民出版社,1977).

Michael Mulkay, 2001, *Science and Knowledge Sociology* (马尔凯:《科学与知识社会学》,林聚任等译,北京:东方出版社,2001).

Joseph Needham, 1999, *History of Ancient Chinese Scientific Thought* (李约瑟:《中国古代科学思想史》,陈立夫等译,南昌:江西人民出版社,1999).

Ni, L.K., 2005, Again Misunderstood the Transzendental (倪梁康:《再次被误解的transzendental》,载《世界哲学》,2005(5): 97–8, 106).

Nie, J.B., Tu Wu, G.Z., and Li, L., 2005, Human Experiments of Japanese Invaders and Their Challenges to Contemporary Medical Ethics (聂精保、土屋贵志、李伦：侵华日军的人体实验及其对当代医学伦理的挑战，《医学与哲学》，2005(6): 35–8).

Pan, N.Y., 2001, Native Anthropology: Rediscovering Knowledge (潘年英：《本土人类学：重新发现的知识》，载《贵州师范大学学报》(社会科学版)，2001(2): 53–4, 60).

Pei, S.J., 1988, Ethnic Botany and Plant Resources Development (裴盛基：《民族植物学与植物资源开发》，载《云南植物研究》(增刊)，1988, Suppl(I): 135–44).

Pei, S.J., and Long, C.L., 1998, *Application of National Botany* (裴盛基、龙春林：《应用民族植物学》，昆明：云南科技出版社，1998).

Andrew Pickering, 2004, *The Mangle of Practice* (皮克林，A.：《实践的冲撞——时间、力量与科学》，邢冬梅译，南京：南京大学出版社，2004).

Karl Popper, 1999, *The Logic of Scientific Discovery* (波普尔：《科学发现的逻辑》，查汝强、邱仁宗译，沈阳：沈阳出版社，1999).

Hilary Whitehall Putnam, 2006, *The Collapse of Fact/Value Dichotomy* (普特南：《事实与价值二分法的崩溃》，应奇译，北京：东方出版社，2006).

Qing, X.T., and Zhan, S.C., 1999, *Taoist Culture New Code* (卿希泰、詹石窗：《道教文化新典》，上海：上海文艺出版社，1999).

Qiu, Hui., 2002, The Scientific View of Practice (邱慧：《实践的科学观》，载《自然辩证法研究》，2002(2): 19–22).

Ren, Y.F., 2007, An Interpretation of Rouse 's Scientific and Practical Hermeneutics (任玉凤：《劳斯的科学实践解释学思想解读》，载《内蒙古大学学报》(哲学社会科学版)，2007(4): 117–22).

Alexander Rosenberg, 2004, *Philosophy of Science* (罗森堡，A.：《科学哲学》，刘华杰译，上海科技教育出版社，2004).

William David Ross, 1997, *Aristotle* (罗斯．W.D.：《亚里士多德》，王路译．北京：商务印书馆，1997).

Richard Rorty, 2003, *Philosophy and the Mirror of Nature* (罗蒂：《哲学与自然之镜》，李幼蒸译，北京：商务印书馆，2003).

Joseph Rouse, 2004, *Knowledge and Power – Toward a Political Philosophy of Science* (劳斯：《知识与权力——走向科学的政治哲学》，盛晓明等译，北京：北京大学出版社，2004).

Oda Sachiko (小田幸子), 1994, A Pictorial Record of Japanese Clocks Preserved at the Seiko Institute of Horology (セイコー時計資料館蔵和時計図録). Tokyo: Seiko Institute of Horology (東京：セイコー時計資料館).

Carl Sagan, 1998, *The Demon-Haunted World-Science As a Candle in the Dark* (萨根：《魔鬼出没的世界：科学，照亮黑暗的蜡烛》，李大光译，长春：吉林人民出版社，1988).

Sheng, X.M., 2000, The Construction of Local Knowledge (盛晓明：《地方性知识的构造》，载《哲学研究》，2000(12): 36–44).

Sheng, X.M., 2003, From Scientific Ssocial Studies to Scientific Cultural Studies (盛晓明：《从科学的社会研究到科学的文化研究》，载《自然辩证法研究》，2003(2): 14–18, 47).

Alan Sheridan, 1997, *Searle and Foucault on Truth* (谢里登：《求真意志——米歇尔·福柯的心路历程》，尚志英、许林译，上海：上海人民出版社，1997).

Shu, W.G., 1982, Thought Experiment Also Has the Nature of Practice (舒炜光：《思想实验固有实践本性》，载《社会科学战线》，1982(4): 1–10).

Peter Singer, 2004, *Animal Liberation* (辛格: 《动物解放》, 祖述宪译, 青岛: 青岛出版社, 2004).

Sergio Sismondo, 2007, *An Introduction to Science and Technology Studies* (西斯蒙多: 《科学技术学导论》, 许为民等译, 上海: 上海科技教育出版社, 2007).

Hashimoto Takehiko (橋本毅彦), 2003, 'On the Accuracy of Japanese Clock' (和時計の精度をめぐつて), National Museum of Nature and Science News (国立科学博物館ニュース), 2003(414).

Tian, D.H., 2005, *The Yellow Emperor's Su Wen Essay* (田代华整理: 《黄帝内经素问》, 北京: 人民卫生出版社, 2005).

Tian, D.H., and Liu, G.S., 2005, *Lingshu Scripture* (田代华, 刘更生整理: 《灵枢经》, 北京: 人民卫生出版社, 2005).

R.A. Uritam, 1983, Ancient Chinese Physics and Nature (尤利达, R. A.: 《中国古代的物理学和自然观》, 载《科学史译丛》, 1983(3)).

Waldrop, 1997, *Complex* (沃德罗普: 《复杂》, 陈玲译, 北京: 三联书店, 1997).

Wang, D., 1995, *Grassland Management* (王栋: 《草原管理学》, 南京: 畜牧兽医图书出版社, 1995).

Wang, Q.H., 1992, *Research on Feng Shui Theory* (王其亨主编: 《风水理论研究》, 天津: 天津大学出版社, 1992).

Wang, Y.F., 2005, A Study on the Scientific Construction of Pickering (王延锋: 《皮克林科学建构论研究》, 上海交通大学博士生论文, 2005年9月).

N. Wiener, 1989, *People's Usefulness: Cybernetics and Society* (维纳, N.: 《人有人的用处: 控制论和社会》, 陈步译, 北京: 商务印书馆, 1989).

Raymond Williams, 2005, *Keywords: Vocabulary of Cultural and Social* (威廉斯, R: 《关键词: 文化与社会的词汇》, 刘建基译, 北京: 生活·读书·新知三联书店, 2005).

Wu Tong, 2005, A Review of the Development of Philosophy of Science Practice (吴彤: 《科学实践哲学发展述评》, 载《哲学动态》, 2005(5): 40–3).

Wu Tong, 2006a, Criticism on the Topic of 'Observation / Experimental Load' (吴彤: 《"观察/实验负载理论"论题批判》, 载《清华大学学报》(哲学社会科学版), 2006a(1): 127–31).

Wu Tong, 2006b, Scientific Practice in the View of Scientific Practice Philosophy -On Rouse and others' Scientific Practice View (吴彤: 《科学实践哲学视野中的科学实践——兼评劳斯等人的科学实践观》, 载《哲学研究》, 2006b(6): 85–91).

Wu Tong, 2006c, Focus on the Scientific Philosophy of Practice (吴彤: 《聚焦实践的科学哲学》, 载《科学实践哲学的新视野》, 呼和浩特: 内蒙古人民出版社, 2006).

Wu Tong, 2007, Scientific Research Begins with Opportunities, Problems or Observations (吴彤: 《科学研究始于机会, 还是始于问题或观察》, 载《哲学研究》, 2007(1): 98–104).

Wu, S.X., Qi, G.R., and Bian, Y.J., 1983, *Time Measurement* (吴守贤、漆贯荣、边玉敬: 《时间测量》, 北京: 科学出版社, 1983).

Bruno Latour Woolgar, 2004, *Laboratory Life: The Construction Process of Scientific Facts* (拉图尔, B.、伍尔加, S.: 《实验室生活: 科学事实的建构过程》, 张伯霖、刁小英译, 北京: 东方出版社, 2004).

Xu, C.F., 2004, *On Aristotle's Concept of Practice* (徐长福: 《论亚里士多德的实践概念》, 载《吉林大学社会科学学报》, 2004(1): 56–63).

Yamada Keiji, 1986, Spatial·Classification·Category (山田庆儿: 《空间·分类·范畴》, 载《日本学者论中国哲学史》, 中华书局, 1986).

Yan, J.M., 2006, *History of Ancient Chinese Medicine* (严健民:《远古中国医学史》, 北京: 中医古籍出版社, 2006).

Ye, S.X., 2001, Local Knowledge (叶舒宪:《地方性知识》, 载《读书》, 2001(5): 121–5).

Yu, G.Y., 1988, *Marx, Engels, Lenin on the Dialectics of Nature and Science and Technology* (于光远等编:《马克思恩格斯列宁论自然辩证法与科学技术》, 科学出版社, 1988).

Yu, W.J., 2001, *Practical Hermeneutics: Reinterpretation of Marxist Philosophy and General Philosophical Theory* (俞吾金:《实践诠释学: 重新解读马克思哲学与一般哲学理论》, 昆明: 云南人民出版社, 2001).

Zhang, R.L., 2005, Practical Philosophy and Its Practical Concept (张汝伦:《作为第一哲学的实践哲学及其实践概念》, 载《复旦学报》(社会科学版), 2005(5): 155–63).

Zhang, X.L., 2001, Providing a Sanctuary for Ancient Chinese Endangered Culture – Proposal for the Establishment of a Confucian Cultural Reserve (张祥龙:《给中国古代濒危文化一个避难所——成立儒家文化保护区的建议》, 载《中华读书报》, 2001(8): 15).

Zhang, Z.C., 2002, *Notes of the Yellow Emperor* ([清]张志聪集注:《黄帝内经集注》, 方春阳等点校, 杭州: 浙江古籍出版社, 2002).

Zhao, J.S., 2003, *Theoretical Interpretation of Acupuncture and Moxibustion* (赵京生:《针灸经典理论阐释》, 上海: 上海中医药大学出版社, 2003).

Zhao, X.J., and Zhu, X.C., 2004, Investigation of SARS Virus Leakage (赵小剑、朱晓超: SARS病毒泄漏调查, 载《财经》, 2004(10). 2004-5-20).

Zhen, Z.Y., 1984, *History of Chinese Medicine* (甄志亚等:《中国医学史》, 上海: 上海科学技术出版社, 1984).

Zhu, Q.S., 2005, The Scientific Content and the Approach of Reformation for Chinese Medicine (朱清时:《中医学的科学内涵与改革思路》, 载《自然杂志》, 2005, 27(5): 249–50).

Zuo, Y.H., 2004, *Academic Division and the Establishment of Modern Chinese Knowledge System* (左玉河:《从四部之学到七科之学——学术分科与近代中国知识系统之创建》, 上海: 上海书店出版社, 2004).

Index

Abbe, Ernst 181
Absolutism 92, 169–171
Ackerman, Robert 236
Allium plant 283–285
Almeder, Robert 136
Anthropology, local knowledge and 82–89, 93
Aobao 286
Apel, Karl-Otto 150, 156
Arab academic community, power and 117–119
Architecture, local knowledge and 286, 290
Aristotle: overview 3, 13; concept of practice and 13–16, 18–19; physics and 201–202
Arthur, W. Brian 218
Atomism 258
Ayer, Alfred J. 162

Bacon, Francis 18, 106, 201–202, 229
Baird, David 78, 176–177, 186–188, 189
Barlow, Peter 188
Barnes, S. Barry 140, 146
Batens, Diderik 302
Being-in-the-world 90, 133
Bergman, Gustav 180
Bernal, John Desmond 22
Bloor, David 52, 140
Bourdieu, Pierre 61, 64, 295
Boyle, Robert 235
Brandom, Robert 66, 70–75, 148, 152–155, 158
Brown, Robert 167, 226
Brownian movement 226
Buchwald, Jed Z. 227
Bukharin, Mikhail 36
Bunge, Mario 268
Butterfield, Herbert 23

Cai Yuanpei 124
Callon, Michael 26, 27
Carnap, Rudolf 128, 162, 305
Carothers, James M. 217
Cartwright, Nancy: overview 11; local knowledge and 91–92, 101–102; new empiricism and 194–207; new experimentalism and 193, 303; nomological machines and 280; phenomenology and 305, 307–309; relativism and 170; replicability of experiment and 239; research and 40–41
Causality, philosophy of scientific experimentation (PSE) and 176
Ceteris paribus laws 101–102, 198–200, 206
Chalmers, Alan Francis 172, 180, 229
Character of First Philosophy 142
Checkland, Peter 195–196, 300
Chen Jaiying 36
Chen Shan 283
China: Chinese Communist Party (CCP) 122–126; Cultural Revolution 49, 124; *fengshui* (*see* Fengshui); modern Chinese society, social knowledge and power in 120–126; PRC, knowledge and power in 122–126; traditional Chinese medicine (TCM) (*see* Traditional Chinese medicine (TCM)); traditional Chinese society, knowledge and power in 120–122
Chuang-tzu 160
Church, Alonzo 162
Churchland, Paul 2
Clark, George Norman 23
Clocks 100–101
Cohen, Morris R. 137

Collins, Harry M. 232–235, 238
Complexity Science 232, 254, 260
Complexity study 216–220
Computer simulation 177
Concept of practice 13–36; Aristotle on 13–16, 18–19; constructionism and 24–25; framework for 58; hermeneutics and 28–35 (*see also* Hermeneutics, PSP and); isomorphism of practical activity 24–25; Kant on 16–17; Marx on 17–21; modelling process 59; objectivity versus relativity of knowledge 58–59; origins of 13–17; practical thoughts embedded in social research of science 21–24; scientific practice versus Marx's practice 21–24; SSK and 25–28
Conceptual practice 52–59; bridging 53, 57–58; filling 53, 57–58; transcription 53, 57–58
Confucianism 121
Constructivism 24, 27, 147
Coulomb's law 93, 199–200, 202–203
Cowan, George 216–217
Crapper, Thomas 193
Crick, Francis 95, 176, 186–187
Cultural practice 95
Cultural Revolution 49, 124

Danxun Li 297
'Dappled world' 194–196, 203
Dasein 35, 80, 133
Davenport, Thomas 176
Davidson, Donald 149–150, 154–155, 158
Davy, Humphry 167, 185, 227
Demarcation of science 267–268
Democritus 258
Dewey, John 137
Dilthey, Wilhelm 3, 5
Discursive practice: overview 51–52; discursive nature of scientific practice 77–79; naturalism and 158
DNA 95, 176, 186–187, 221
Downes, Stephen 138, 141
Dreyfus, Hubert 70–73, 75–76
Duhem-Quine thesis 162, 240

Ecological knowledge 283–285
Eddington, Arthur 192
Einstein, Albert 48, 91, 174–175, 192
Eitel, Ernst J. 265
Electromagnetic motor 176, 188–189
Embodied cognition, PSP and 2, 10

Empiricism 5
Engels, Friedrich 21, 23
Ephemeris Time (ET) 99
Epistemic things 133
Epistemology: Kantian epistemology 147–148, 150, 158; local knowledge, epistemological significance of 92–93; naturalistic epistemology 138, 140; philosophy of scientific experimentation (PSE) epistemological significance of 176; scientific instruments, epistemological classification of 182–185; traditional epistemology 142–143
Ethnobotany 281–294; architecture and 286, 290; aspects of 292; as combination of natural and social sciences 291–294; defined 84–85; ecological knowledge and 283–285; folksongs, natural knowledge embedded in 282–283; geographical knowledge and 286, 290; hermeneutics and 293–294; medical knowledge and 286; Mongolian local knowledge and 283–285, 287–289; problems of 290–291; significance of 290–291
Expectancy effect 234
Experimental practice: overview 44–45; new experimentalism, PSP and (*see* New experimentalism, PSP and); observation versus 173; philosophy of scientific experimentation (PSE) (*see* Philosophy of scientific experimentation (PSE)); replicability of experiment in 238–239 (*see also* Replicability of experiment); theory and 190, 229–230; types of experiment 178
Experimenter's regress 235

Faraday, Michael 40, 188–189, 191, 227, 239
Feminism 67, 142, 208
Fengshui 264–280; daily life, relationship to 271; defined 264; demarcation of science and 267–268; experience and 267; 'four seasons and five directions' and 276; 'harmony of yin and yang' 277; *I Ching* and 277–278; increasing interest in 265; Liqi School 276, 278; as local knowledge 270–273; mysteries of 273; normative framework of practice 274–278; observing 'sand pattern' 274; optimal layout and sites 272; as practical knowledge 269–270;

Index 323

practices of 264–267; in PSP 279–280; *Qi* theory and 273, 277–279; scientific explanation of 273; selecting 'dens' 275; 'seven luminaries and nine stars' and 276; surveying 'dragon' 274; theories of 264–267; 'unity of heaven and human' 277; viewing 'flow of water' 275; world view of nature, relationship to 271–273; *Wu-Xing* theory and 276, 279; Xingshi School 276, 278; *Yin-Yang* theory and 250–255, 273, 275–277, 279
Feyerabend, Paul K. 8, 106–108, 163, 267
Fine, Kit 268
Five Zang 244
Fleck, Ludwik 97
Folksongs, local knowledge and 282–283
Food and Drug Administration (US) 263
Foucault, Michel 20, 68, 105, 107–112, 114–116
Foundationalism 170, 204–205
Frankfurt School 106
Franklin, Allan 2, 190, 236
Frege, Gottlob 3, 128, 137, 147, 304–305
Fujimura, Joan H. 26, 66
Fuller, Steve 138, 141

Gadamer, Hans-Georg 15–16, 28
Gaillard, Mary 215
Galileo 48, 50, 167, 192, 226–227, 307
Galison, Peter 2, 191, 236, 301–302
Garfinkel, Harold 26
Geertz, Clifford 82–89
Geographical knowledge 286, 290
Giere, Ronald 2, 138–139, 146
Gilbert, G. Nigel 27
Globalization 83
Gooding, David: overview 2, 11; conceptual practice and 53; new experimentalism and 191–192, 227, 302; replicability of experiment and 236, 238–239
Gravitation 91, 199, 202
Grazing 282–283
Griffin, David 232
Guo Jianming 121

Habermas, Jürgen 155–156
Hacking, Ian: overview 2; conceptual practice and 53; experimental practice and 44–45; 'experiment has its own life' 224, 226; local knowledge and 90–91; naturalism and 151; new experimentalism and 190–193, 300–304; philosophy of scientific experimentation (PSE) and 173–174; replicability of experiment and 236, 238–239; research and 40–41; scientific instruments and 180–182; scientific practice and 37
Haldane, J.B.S. 23
Hall, Edwin H. 44–45
Hall Effect 44–45
Hamilton, William Rowan 53, 55–58
Hanson, Norwood Russell 90, 173, 224
Harré, Rom 182–185, 308
Harshberger, John William 84–85
Haugeland, John 134
Hayes, Steven C. 193
Heelan, Patrick A. 305–306
Hegel, G.W.F. 160
Heidegger, Martin: overview 2–3, 5; hermeneutics and 28–36; laboratory practice and 47; local knowledge and 88–91; normativity and 129, 132–133; opportunity as starting point of research and 211; phenomenology and 304–305; 'poetic habitation' 271; relativism and 166; research and 41; scientific instruments and 182; scientific practice and 39, 80
Heidelberger, Michael 177
Hermeneutics, PSP and 28–35; overview 3–4, 304–310; different senses of hermeneutics 28–29; ethnobotany and 293–294; hermeneutic nature of scientific practice 62–81; new experimentalism versus 9–10; practical hermeneutics 18, 29–34; theoretical hermeneutics 31–34
Herschel, Caroline 227
Herschel, William 227
Hertz, Heinrich 191, 227
Hesse, Marry 9, 39
Hessen, Boris 22–23, 36
High-energy physics 92, 179
Hippocrates 13
Historicism 1, 106
Hobbes, Thomas 235
Hogben, Lancelot 23
Holland, John 45, 218–219
Hook, Sidney 137
Huang Di Nei Jing 244–245, 249–250, 260, 263
Husserl, Edmund 28, 128, 147, 304–305, 307–310

I Ching 278
Indexicality of knowledge 89

Insect flight measurement 220–221
International Atomic Time (TAI) 99
Interpretive anthropology 85

Jiahua Wei 297

Kant, Immanuel: concept of practice and 16–17; constructivism and 147–148; normativity and 147–148
Kantian epistemology 147–148, 150, 158
Keller, Evelyn Fox 2, 177
Kelvin, Lord 40
Kertész, András 142–143
Kitcher, Philip 127
Knorr-Cetina, Karin D.: overview 2; concept of practice and 24–26; laboratory practice and 46, 174; local knowledge and 93–96; naturalism and 140; opportunity as starting point of research and 214–215; research and 41
Knowledge: architecture and 286, 290; ecological knowledge 283–285; folksongs, natural knowledge embedded in 282–283; geographical knowledge 286, 290; local knowledge (see Local knowledge); measuring knowledge and practice 189; medical knowledge 286; model knowledge and practice 186–187; of nature and law 201–203, 207; objectivity versus relativity 58–59; power and (see Power, knowledge and); relativity of 91; SSK (see Sociologists of scientific knowledge (SSK)); standardization of 96–99; universal knowledge (see Universal knowledge); working knowledge (see Working knowledge)
Kripke, Saul 128
Kuhn, Thomas: overview 2, 5, 8, 12; concept of practice and 27; demarcation of science and 267; hermeneutics and 28, 31; local knowledge and 88; naturalism and 127, 138, 143, 146, 147; opportunity as starting point of research and 222; power and 106–107, 109; relativism and 163, 165–167; replicability of experiment and 231; scientific practice and 37–40, 44; thought experiments and 48–49
Kuhnian historicism 138

Laboratory practice 46–48 see also Research; artificial research environment, constructing 47–48; constructing phenomenal micro-world 46; highlighting 46; intervention 46–47; isolation 46; local knowledge and 95–96; manipulation 46–47; phenomenology and 307–310; philosophy of scientific experimentation (PSE) and 174–175; power and 112–114, 118, 120; tracking phenomenal micro-world 47
Lakatos, Imre 9, 163
Lange, Rainer 177
Lao Tzu 79
Latour, Bruno: overview 2; concept of practice and 25; conceptual practice and 56; discursive practice and 51–52; laboratory practice and 47; local knowledge and 95–98; opportunity as starting point of research and 212–214; philosophy of scientific experimentation (PSE) and 193; replicability of experiment and 239; research and 41
Laudan, Larry 9, 127, 138, 164, 267–268
Lebenswelt 309
Leibapriori 150–151, 156
Liang Shuming 245, 255
Lihong Liu 244
Literary technology 78
Liu Huajie 170
Livingston, Eric 26
Local knowledge 82–104; absolutism and 92–93; agents of 86; anthropology and 82–89, 93; architecture and 286, 290; *ceteris paribus* laws and 101–102; conceptual background of 82–84; ecological knowledge 283–285; epistemological view of 88; ethnobotany (see Ethnobotany); *fengshui* (see Fengshui); folksongs, natural knowledge embedded in 282–283; Geertz's notion of 84–87; geographical knowledge 286, 290; indexicality of 93–95; laboratory practice and 95–96; medical knowledge 286; modern knowledge, in contrast with 85–86; Mongolian local knowledge (see Mongolian local knowledge); problems with 86–87; from PSP perspective 102, 281–282; science as universal knowledge 95–102;

scientific knowledge as 98, 194; scientific practice and 95; significance in PSP 87–95; traditional Chinese medicine (TCM) (*see* Traditional Chinese medicine (TCM)); universal knowledge versus 6–7, 10–11, 92–93, 197, 206; Western knowledge, in contrast with 85
Logical positivism 158, 161–162, 173, 267
Logocentrism 86
Lorentz, Hendrik 40
Lynch, Michel 2, 26

MacIntyre, Alasdair 30, 160
Maffie, James 146
Marx, Karl: Aristotle compared 18–19; concept of practice and 17–21; constructionism and 24; hermeneutics and 19–21; scientific practice versus 21–24
Materialism 232
Material nature of scientific practice 77–79
Maxwell, Grover 180
Maxwell, James Clark 40
Mayo, Deborah G. 2, 177, 190, 192, 227, 236
McDowell, John 152, 154
Medical knowledge 286
Merton, Robert K. 9, 23, 36, 106
Michelson-Morley experiment 210
Microscopes 180–182
Modality 130–134
Modelling process 59, 177
Modernism 7–9
Mongolian local knowledge: architecture and 286, 290; botanical knowledge 283–285, 287–289; ecological knowledge 283–285; folksongs, natural knowledge embedded in 282–283; geographical knowledge 286, 290; medical knowledge 286
Morgan, Mary 177
Morpurgo, Giacomo 239
Mulkay, Michael 27, 105

Naturalism: overview 1; constructivism and 147–148; continuity of essences 139; continuity of methodology 139; continuity of statuses 138–139; discursive practice and 158; historical issues 137–139; Kuhnian historicism 138; naturalistic epistemology 138, 140; normativity and 127–128, 142–148; PSP and 139–142, 152–155; quasi-transcendental philosophy PSP as, 147, 155–158; SSK and 140; traditional epistemology and 142–143; transcendental normativity, critique of 148–152, 157
Naturalistic epistemology 138, 140
Needham, Joseph 23
Neptune 199
Neurath, Otto 129, 195
New empiricism 194–207; overview 300–304; *ceteris paribus* laws and 196–200, 206; concrete versus abstract 203–204; 'dappled world' and 194–196, 203; knowledge of nature and law 201–203, 207; plural local realism and 194–196, 205–206; power and knowledge and 206–207; social construction and 204–205; as supplanting PSP 205–207
New experimentalism, PSP and 224–230; overview 2, 300–304; achievements of 228–229; criticism of 228–229; experiment, relation with theory 190, 229–230; 'experiment has its own life' 225–228; hermeneutics versus 9–10; observation versus theory and 228–229; realism and 190–191; relativism and 167; replicability of experiment and 236–241; shortcomings of 228–229; SSK versus 240–241; strategy for criticism 225–228; theory and 228–229; 'theory-ladenness' of observation and 190, 192, 224–228
Newton, Isaac 91, 197–199, 201–202, 205, 226
Newton's laws 93, 197
Nietzsche, Friedrich 20
Nietzsche commitment 129
Nomological machines 199–200, 280, 307, 309
Normativity of scientific practice 127–136; overview 134–135; development of 128–130; historical background 128–130; importance in scientific practice 130–134; modality and 130–134; naturalism and 127–128, 142–148; transcendental normativity 148–152, 157

326 *Index*

Observation: experiment versus 173; new experimentalism and 228–229; relationship with question and opportunity 212; as starting point of research, opportunity versus 209–215; 'theory-ladenness' of 172–174, 189–190, 224–228; theory versus 228–229
Ontology: philosophy of scientific experimentation (PSE), ontological significance of 175–176; scientific instruments, ontological classification of 182–185
Opium War 122
Opportunity as starting point of research: cases of 215–222; complexity study 216–220; insect flight measurement 220–221; observation as starting point versus 209; relationship with question and observation 212; significance of 222; solar neutrino experiment 215–216; SSK and 208, 212, 220

Paradigm 164–168
Peirce, Charles Sanders 93, 150–151, 156
Perrin, Jean Baptiste 226
Phenomenology: laboratory research and 307–310; PSP and 304–310
Philosophy of scientific experimentation (PSE) 172–193; causality and 176; computer simulation, research on 177; epistemological significance of 176; instruments and (*see* Scientific instruments); laboratory practice and 174–175; material realization and 175–176; modelling process, research on 177; new experimentalism (*see* New experimentalism, PSP and); ontological significance of 175–176; relationship between science and technology 176–177; relativism and 173; research themes 175–185; significance of 172–175; theory, role of 177; working knowledge, theory of 186–189 (*see also* Working knowledge)
Philosophy of Scientific Practice (PSP) *see specific topic*
Pickering, Andrew: overview 2; concept of practice and 26–27; conceptual practice and 53, 55–59; local knowledge and 92, 96; naturalism and 140; opportunity as starting point of research and 214–215; replicability of experiment and 232; scientific practice and 64
Pinch, Trevor 209, 215–216, 302
Plato 13, 21
Plural local realism 194–196, 205–206
'Poetic habitation' 271
Poiesis 13–15
Political participation, power and 116–118, 122–126
Popper, Karl: local knowledge and 95; opportunity as starting point of research and 208–211, 221; philosophy of scientific experimentation (PSE) and 176, 179; relativism and 162–163; scientific practice and 44
Positivism 1, 8–9
Postmodern Science 232
Power, knowledge and 105–126; overview 105–109; Arab academic community and 117–119; characteristic presentation of 111–120; characteristics of in practice 109–111; Chinese society, social knowledge and power in 120–126; intervention and 113; laboratory practice and 113–114, 118, 120; new empiricism and 206–207; political participation and 116–118, 122–126; PRC, knowledge and power in 122–126; process of scientific knowledge creation and 112–115; scientific discipline and 115–118; scientific knowledge as constructed by 111–112; scientific practice and 68–70; signs and 113; SSK and 108–109; standardization and 113; tracing and 113; traditional Chinese society, knowledge and power in 120–122
Practical hermeneutics 18, 29–34
Praxis, 13–16
'Present-at-hand' 31–32
Protagoras 160
PSE *see* Philosophy of scientific experimentation (PSE)
PSP (Philosophy of Scientific Practice) *see specific topic*
Putnam, Hilary Whitehall 136, 164

Qi theory 273, 277–279
Quasi-transcendental philosophy, PSP as 147, 155–158
Quaternion 54–57
Question: relationship with observation and opportunity 212; as starting

point of research, opportunity versus 209–215
Quine, Willard Van Orman: concept of practice and 30–31; naturalism and 138–139, 143, 145, 146, 149–150; normativity and 127; research and 39
Quine commitment 129

Radder, Hans 2, 11, 236–237, 239, 302–304
Rapoport, Amos 286
Rationalization 83
'Readiness-to-hand' 31–32
Realism 5–6, 191
Reductivism 232
Relativism 159–171; absolutism versus 169–171; feminism and 208; historical background 160; of knowledge 91; logical positivism and 161–162; new experimentalism and 167; paradigm and 164–166; performance in philosophy of science 159–161; philosophy of scientific experimentation (PSE) and 173; problems of 159–161; PSP and 165–169; SSK and 166; theory-dominated results in 164–165; traditional philosophy of science and 161–165; verifiability and 162–163
Replicability of experiment 231–242; overview 241–242; challenges 231–232; comparison of new experimentalism and SSK 240–241; concept of replicability reconsidered 236–238; in experimental practice 238–239; experimenter's regress 235; fixed theoretical interpretation and 237; of material realization 236–237; new experimentalism and 236–241; as non-local norm 239; problems of 233–234; ranges of 237; of results 237; SSK and 232–235, 240–241; traditional views 231–232; types of 237
Representationalism 210–211
Research *see also* Laboratory practice; artificial research environment, constructing 47–48; on computer simulation 177; as distinct scientific activity 38; existing resources, guided by 41–42; exploring unknown world as purpose of 41; intervening characteristics 41; lack of unified theoretical background 38–40; on modelling process 177; observation as starting point of, opportunity versus 209–215; opportunistic characteristics 41; opportunity as starting point of (*see* Opportunity as starting point of research); practical character of scientific research activities 41–42; question as starting point of, opportunity versus 209–215; starting point of (*see* Starting point of research)
Restivo, Sal 23
Rheinberger, Hans-Jörg 71, 133
Rhetorical force 78
Rigden, John S. 301–302
River research 297–302
Rorty, Robert 268
Rosenthal, Robert 234
Ross, William David 13, 207
Rous, Francis Peyton 71
Rouse, Joseph: overview 2–7, 11–12; concept of practice and 21; cultural studies of science and 280; discursive practice and 158; hermeneutics and 28–36; Husserl and 304–305; knowledge and 269–270; laboratory practice and 46–47; naturalism and 127–135, 140–147, 149, 152–158; new empiricism and 194; new experimentalism and 304; normativity and 127–135; opportunity as starting point of research and 209, 214–216, 221, 223; phenomenology and 304–307, 309; power and 105–106, 108–116; relativism and 166–167; replicability of experiment and 242; research and 41; scientific practice and 37–40, 43, 61–68, 70–77, 79–81, 266; thought experiments and 50–51
Rújula, Alvaro De 215
Russell, Bertrand 128, 137
Rutherford, Ernest 179

Sabour, M'hammed 116–118
Santa Fe Institute 216–219
Savigny, Eike von 140
Schatzki, Theodore R. 140
Scholasticism 201
Science, Technology and Society (STS) 21, 36
Scientific experimentation *see* Philosophy of scientific experimentation (PSE)
Scientific instruments 179–185; epistemological classification of

182–185; ontological classification of 182–185; philosophical significance of 180–182; in traditional philosophy and history of science 179–180; types of 185

Scientific practice *see also specific topic*; agents of 64–67; attention or contact with 295–300; case studies 42–43; conceptual practice 52–59 (*see also* Concept of practice); constitution or becoming of practice 63–70; difference in 67–68; discursive nature of 77–79; discursive practice 51–52, 158; experimental practice 44–45; features of 61–62; hermeneutic nature of 62–83; important issues in 73–74; laboratory practice 46–48 (*see also* Laboratory practice); local knowledge and 96; Marx versus 21–24; material nature of 77–79; meaningful configurations of world and 74–77; nature of 61–62; patterns of practice 72–73; power and 68–70; practical character of scientific research activities 41–42; relation between practice and patterns of practice 70–72; resistance in 67–68; significance of in PSP 37–43; spatiotemporal openness of 79–81; temporal events and processes 63; theory dominance versus 37–41; thought experiments 48–51; types of, 43–59 (*see also* Types of scientific practice)

Scientific rationality 106
Seasonal time 99–101
Secularization 83
Sellars, Wilfred 70
Severe Acute Respiratory Syndrome (SARS) 261
Shaffer, Simon 302
Shanghong Zhang 297
Shapere, Dudley 164
Shapin, Steven 232, 235
Shu Weiguang 49–50
Significance of practice in PSP 37–43
Simon, Herbert A. 2
Singer, Charles 23
Sino-Japanese War of 1894–1895 121
Social constructivism 204–205
Sociologists of scientific knowledge (SSK): overview 2; concept of process and 24–28; naturalism and 140; new experimentalism versus 240–241; opportunity as starting point of research and 208, 212, 220; power and 108–109; relativism and 166; replicability of experiment and 232–235, 240–241

Solar neutrino experiment 215–216
SSK *see* Sociologists of scientific knowledge (SSK)
Standardization: of knowledge 96–99; power and 113
Starting point of research 208–223; overview, 208–209; observation as, opportunity versus 209–215; opportunity as (*see* Opportunity as starting point of research); question as, opportunity versus 209–215
Stephanides, Adam 53, 56–58
Stern-Gerlach apparatus 184
Storer, Norman W. 22
Stuwer, Roger 301–302
Sundial 99, 103

Tang, Wenpei 36
TCM *see* Traditional Chinese medicine (TCM)
Thagard, Paul 138, 268
Theoretical hermeneutics 31–34
Theory: experimental practice and 190; experiment and 229–230; new experimentalism and 228–229; observation versus 228–229; philosophy of scientific experimentation (PSE), role in 177
'Theory-ladenness' of observation 172–174, 189–190, 224–228
Thought experiments 48–51
Thyrotropin releasing hormone (TRH) 97–98
Time measurement 98–101
Toulmin, Stephen 8
Traditional Chinese medicine (TCM) 243–263; autonomy 250; basic characteristics of 243–249; Complexity Science and 254, 260; cultural diversity, contribution to 254–255; debate regarding scientific status of 255–257; diagnosis 246–247; disease spectrum in 247; Five Zang and 244; historical evolution of 249–250; holistic nature of 258; *Huang Di Nei Jing* 244–245, 249–250, 260, 263; *Jing* and *Luo* 246–247; as local knowledge 261–263; local view of 257–259; Main and collateral

channels 249–250, 255–256, 258–260, 262–263; natural and social environment, effect of 247–248; prior to systemization 249–250; *Qi* theory and 244; rationality of 259–261; locality of 247–249; scientific methodology, contribution to 254; slow development of 253; strongly practical nature of 246–247; systemization 250; treatment 246; uniting metaphysics and physics 244–246; various schools of 248–249, 251; viscera doctrine 247, 249–250, 258, 260, 263; Wang-Wèn-Wèn-Qie method 246; Western science and 250, 252–253; *Wu-Xing* theory and 250–255; *Yin-Yang* theory and 250–255
Traditional epistemology 142–143
Traditional philosophy of science versus PSP 4–9; modernism and 7–9; neglect of scientific practice 4–5; realism and 5–6; universal versus local knowledge 6–7
Transcendence 142
Transcendental constructivism 148, 150, 152, 156–157
Transcendentalism 5
Transcendental normativity 148–152, 157
Traweek, Sharon 26
Tsinghua University 220, 296–300
Types of scientific practice 43–59; conceptual practice 52–59 (*see also* Concept of practice); discursive practice 51–52, 77–79, 158; experimental practice 44–45; laboratory practice 46–48 (*see also* Laboratory practice); scientific practice 43–44; thought experiments 48–51

Universalism 82–83, 92–93, 96, 199, 204–205
Universal knowledge: deconstruction of 89–90; local knowledge versus 7, 11–12, 92–93, 197, 206; scientific knowledge as 95–103, 96–97
Universal Time (UT) 99

Unsolved problems in PSP 9–11; hermeneutics versus new experimentalism 10–11; theory versus experiment 10; universal versus local knowledge 11
Uranus 199

van Bendegem, Jean Paul 302
van der Waals law 92
Van Gogh, Vincent 306
Verifiability 162–163
Virology 71

Wang, Dong 284
Wang, Hailong 84
Wang, Qingjie 36
Wardrop, John Glen 216, 220
Wasserman reaction 97
Watson, James 95, 176, 186–187
Weber, Max 120
Wilson Cloud Chamber 184
Winch, Peter 70, 73
Wittgenstein, Ludwig: concept of practice and 20, 27–28; conceptual practice and 52; scientific practice and 79
Woodbridge, Frederick James Eugene 137
Woolgar, Steve: overview 2; concept of practice and 25–26; discursive practice and 51–52; laboratory practice and 47; local knowledge and 95–98; opportunity as starting point of research and 212–214; research and 41
Working knowledge: measuring knowledge and practice 189; model knowledge and practice 186–187; practice and 187, 189; theory of 185–189
Wu-Xing theory 250–255, 276

Xu, Changfu 15

Yan, Fu 255
Yin-Yang theory 250–255, 272, 275–277, 279
Yurts 286, 290
Yu, Wujin 19–21, 36

Zhang, Gongyao 256

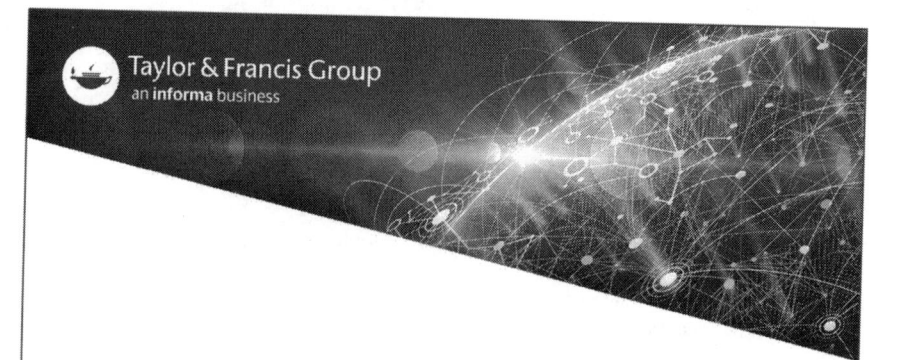

Taylor & Francis eBooks

www.taylorfrancis.com

A single destination for eBooks from Taylor & Francis with increased functionality and an improved user experience to meet the needs of our customers.

90,000+ eBooks of award-winning academic content in Humanities, Social Science, Science, Technology, Engineering, and Medical written by a global network of editors and authors.

TAYLOR & FRANCIS EBOOKS OFFERS:

- A streamlined experience for our library customers
- A single point of discovery for all of our eBook content
- Improved search and discovery of content at both book and chapter level

REQUEST A FREE TRIAL
support@taylorfrancis.com